Masters of the
Lost Land

Masters of the Lost Land

The Untold Story of the Amazon
and the Violent Fight for the
World's Last Frontier

Heriberto Araujo

MARINER BOOKS
New York   Boston

HarperCollins books may be purchased for educational, business, or sales promotional use. For information, please email the Special Markets Department at SPsales@harpercollins.com.

FIRST EDITION

Designed by Emily Snyder
Title page photograph © Photo Smoothies/Shutterstock
Map by Nick Springer

Library of Congress Cataloging-in-Publication Data has been applied for.

ISBN 978-0-06-302426-7

23 24 25 26 27   FRI   10 9 8 7 6 5 4 3 2 1

To Mrs. Butterfly and Ms. Little Butterfly

Contents

Map   viii

Preface   xi

Part I: It's All About Land

1: The Escape   3

2: The Criminal Syndicate   21

3: Terror on the Nut Road   39

4: The Chainsaw Murder   59

Part II: Rise and Fall

5: The Boomtown   79

6: Early Challenges   88

7: Crickets and Cattle   95

8: No Longer Meek   106

9: Hunting Souza   120

10: Nowhere to Hide   132

11: Nothing Shining in Eldorado   140

12: Death and Salvation   148

Part III: After He's Gone

13: An Unusual Case · 161
14: The Evidence Man · 169
15: A Cause Larger Than Death · 184
16: The Law of the Gun · 192
17: Land or We Burn the Jungle · 202
18: Amazonian Justice · 210
19: Sink or Swim · 225

Part IV: The Downfall

20: The Widow Must Fall · 235
21: "Load the Trucks" · 245
22: She Is Out · 258
23: The Trial · 268
24: A Certain Sense of Justice · 282
Epilogue · 297

Acknowledgments · 311
A Note on Sources · 315
Notes · 319
Index · 387

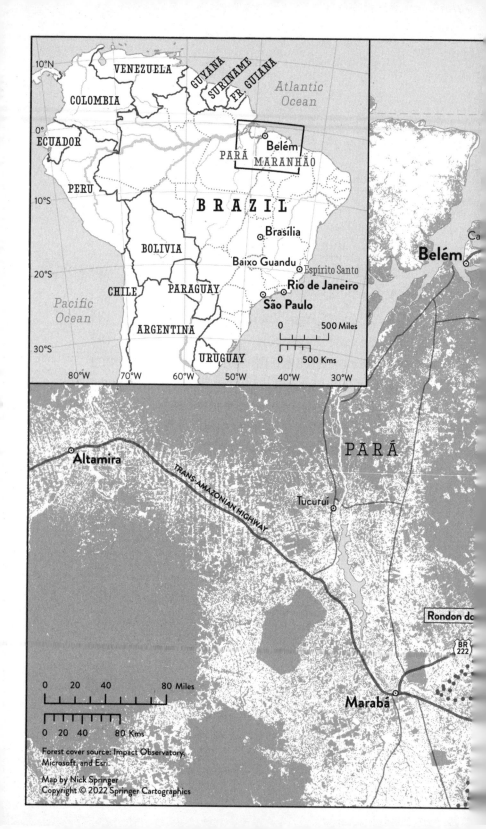

Map labels visible:

10°N, VENEZUELA, GUYANA, SURINAME, FR. GUIANA, COLOMBIA, Atlantic Ocean, ECUADOR, 0°, PARÁ, Belém, MARANHÃO, PERU, BRAZIL, 10°S, Brasília, BOLIVIA, Baixo Guandu, Espírito Santo, 20°S, PARAGUAY, Rio de Janeiro, CHILE, São Paulo, Pacific Ocean, ARGENTINA, URUGUAY, 30°S, 80°W, 70°W, 60°W, 50°W, 40°W, 30°W, 0 500 Miles, 0 500 Kms

Belém, PARÁ, Altamira, TRANS-AMAZONIAN HIGHWAY, Tucuruí, Rondon do, BR 222, Marabá, 0 20 40 80 Miles, 0 20 40 80 Kms, Ca

Forest cover source: Impact Observatory, Microsoft, and Esri.

Map by Nick Springer
Copyright © 2022 Springer Cartographics

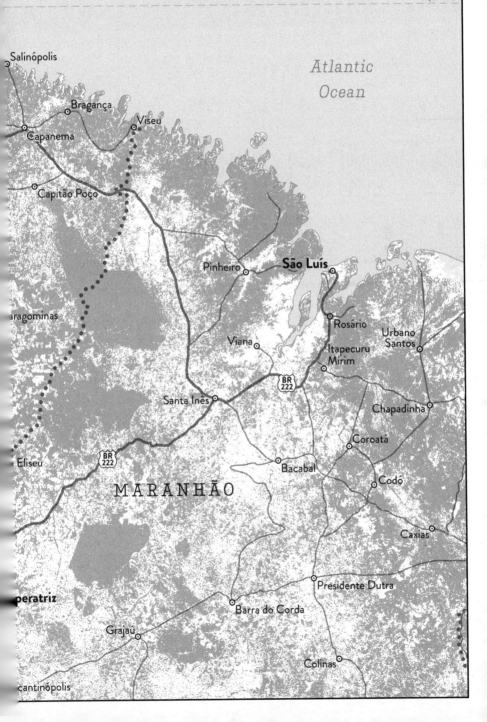

Preface

THE stories of Rondon do Pará had prepared me for a dodgy, crime-ridden place, but when I first visited the little Brazilian town on the eastern edge of the Amazon, it didn't look particularly threatening to me.

I'd been brought to Rondon do Pará—the seat of a municipality of the same name encompassing some 3,200 square miles in Pará state—by a tip from a researcher who had spent years investigating organized criminality and deforestation.[1] The issue of violence in the region had piqued my investigative interest since 2013, when I'd moved to Brazil from China, my former base as a journalist. The fate of the Amazon regularly captured media attention, often due to the appalling rates of environmental destruction, a story that became even more urgent after President Jair Messias Bolsonaro took office. But rather than the headline-grabbing man-versus-nature tale, what fascinated me was understanding why and when the jungle had become not only a front line of climate change but also a place of human contest. Criminality plagues many areas of Brazil, and the state presence remains feeble in the interior of the continent-sized country. But even within this context—which is no grimmer than in other vast resource-rich nations with weak institutions—the pace of homicides in Pará is staggering.

In the twenty-first century, more than 2,000 people have been killed worldwide for defending their lands or the environment—a murder rate on the rise, at least on record.[2] Brazil accounts for about a third of these homicides, with most of them in the Amazon. Pará sits at the top of the list of the most lethal states, and murder is just the most extreme tactic used to silence campaigners. Environmental and land defenders also suffer death threats and nonlethal physical attacks, and women activists face sexual assault.[3] The stories of these victims rarely reach international audiences, and when they do, it's because reporters retrace the lives of the victims and their dangerous careers as grassroots activists that exposed them to violence or death.

I wanted to get to the roots of these conflicts to capture, in a single narrative, the factors that have made the largest rainforest on Earth the world's most dangerous place for environmental and land activists. Pará, Brazil's second-largest state, covering a land mass almost three times that of California, seemed like an obvious area to investigate. In addition to being dangerous for campaigners, it is also a place where the rule of law is said to be plagued by corruption and impunity, imparting a sense of lawlessness in cattle, wood, mineral, and agrocommodities production on many of the state's frontiers. These areas are connected to the world through supply chains reaching markets as far away as China, Russia, Spain, and the United States.

My source suggested that I do my research in southeastern Pará, one of the most developed areas and home to tens of thousands of migrants who have moved from throughout Brazil to cash in on the riches of both natural resources and the seemingly endless availability of land. Marabá, a sprawling city built up along the Tocantins River, is the regional hub, where direct flights land from Brasília, but my source advised me to focus on a lesser-known town: Rondon do Pará, or Rondon, as people call it.

For decades, he said, Rondon had been home to criminal organizations that had plundered the region's rainforest, often resorting to hired assassins to silence opponents. He described the town, which is close to the border of Maranhão state, as a nest of gunmen, or *pistoleiros*, with ties not only to powerful rural elites but also to law enforcement and politicians in the Brazilian congress. At the top of this entwined and murky system of power, according to my source, were large landowners in the region. They were *"os donos do lugar"*—the masters of the place.

"It's a risky town for a reporter," the source told me, "but I think you should go and contact the people of the local rural workers' union. They have lots of stories to tell."

RONDON HAS A history typical of its time and place. In the late 1960s, when the Brazilian federal government decided to build a grid of highways penetrating the hinterlands, settlers moved into the lands of an Indigenous tribe, triggering clashes in what is today Rondon but was at the time a forest so dense and alive that, according to pioneers, a monkey could cross it from branch to branch without ever setting foot on the ground.

The tribe was removed by the government and transferred to a reservation, and colonists began to carve out gardens, farms, and cattle ranches. Pioneers recalled arriving in the area on foot and settling in the absolute wilderness, often near streams and rivers. As they waited for their new gardens to provide a first harvest of rice and manioc, they survived on a diet of river turtle eggs, fish, and roasted armadillos. Some died from malaria or snakebite; others aged rapidly due to the great hardships they endured, their hands wrinkled and bony from clearing areas and building wooden cabins with rusty axes. The extreme loneliness of the settlers was eased only by the sense of opportunity.

As the colonization process progressed, with trails hacked through the wilderness, word spread that the region was open for development. Hordes of migrants and entire families of sharecroppers and landless peasants from Maranhão, one of Brazil's poorest states, flocked into the region to claim a plot. Wealthy ranchers, dealers in wood, and land sharks also moved in, traveling in small planes that flew for hours over a carpet of green before landing on tiny and bumpy airstrips. Covered in sand and dotted with weeds, these were located near Marabá or Paragominas, another outpost that had developed earlier than Rondon. The men in the planes came from rich states in the southeast of Brazil like Minas Gerais or Espírito Santo. Soon these rural entrepreneurs, often referred to as *fazendeiros*, formed a rural elite influencing politics and shaping the future development of Pará.

Although the Amazon is often imagined as a natural continent of expansive fertility—around 10 percent of all living species on Earth are found there, and the basin, sustained by the mighty Amazon River and over a thousand tributaries, holds about a fifth of the planet's total supply of freshwater—Rondon isn't pristine.[4] The early roads penetrating the interior allowed settlers to bring in power tools and undertake a frantic process of logging and clearing, especially during the 1970s and 1980s, when Rondon grew and prospered as a logging boomtown before transitioning to agribusiness. The expanding human presence could be seen in photographs taken by satellites. In these aerial images, the human footprint reaching into the wilderness resembled the skeleton of a fish, with state and federal roads forming the spine and thin trails opened by pioneers as the ribs.

Today, these fishbone patterns are visible on the jungle's most pristine frontiers, mostly in the central area and on the western border of the Brazilian Amazon, but they no longer exist in Rondon.[5] Most of the original vegetation has been either burned or otherwise altered, and I witnessed this during my first journey. A smooth two-hour drive on a decent asphalt highway connecting Rondon to Marabá exposed a landscape dominated by rolling pastures and a growing number of soybean fields. Some islands of towering vegetation still abut farmland, but those scattered patches of forest were the object of logging. Even most of the castanheira trees—a species of nut tree that had once provided critical income to gatherers of the nutritious Brazil nut—had been logged, despite a national ban instituted to protect this beautiful and massive tree that can live as long as four centuries.[6] While I had vaguely hoped to see a wild jaguar—a beast formerly so common in these forests that pioneers, unafraid, had even domesticated some specimens and treated them like pets—I was disappointed; the sole animal in sight was the humpbacked, floppy-eared, glossy white Nelore cow, the ultimate conqueror of the frontier.[7]

More disappointment stemming from early misconceptions awaited me in Rondon. Throughout my years as a Brazil-based journalist, I had visited many far-flung hamlets, and I had expected Rondon to be one such wild place. I had been in villages where local criminals were so infamous that heavily armed squads of federal officers sent to fight loggers or gold prospectors stayed in their hotels when they were off duty for fear of being attacked.[8] I'd visited outposts where environmental offenders had no qualms about vandalizing pickups or helicopters used by law enforcement for anti-deforestation raids and where envelopes full of bullets were delivered to visiting journalists in their hotel rooms to drive them away.[9]

But Rondon didn't resemble any of those backward spots. The city had a rural but modern atmosphere. The highway that had brought me there—the BR-222, formerly called the Nut Road due to the once-ubiquitous castanheira trees—went straight through the town, bisecting it from west to east. The southern part was more urbanized and populated despite its sloping terrain, as the development on the northern area was bounded by a river. Most streets were paved, although saffron-colored dust was everywhere. The architecture was simple and unpretentious. There were some large Pentecostal and Catholic churches and a few three- and four-story buildings, but the structures were mostly unsophisticated one-story cabins with white, yellow, green, and pink facades. Some were encircled by high walls overgrown with the shining leaves of mango trees.

The city had a small university campus and broadband internet service, cable TV, and 4G phone signal. Early in the morning and late in the afternoon, when the heat was less intense, the roadside and Rondon's central square, Praça da Paz or Square of Peace, bustled with activity. Supermarkets, furniture shops, pharmacies, and agricultural supply stores welcomed customers and advertised their promotions over loudspeakers. Smiling and chatty uniformed students flirted in cafeterias offering bowls of frozen açai berries. Pickup drivers refilled their mud-encrusted vehicles' tanks at gas stations. At night, the terraces of bars, pizzerias, and barbecue restaurants serving beef cuts from the nearby fazendas became overcrowded by families in shorts and flip-flops. They drank chilled beers as they listened to sertanejo, Brazil's answer to country music, and watched football matches on LCD televisions. People seemed relaxed, and I didn't see anyone armed or displaying an aggressive attitude. The only noticeable lawbreakers in that town lacking a single traffic light were motorbike drivers not wearing helmets.

This sense of joviality was also on display during the most important event of the year, the annual livestock convention and exposition, called ExpoRondon. It had been founded by the fazendeiros forty years prior and had grown to become a magnet for tourists from across southern Pará. Caravans of cars and trucks rushed to the city to enjoy a week-long program of festivities, including rodeo shows and concerts of renowned sertanejo artists. Hundreds of the town's residents put on their cowboy clothes to participate, on horseback, in a spectacular parade through town led by a large truck on which former mayor Shirley Cristina de Barros Malcher launched her speeches to the happy crowd. Although less exuberant and sensual, some of the scenes I witnessed during ExpoRondon reminded me of the Carnival festivities in Ipanema, the beachfront neighborhood where I lived in Rio de Janeiro. My research would soon lead me to scrutinize the story of Shirley Cristina and her father, Josélio de Barros Carneiro.

Life in the interior, access to which involved hours of travel along potholed dirt roads, was less showy. On deserted roads that cut into pastures and vegetation, I passed logging trucks with beds piled up with freshly cut trees, the vehicles driven by shirtless men wearing gold necklaces. I also encountered humble family farmers transporting milk, crops, and fruit on converted motorcycles and, occasionally, I saw fully equipped pickups with tinted windows, engines rumbling, sending up clouds of orange dust as they sped along.

In that hinterland of infinite blue sky, social inequalities were obvious. Peasant settlements were dominated by low-income communities where large families continued to live in the same rickety wooden huts built at the time of colonization, with hammocks serving as their beds. Some even lacked electricity, their inhabitants waking up at dawn to work the fields and going to bed early at night. They relied on their gardens to produce beans, manioc, rice, and fruit that fed entire families; they took away the surplus to Rondon to sell in markets. Their sun-weathered faces and shabby clothes were a testament to their hard work and lives, yet I could see pride and a sense of accomplishment in those people.

Not far from there lay soybean plantations and massive fazendas producing cattle. Long twisted strands of barbed wire and wooden fences paralleled the roads crisscrossing the interior, enclosing massive spreads that looked to me like insular, inaccessible places. They conveyed a sense of possibility, of a frontier still in the making. They also showed that, despite the Amazon's centuries of resistance to human domination, nature had finally bowed to the will of global capitalism. This dramatic shift in the jungle's history began in the 1960s, when a new development model was planned and implemented that, as a researcher wrote, valued "commodity production and large land holdings, and devalued nature, devalued forest . . . [and] the people who already live there."[10] The echoes of those momentous policies are still shaping the Amazon.

ONE MORNING, FOLLOWING my source's lead, I drove to the headquarters of the rural workers' union, known in Portuguese as Sindicato dos Trabalhadores e Trabalhadoras Rurais de Rondon do Pará. It was an old and rather battered two-story building located downtown, with a peeling turquoise facade and metal security bars over the windows. I entered and met a female receptionist, who immediately recognized my foreign accent and asked, even before I could say my name, if I was the reporter who had been calling over the last few days. I said I was, and she asked me to sit. The room where I waited was plastered with photographs of union members working in bountiful gardens. Slogans on posters urged peasants to produce in a sustainable, pesticide-free fashion and to respect the forest.

Fifteen minutes of silence followed. Then I saw a black pickup pull up by the entrance. A man and two bespectacled women, one visibly younger, stepped down from the vehicle and entered the Sindicato. They all looked at me, the man scrutinizing me closely, and the eldest of the two women walked over. In her fifties, she was short and slightly overweight, her black hair pulled back in a ponytail. She had dark-brown eyes and wore a red blouse and a knee-length cream-colored skirt.

"*Bom dia,*" she said, speaking slowly while she softly shook my hand. "I am Maria Joel Dias da Costa."

We were ushered into a large, empty room. Sunbeams filtered through the window blinds, but the room remained poorly illuminated. The man, I was informed, was a police officer who escorted Maria Joel around the clock. He remained outside by the door, which was left ajar. I sat face-to-face with Maria Joel at a small wooden table, and the other woman, who had introduced herself as Joélima, Maria Joel's middle daughter, sat by her mother's side. As I opened my notebook and checked the batteries of my voice recorder, Maria Joel lethargically mopped a faint dew of sweat from her brow and then asked me in her tiny, barely audible voice what I wanted to know about her. Some in Rondon considered her a heroine, and I had expected to hear a testimony of unflinching persistence and success, but the first impression I had was one of gravity. I could tell she was preparing to deliver an account that would dig up unpleasant memories for her. It was noon on May 29, 2017, and though I was unaware of it at the time, this tranquil, fragile-seeming woman was one of the most courageous individuals I would ever meet. Perhaps I had been expecting to hear a romantic and hokey narrative of heroes and villains, but hers, in truth, was a testimony of change, adaptation, and survival.

Maria Joel and I would repeat that scene many times in the coming years, and her story had a real impact on my view of the Amazon and the grassroots movements that campaign for both environmental protection and social justice. While I had wanted to tell a story explaining why the rainforest is today a profoundly contested place, Maria Joel's account would lead me to consider the personal price that individuals are compelled to pay to defend important causes.

Soon after I began working on this book, I realized that the success of the project would rely substantially on Maria Joel's ability to remain alive and continue with our interviews. Her memories were pivotal to the immersive story I envisioned writing. The facts and the historical context were there, waiting to be unearthed in official documents, court files, press clippings, and eyewitness accounts. But if Maria Joel were murdered—a real possibility, considering that her family had been facing death threats for over a quarter of a century—I might never obtain critical details of her saga.

Later, I would understand why Maria Joel had agreed to grant me access to her life and her memories. She wanted to leave a detailed account as a legacy—a way to prevent her family's story from fading into oblivion. That had been the greatest mission of her life. Like the history of Rondon itself, her family saga involved many twists and turns. I made clear to Maria Joel that I'd be pursuing the facts and would not allow any personal empathy I might develop toward her or her children to sway me in telling that story. She agreed.

Ultimately, the greatest challenge I encountered while piecing together her story and that of Rondon was corroborating facts and navigating the waters of propaganda. In that town, two versions of history coexist, often in conflict—a public one, in which pioneers had fostered progress and civilization, and another one, often told in undertones, which depicted those same pioneers as feudal lords threatening the foundations of rule of law and democracy.

In the end, to write this book, I would interview two hundred people, from judges and fazendeiros to Netflix celebrities and top federal and state officials. I would scour a dozen archives scattered throughout five Brazilian states, Europe, and the United States. This investigation would involve research into issues from the historical past to the crucial years of Brazil's contemporary history, right up to the very present. Because this story, which begins back in the 1960s, still isn't fully settled as of this writing in 2022.

While undertaking this research, I often thought about a statement by the American reporter and author Barbara Demick, who has written books on places as difficult to access as Tibet and North Korea: "Journalists are contrarian creatures. If we are told we cannot go somewhere, then predictably we try to go."[11] When controversial eyewitnesses closed their doors to me, I felt myself bewitched by curiosity and some sense of duty to try to achieve "the best version of the truth you could come up with," as Carl Bernstein has described the essence of investigative journalism.[12] Ultimately, I devoted four years to writing a story that I believe epitomizes why the Amazon, a strategic region in our era of climate crisis, has been blackened over the last half century by the dual flames of fire and lawlessness.

Part I

It's All About Land

1

The Escape

SINCE first setting foot on the Te-Chaga-U ranch, Gil Bonifácio Carvalho Neto had felt a growing sense of dread—but it was only after uncovering a hidden clearing in the jungle that he began to truly fear for his life. The slender twenty-five-year-old laborer had pushed through the brambly vegetation to investigate a strange cloud of dark smoke rising above the canopy.[1] Expecting to find a brush fire, Gil had instead stumbled across an apparent killing field. He should never have discovered that place. But now he could never unsee it.

It was mid-1994, and everything had happened almost by chance. Gil was in a garden, harvesting manioc, when he was approached by Manuel, a cowhand at the sprawling estate where both men lived and worked.[2] Manuel was anxious, his tone marked by a kind of nervous exhilaration as he explained his sighting of a plume billowing above the nearby jungle. At that, Gil, his body sweaty from the searing heat, peered up to see the smoke rising through the treetops. He was hesitant to venture beyond the boundaries of his work; people murmured of strange occurrences on that fazenda tucked into the wilderness spanning the municipalities of Rondon do Pará and Paragominas.[3]

However, intrigued, Gil and Manuel finally decided to have a look. They zigzagged across the ranch, along pathways and dirt roads and past patches of jungle and grassy pastures occupied by Nelore cows.[4] As they approached the hidden bonfire, they became transfixed by the scene before their eyes. In the flames, Gil recognized fragments of charred rubber, steel cords, and plastic bags intermixed with what looked like scattered pieces of broken flutes. Shafts of morning sunlight through the thicket illuminated a massive tree toppled on a bed of leaves. Its bark and heartwood, scorched by the fire, had been carved away.

The lack of light made it difficult to decipher the contents of the embers. Gil and Manuel had likely expected to find some brush set ablaze to clear the land, a typical form of the slash-and-burn agriculture common in this region, where environmental destruction was approaching "the worst levels ever," as the press wrote in those years.[5] But Gil finally realized that the remains in the fire were human.[6] Immediately, he thought of Ceará, a laborer who had recently vanished from the fazenda.[7]

Ceará was a compact man in his twenties who, like Gil, had migrated to Pará in search of a better future.[8] The military government had launched a series of programs in the 1960s to colonize the Amazon, and since then the sparsely populated state of Pará had attracted waves of would-be settlers and fortune hunters. Ceará and Gil had initially chosen Paragominas—at the time described as "one of the world's logging capitals"—to pursue economic betterment.[9] The men had probably expected to find good-paying jobs in one of the sawmills springing up there, but then a seemingly better opportunity came up.

One day Gil met a recruiter named Chico. He worked for a cattleman named Josélio de Barros Carneiro, who was often in need of workers for his estates. Chico hired seasonal farmhands he found in the hotels of Paragominas, where transient migrants slept while looking for jobs.[10] In those dingy places, Chico offered employment on Josélio's ranches, and to entice laborers to accept the deal, he himself paid their unsettled bills at the inns.[11] The day Gil met Chico, the recruiter said that he wanted him to work on a 10,700-acre fazenda called Te-Chaga-U, which means "longing" in the Guarani language and is pronounced *Ti-shaga-u*.[12]

In early 1994, Gil was driven to the estate, located some 100 miles from Rondon, along an unpaved, rutted road punching into the jungle.[13] It was during that trip that Gil met Ceará. Soon thereafter, both laborers found themselves facing Josélio himself, a pale-skinned, five-foot-seven-inch man in his late fifties with intense blue eyes, chestnut-colored hair, and a deep voice.[14] He didn't live on the ranch, which was run by his overseers, but one of his instructions, according to Gil's account, was that they were not to leave the property without his approval. "Don't try to leave the fazenda," Josélio said, his accent revealing he hadn't been raised in Pará.[15] Apparently, Gil couldn't even go to nearby Rondon.[16] Although Josélio never mentioned how much they would be paid or when, Gil and Ceará worked clearing land for about two months. It was an exhausting, dangerous job, and it was carried out without protective gear. Sometime in April 1994, when both laborers were already unhappy about their situation, Ceará was called to have a talk with the boss.

As Gil stared at the embers near the log that day, he recalled having witnessed Ceará entering the fazenda's headquarters early in the morning, when other farmhands were having coffee before work. Chico, the recruiter, was also there. Moments later, Ceará had emerged from the white one-story house accompanied by Josélio and the managers of the Te-Chaga-U, a pair of menacing men nicknamed Souza and Rai, according to Gil.[17] The group, he said, had jumped into a vehicle and taken off toward the woods, apparently with the intention of repairing a remote fence. That was the last time Gil had seen Ceará. Now he believed that Ceará's body had been burned along with some old tires.[18]

"We can't talk about this to anyone," urged Manuel, the cowhand, who had also realized that there were human remains in the cinders. "If Josélio or his hitmen know that we have been here, they'll kill us."[19]

Gil recalled the grim accounts other employees had shared with him in confidence. Colleagues at the Te-Chaga-U said that Josélio was a murderer who had ordered many hits. "People were scared to death," one laborer would later say. "When they spotted Josélio coming to the ranch, employees ran to hide from his sight and pretend they were working."[20] Gil also remembered the words of the stocky, bewhiskered Souza. On one occasion, according to Gil, the overseer had said that "the most beautiful thing in life to watch is the fearful grimace of a laborer who is about to die."[21] Souza had also ordered Gil and several others to load a pickup with beat-up tires and said with a sneer, "Who knows if they will be used with you!"[22]

As he remembered some of these events, Gil understood that he was in great trouble.[23] In a panic, he struggled to figure out what he should do next. This place was cut off from the world, walled in by the rainforest and lacking any public transportation. Born to a poor family in Maranhão, the state bordering Pará to the east, Gil felt more vulnerable than ever before in his life.

He decided to seek help from Luiz Bezerra Cavalcante, a thirty-nine-year-old man with black hair and a slight build, and his wife, Sueny Feitosa Cavalcante.[24] Also migrants from Maranhão, they lived on the estate with their ten children. They weren't friends, but Gil had toiled in the fields with Luiz, and his wife was kind and sometimes invited him to join them for dinner in the shack that Josélio lent to the family. It was a tumbledown cabin without tap water, electricity, or beds, located within the ranch's boundaries.

Moving quickly away both from the apparent remains of Ceará and from Manuel, whom he didn't fully trust, Gil headed toward the Cavalcante home.

AFTER A HIKE away from the fire, Gil reached the Cavalcante home and told them what he had just seen. Oddly, Luiz and Sueny, who had been living on the ranch for almost four years, didn't look entirely shocked at the story. "You know, we've heard so many things about him," Luiz admitted, his wife nodding silently.[25]

Luiz and Sueny had heard accounts indicating that other workers like Ceará had vanished after meetings with Josélio, their bodies apparently burned up along with tires.[26] The wife of an employee had once told Sueny that she had seen a corpse being transported in the trunk of a Volkswagen owned by Josélio's family. When Sueny had asked around to learn more, Souza had approached her and "threatened to burn her tongue and that of her husband if she mentioned that issue [with people] in the fazenda."[27]

Luiz had also heard whispers that gunmen allegedly employed by

Josélio had executed rival ranchers in nearby towns, but he had no proof of this.[28] Some stories were published by the press, but Josélio rebuffed them, accusing his foes of "fake news." "I vehemently deny my name being mentioned in the context of such a sordid crime," he wrote to the *Diário of Pará* in 1990, in response to an article stating that Josélio hid pistoleiros on his estates.[29] "For forty years, I've been a rural entrepreneur, and I've spent half of this time in Pará, where I carry out large-scale stockbreeding and agriculture." Josélio liked to be called a fazendeiro, the Portuguese word for a rancher or an owner of a fazenda.

What Luiz and Sueny did know for a fact, because they had experienced it firsthand, were the harsh working conditions on the Te-Chaga-U. In four years, Luiz had "never bought a pair of trousers for himself or a dress for his wife" because of his limited access to money.[30] Initially he'd had a work contract and received some payment, but for two years now he claimed that he'd not seen a single cent. When he approached Josélio for payment, the answer was invariably that Luiz was "still indebted" because his wife had bought food and supplies on credit in the sole shop where the family could purchase them—a small grocery store within the fazenda's boundaries where, in Sueny's words, "the price of merchandise suddenly doubled from morning to night." Black beans, rice, and cooking oil were some of the few foodstuffs they could find. Sueny would later report that meat was unavailable to them, despite the fact that the ranch was home to large herds of cattle raised for profit.

"When Josélio gave us some beef, it was only the bones. He treated us like dogs," she would proclaim.[31]

After Sueny heard Gil's hair-raising story about Ceará, the feisty thirty-one-year-old woman decided to escape with her children. Both Luiz and Gil agreed, but the crucial question was how. The common-sense solution would be to simply announce to Josélio that they wanted to quit and be provided with transportation to Rondon. But Luiz knew from his own experience that walking off wasn't that simple. Josélio had thumbed his nose at the request every time he or his wife had attempted to visit the town. "To do what?" the fazendeiro used to reply, according to Luiz's account. "You've got everything you need right here. If you need something, you can buy it at the grocery store."[32] Not even when someone in their family needed medical treatment—for instance, when Sueny was pregnant and, in her own words, "really needed a checkup"—were they brought to town.[33] Occasionally, Luiz admitted, they had no option but "to disobey Josélio and leave the ranch for a few days, because someone in the family really needed to see a doctor" or because they had to deal with urgent paperwork. Then the problem became transportation. "People [in the area] refused to give us passage, because they knew we were workers of Josélio, and they were afraid of him," Luiz recalled.

The issue of how to escape the fazenda remained unresolved until a few days later, when Sueny saw an opportunity. Alone, she managed to jump on a passing car and traveled to Rondon. Once there, she didn't go to the police, visiting instead the rural workers' union, the Sindicato dos Trabalhadores Rurais de Rondon do Pará. She was met by the president of the union, José Dutra da Costa, a short, paunchy man in his late thirties with curly ink-colored hair and glittering black eyes. Everyone knew him by the nickname Dezinho (pronounced *De-sí-nio*). His wife, Maria Joel, was a shy and slim woman with beautiful black hair.

Founded in 1982, the Sindicato—as it was popularly called—was a small union that originally helped farmworkers and subsistence farmers obtain retirement pensions and other financial benefits from the government. Things had radically changed after the outspoken Dezinho was elected president in 1993. He had made it clear that his objective would be transforming the Sindicato into an organization advocating for agrarian reform and defending the rights of rural workers in Rondon. He had two priorities: eradicating debt bondage, or "slave labor," as he called it, and fighting an array of land-grabbing schemes called *grilagem* in Portuguese. For this, Dezinho had established personal links with members of the Federation of Agricultural Workers of the State of Pará (FETAGRI), with which the Sindicato was affiliated. With its headquarters in Belém, the state capital, FETAGRI was a larger peasant association with privileged access to politicians of the left-leaning Workers' Party, or Partido dos Trabalhadores. Founded by Luiz Inácio Lula da Silva, the Workers' Party was rapidly expanding its influence across Brazil.

Dezinho's involvement in the movement had made him a marked man. The same year he became president of the Sindicato, his name appeared on a list of "people to be eliminated" that apparently circulated among Rondon's underworld and was published by the press.[34] Without delay, Dezinho and Maria Joel enacted draconian security measures to prevent a gunman from killing any of their four children: three girls and one boy aged five to eleven. Their operating principle was that their home was the safest place, and thus the children were outright forbidden, almost overnight, from participating in activities like riding bikes in the neighborhood or swimming in one of the town's nearby streams, where other kids enjoyed jumping into the water from overhanging palms. Their children protested the strict limitations to their social lives, but Maria Joel, who worried about the consequences of her husband's activism, ignored her children's complaints.

When Sueny sat with Dezinho at the Sindicato's office and revealed her address, the union leader's head immediately shot up at the familiar name. "How is life in the Te-Chaga-U fazenda?" he asked, looking Sueny straight in the eyes.

Sueny understood that Dezinho, without saying it straight out, was signaling to her that he knew what happened on the ranch and wanted to know more. By that time, the activist already believed that Josélio had committed murders that had never been investigated because one of Josélio's daughters, Josélia Leontina de Barros Lopes, was a prosecutor in the Rondon do Pará district attorney's office.[35] Ambitious and courageous, Dezinho was eager to confront the rancher, but he had never been able to gather enough reliable evidence to bring him to justice. That day, Sueny trusted Dezinho and told him everything about the plight her family had been facing, also mentioning Gil and his unsettling discovery.

The account led Dezinho to believe that he had at last found his opportunity to prosecute Josélio. "We can help you leave this place," the activist replied, promising Sueny to find a safe place for her family and for Gil once they had managed to flee the fazenda. Dezinho also said that, once out, they would have support from the Sindicato to rebuild their lives. Sueny agreed and returned to the Te-Chaga-U with the news.

About a month after Gil had found the human remains, Sueny made her move.[36] The family persuaded one of the ranch's drivers to transport Sueny and the children to Rondon during one of Josélio's absences. The man was deeply reluctant, but eventually they managed to convince him. The plan was that Luiz would leave afterward, once he had "paid with work the debts owed to Josélio," because, in Sueny's words, "no one left the Te-Chaga-U without paying the debts."

The cargo bed of a pickup was packed with the ten kids as well as some of the belongings they had gathered over the years—pots, blankets, clothes, shoes. The truck slowly pulled away over the dirt road, engulfing its passengers in a plume of dust. Sitting in the passenger seat, Sueny wished Luiz, the man she had married at sixteen, had been with her, aware that there was a real chance she would never see him again.

He won't get out of there alive, she thought as the familiar sights of the Te-Chaga-U slowly disappeared behind them.

SUENY'S FEARS OF never reuniting with Luiz were not unfounded. During this period, international organizations and the media painted a gloomy picture of what awaited migrant laborers in that area of the Amazon. The Inter-American Commission on Human Rights, a body of the Organization of American States (OAS) that monitors rights in the American hemisphere, described Brazil's hinterland as struck by chronic violations of human rights, like "semi-slavery" and "impunity and judicial inaction."[37] In a comprehensive report, the commission wrote that farmworkers in Pará were exploited, enduring "conditions of servitude," and couldn't "leave the hacienda without paying those debts. When at times they attempt to do so, the contractors' hatchet men stop them with the help of firearms, which they will shoot if the threat does not suffice. Since most of the haciendas are in isolated locations, such attempts to get away are fraught with difficulties and danger."

Despite this, Luiz and Gil did manage to leave the Te-Chaga-U. Never mentioning Ceará or the burned body (or *bodies*, because at that point it wasn't clear how many there were in the woods), Luiz approached Josélio and said he wanted to "move on" with his life. After much discussion, the boss agreed, but only once Luiz had settled his debts, which was accomplished some weeks later.[38] Afterward, Luiz reached Rondon with the help of an acquaintance who picked him up at a specific location near the ranch. Gil accompanied him, and they joined Sueny and the children in a cabin on the outskirts of town. The context of that departure and the reasons it unfolded without further trouble would be revealed only later.

Having managed to escape without raising suspicions, Gil and Luiz went to meet Dezinho. If for the beleaguered laborers, the escape was the end of a painful experience, for the activist, it was the beginning of a possible legal crusade against Josélio, so Dezinho devoted his attention to getting full testimonies from the farmhands. Both men revealed all they had witnessed and said they would be willing to testify in court. Gil was so determined to expose what he believed was the homicide of Ceará that he even agreed to personally lead the police to unearth the human remains.

Also participating in the meeting was Dezinho's special advisor, José Soares de Brito. Brito was a grassroots activist who had previously attempted—and failed—to be elected Rondon's mayor as a candidate of the Workers' Party. Brito was also a central figure in the ongoing transformation of the Sindicato, and, for this, he was a marked man. Some years prior, his house had been the target of arson while he, his wife, and their son were sleeping inside. When he'd reported the crime, the town's chief of police had replied that he "lacked the means" to investigate.[39] Officers, the press wrote, had refused to collect evidence from the burned cabin or to interrogate a single suspect.[40] Brito attributed that attack to Josélio and other local ranchers opposing the Sindicato. The fazendeiro denied any wrongdoing.

After hearing the account of life on the Te-Chaga-U from Gil and Luiz, Dezinho contacted colleagues at FETAGRI. With their help, he reached José Geraldo Torres da Silva, a member of the state legislature of Pará known as Zé Geraldo.[41] A former smallholder, the burly man with a husky voice and meaty hands was a member of the Workers' Party and a friend to the causes of peasants. He agreed to help and, soon afterward, met with the state's secretary of public security, Paulo Sette Câmara. Zé Geraldo explained that escaped eyewitnesses had revealed the existence of debt bondage, multiple death threats, and at least one murder on a fazenda near Rondon. He urged Secretary Câmara to take measures and send Belém-based officers to raid the Te-Chaga-U without delay.

The possibility of Pará's top cop taking swift action quickly faded. Secretary Câmara, Zé Geraldo would recall, was not that enthusiastic about the idea of sending in the police. "Everyone talks about clandestine cemeteries in estates [of Pará], but nobody has proof of it," Zé Geraldo recalls him saying at their first meeting.[42]

Regardless of Secretary Câmara's lukewarm response to the case, the wider context didn't produce the sense of urgency the Sindicato had expected. At the time, Brazilian civil servants tended not to consider debt bondage a criminal offense, much less a human rights abuse. The International Labor Organization (ILO), a United Nations agency that monitors work standards across the planet, would write that as late as in 1992, and despite the evidence confirming multiple cases in Pará and other Amazonian states, "the representative of the Government of Brazil denied that forced labor existed in the country, stating that the cases mentioned [by civil society at an ILO conference] merely constituted violations of labor legislation."[43]

The ingrained mentality of outright denial seemed to be founded in the understanding of debt bondage as part of a system of exchange, credit, and dependence known as *aviamento*.[44] Although the foundations of this system date back to when Brazil was a colony of Portugal, aviamento had reached its height in the Amazon during the rubber boom of the late nineteenth and early twentieth century, when tappers scattered throughout the jungle supplied export houses with latex—the basis of wild rubber, then a strategic raw material for a world undergoing an industrial revolution.[45] In a context marked by profound isolation and remoteness, tappers exchanged their produce—extracted from *Hevea brasiliensis,* or rubber trees— with bosses, patrons, and passing traders who sold them food and supplies "at grossly inflated prices," thus forcing them to live in a perpetual state of indebtedness.[46]

Human rights organizations maintained, however, that behind contemporary debt bondage was the shadow of Brazil's enduring legacy as the last country in all the Americas to abolish the slave system in 1888.[47] "Slavery remains a current practice in Brazil, 500 years after Columbus," decried the London-based group Anti-Slavery International.[48] Estimates indicated that at least 18,000 people nationwide were subjected to slave-like working conditions at the time, although the number was likely much higher because the abuses were largely underreported.[49] About 80 percent of all known complaints originated in Pará, and the state also accounted for the highest number of murders associated with disputes over the control of land, especially as a consequence of land-grabbing schemes that made fazendeiros, smallholders, migrant peasants, and Indigenous tribes jockey for vast areas.[50]

By the time Zé Geraldo approached Secretary Câmara, some things had nevertheless begun to improve, thanks to President Fernando Henrique Cardoso and his policies. "The existence of slave labor in the country was officially recognized," according to the ILO, and tougher legislation was passed to ban the practice, although the UN agency accused civil servants of pursuing only a halfhearted campaign to punish the offenders and make the law prevail.[51]

This complex social and political context explained why Secretary Câmara initially didn't pay much attention to the case of the Te-Chaga-U. The drama that Zé Geraldo described to him was too common to justify deploying special agents to the interior. However, Zé Geraldo rethought his strategy and realized that one element unexpectedly played in his favor: the fact that a daughter of Josélio, the twenty-nine-year-old Josélia Leontina, had been the district attorney in Rondon since July of 1993. She was also a shareholder of the

Te-Chaga-U ranch, which made the conflict of interest too blatant for the top authorities to ignore, Zé Geraldo argued.[52]

After some months of back-and-forth and a great deal of pressure, Secretary Câmara agreed to send a secretive fact-finding mission to the fazenda. The date he communicated his decision to Zé Geraldo—May 25, 1995—was by no means a coincidence.[53] The regional newspaper *A Provincia do Pará* published a story that day in which Josélio was accused as one of the suspected instigators in the recent double homicide of a fazendeiro and his bodyguard near Paragominas.[54] The motive appeared to be linked to Souza, the overseer of the Te-Chaga-U.[55]

A month later, on June 25, a group of five law enforcement agents finally traveled the 310 miles from Belém to Rondon in a couple of nondescript white vans. At the head of the operation to find whatever remained of the alleged murder site found by Gil a year earlier was detective João Nazareno Nascimento Moraes, an experienced agent who had already heard of Josélio's bad reputation.[56] Zé Geraldo, members of FETAGRI, and an activist lawyer also traveled with them to ensure that the mission was carried out properly and to a satisfactory end, because cases often went cold after the police received bribes.

Sueny, Luiz, and Gil had already rebuilt their lives but had remained in the region, the two laborers making a living as construction workers. When Moraes approached them, it took some time to convince them to participate in the operation, no matter what they had previously said to Dezinho. "Luiz only agreed the day after we arrived, in the afternoon," recalled Zé Geraldo. "He said: 'We leave from here with you all, and the very same day, because if we stay a single night in Rondon, my entire family will die.'" The state legislator gave his word that he would make sure this happened. It was agreed that Gil would join the task force going to the Te-Chaga-U.

In the early morning of June 27, a day and a half after law enforcement agents arrived in the town, the operation was launched.[57] Gil was a bundle of nerves when the two vans carrying the police officers, Zé Geraldo, and Dezinho, who had joined the group, entered the boundaries of the fazenda. The ski mask he wore to conceal his identity was soaked in sweat, despite the early morning chill.[58]

"There," Gil announced once he had identified the area where he had seen the fire. Agents holding sawed-off shotguns followed Gil's directions to reach the spot—a wooded area a hundred meters or so from the road they were driving on and where the light at dawn struggled to penetrate the wilderness.[59] Through the eyeholes in his balaclava, Gil immediately recognized the bulky overturned tree trunk, which helped him to identify the place where he had seen the rubber and the bones.

Detective Moraes ordered his men to dig. Without much effort, the police confirmed Gil's allegations. Officers unearthed carbonized rubber and plastic bags, bone fragments, and about six kilograms of circular bands of metal, steel cords from inside the tires that had survived the flames.[60] The agents piled the evidence into a large white plastic bag similar to a human remains pouch and sealed it.[61] The air filled with tension.

"That's enough," Moraes said, stepping back, sure that the findings would be sufficient to file for an arrest warrant from Rondon's judge. Now he took a moment to weigh his options. Should he attempt to arrest Josélio on the fazenda? Previous police operations had confirmed that, although the sale of guns in the country was restricted and, overall, banned to civilians, estates in that area of Pará were filled with weapons, often provided by mobsters or corrupt police officers.[62] Dezinho and Brito would later claim to have witnessed Josélio being supplied with arms and ammunition at Rondon's police station.[63]

After a few minutes of hesitation, Detective Moraes decided to visit the ranch's headquarters. Once there, he found only some workers, who told the police that Josélio had been away for "about sixty days."[64] Zé Geraldo suspected that despite his team's precautions, the fazendeiro must have been informed of the mission as it was being planned in Belém and fled the scene.

When pressed, ten farmworkers told the group of their living conditions. One stated that he and his wife had been employed for a year but had never received payment. "The 'law' imposed by the boss to his employees is that they absolutely can't leave the ranch unless they complete a year [of work]," the man declared.[65] Another explained that he had been working for four years but had received only food and shelter in return, and a cowhand claimed he had done unpaid work for nine months.[66]

"You are free to go," Moraes told the ranch's staff. "You are no longer forced to remain here. I'll send a vehicle tomorrow to pick up any of you who want to leave."[67] None of them accepted the offer. According to Zé Geraldo, they were too poor to flee without a viable alternative. Luiz, Sueny, and Gil provided statements, Gil signing the police transcription of his deposition with his thumbprint, because he was illiterate.

Back in Rondon, Moraes asked the local judge, Ana Lúcia Bentes Lynch, to immediately issue a warrant to arrest Josélio wherever he was caught and thus put an end to the "savagery" found on the ranch, the detective wrote.[68] (A scientific analysis of the evidence collected at the scene later confirmed the presence of, among other contents, carbonized vertebrae, shoulder blades, ribs, tibia, clavicle, molars, and skull fragments belonging to at least two adults.[69]) But Judge Lynch disagreed, denying Moraes's requests, arguing that the probe wasn't airtight and lacked "elements proving the existence of the crime."[70] According to the judge, there was no "nexus . . . between the disappearance of Ceará and the human bones found in the fazenda."[71] Only days after unearthing the remains, the case began to collapse.

In response, the Sindicato and FETAGRI turned to the media to pressure the authorities. Regional newspapers published a series of articles denouncing the discovery of both "enslaved immigrant workers" and a "clandestine cemetery" on the Te-Chaga-U.[72] Gil and Luiz posed for photographs with their faces hidden, and the story was picked up by the influential *Folha de S. Paulo*, a paper read daily by the political and economic elite.[73] Far from an isolated case, the media speculated on the possibility that as many as twenty ranches in the interior of Rondon may each have a "clandestine cemetery" to dispose of bodies of murdered farmworkers.[74]

As the public pressure mounted, Secretary Câmara announced that he would do all that was necessary to punish the criminals.[75] Coincidentally, President Cardoso spoke about slave labor the same day that the remains were found in Josélio's fazenda. Although he probably wasn't aware of the raid and was referring to the problem in general terms, Cardoso announced to the nation that he would set up a special task force to combat modern slavery across the country.[76] "This needs to stop," he declared in his weekly radio address *Palavra do Presidente* ("Word from the President"), aired on June 27.[77]

Two days later, the Barros family publicly responded to the accusations. Another of Josélio's daughters, Shirley Cristina de Barros, also a lawyer and shareholder of the Te-Chaga-U, sent a fax to the press decrying the charges against her father as "untrue and totally biased."[78] She said that the bones were part of a graveyard built by the previous owner, a man who had buried six family members there prior to Josélio's acquisition of the land. In her narrative, the scandal was simply an error on the part of the police or, worse, a setup to justify the Workers' Party's "pure persecution" of her family. The operation, she argued, was part of a larger plan to force the authorities to seize the fazenda, break it up into smaller plots, and redistribute the land among landless families with links to the Sindicato. In short, everything was the result of a pernicious politically motivated propaganda campaign.

With Josélio's whereabouts still unknown, Moraes returned to Rondon to investigate the story of the preexisting graveyard. He found Alfonso Dias Soares, the former owner, and the man admitted that he had buried his family there but denied that the evidence collected was the bones of his kin. "My relatives were buried seven palms deep," he said.[79] (A palm is about 22 centimeters and is a unit of measurement used in Brazil.) Eventually, an officer would confirm his narrative by finding the cemetery 300 meters away from the suspected killing field.[80]

More frightening details on the Te-Chaga-U and its owner were disclosed by Dias. Asked by the police why he had sold the land to him, the former owner replied, "I feared that my family and I would be killed. . . . Almost all the landholders in that area sold their properties to Josélio out of fear, because we knew that he had the custom of sawing in the middle [cutting in half] with a chainsaw the peasants who displeased him by not selling the plots."[81]

Josélio emerged to defend himself on July 17, about three weeks after the raid.[82] He reappeared in the coastal state of Alagoas, some 1,200 miles east of Rondon, to claim that he was innocent and deny his involvement in any homicides. In a long statement to the police, he repeated the story that the human remains were of Dias's relatives, but weeks later, he altered his testimony, arguing instead that the bones might have been dumped on his fazenda by "third parties" traveling on roads across his properties.[83] He never fully explained this theory, but the point was that anyone but him could have participated in the murder and the cremation. His daughter Shirley Cristina also attempted to downplay the importance of the issue by declaring to the press, "Human remains can be found anywhere around here."[84]

In the meantime, Josélio's attorney worked to build a defense strategy. In her motions, she depicted Josélio as an exemplary rural entrepreneur who, having been born in Espírito Santo, a coastal state bordering Rio de Janeiro to the north, had settled on the frontier to dedicate "his life to his job."[85] She filed a series of certificates and complimentary statements issued by the police of Pará and by several ranchers' and loggers' associations and also a petition signed by dozens of townspeople in which ordinary citizens asserted, apparently spontaneously, to be "astonished by so much injury, defamation, and calumny" against the Barroses.[86] The accusations of debt bondage were also rebuffed by the statements of current Te-Chaga-U workers, who, in the presence of Shirley Cristina, claimed to enjoy salaries, paid holidays, healthcare benefits, and even commissions on the cattle business.[87] Luiz, Josélio's attorney said, had received his salary, and Gil was indeed a farmhand hired through Chico, the recruiter, but Josélio had never "met him personally."[88]

The Rondon do Pará district attorney's office—where Josélio's daughter worked, though she never signed any official document relating to the case—backed the judge's ruling that human remains were not a sufficient reason to press charges. "The fact that some carbonized human bones were found on this property isn't proof that the person [Ceará] may have been the victim of a crime," reasoned a prosecutor.[89] Moraes continued to investigate for months to try to shore up the probe, although he never managed to demonstrate that the remains were Ceará's, failing to even uncover the laborer's full name. (Ceará probably referred to the man's hometown—the northeastern state of Ceará—as in Amazonian hamlets where migrants were continually coming and going and laborers often called one another by nicknames referring to their home states.)

Ultimately, the crucial piece of evidence—the charred bones—were of no help for Moraes. The evidence was so fragmentary and carbonized that the forensic experts could not gather data such as cause and manner of death.[90] Angered by the lack of support to prosecute Josélio even though human bones were found and testimonies indicated the existence of debt bondage and murders, Moraes began to criticize the "multiple attempts to impede the investigation."[91]

"The freedom of these men," the detective wrote in a report, referring to Josélio and his overseers, "discredits the public institutions responsible for the rule of law and social order."[92] Moraes even quoted Socrates—"Injustice can never be more profitable than justice"—to try to convince the authorities, but to no avail. In 1997, the case was close to dead when the investigation was halted; in May 1999, it was officially shelved due to lack of evidence.

"The indictee Josélio de Barros Carneiro can't remain ad eternum with the issue looming over his head like the sword of Damocles," justified the judge.[93] The district prosecutor supported the motion to dismiss.[94]

By then, Gil, Luiz, and Sueny were no longer living in Rondon. They had been removed from the town immediately following the raid. Luiz later claimed he'd been right about the great risk to his family. Days after moving out, a relative informed him of gunmen lurking around the home where the family had originally planned to stay. Luiz and Sueny ended up spending the subsequent ten years hiding in a remote plot deep in the jungle. Gil, who stayed in touch with Luiz and FETAGRI for a while, eventually lost contact.

The outcome for the organizers of the operation was also bleak. Zé Geraldo claimed to have been hunted for two years by pistoleiros who "showed his picture routinely" to neighbors near his farm in Pará.

According to him, this form of persecution was in retaliation for his crucial role in the case. A FETAGRI activist who had participated in the mission "received multiple death threats after the raid and ran away to Europe, hiding for three months in Italy."[95] Brito and Dezinho, who refused to leave the town, would henceforth face great hardships. Still, the police operation would be a momentous time for the Sindicato and its confrontational strategy for dealing with the local landed class.

But before examining what happened after the events of 1995 and how they changed the lives of both Dezinho and Josélio—considered by many to be the leaders of two opposing local movements embodied by landless peasants and fazendeiros—we must go back in time and answer two crucial questions that haven't been fully addressed during the unfolding of the case: Who was Josélio de Barros, and why was he living in that swath of jungle?

2

The Criminal Syndicate

AT about eleven a.m. on February 26, 1967, almost thirty years before the Te-Chaga-U raid, three young men with pencil mustaches approached a downtown bar in the seaside town of Nova Almeida, a resort destination in Espírito Santo. They were armed and prepared to commit a murder that would have irreversible consequences for Josélio, then a good-looking thirty-two-year-old man who helped his father, José de Barros, with his cattle business.[1]

It was Sunday, and summer was at its peak in the southern hemisphere. The tiny town of Nova Almeida, founded in the sixteenth century by Jesuit priests, was bustling as the mustached men prepared to launch a furious attack on the bar.[2] Families in flip-flops walked leisurely while shirtless, noisy children ate ice cream and ran around on the sidewalks. The Praia Grande, or Big Beach—located less than a mile from the bar and famous for its flat, hot waters—was packed with couples drinking chilled coconut water and tanning under the tropical sun.

The tourists unnerved Josélio, the mastermind of the assault. He had asked the three men to "take all precautions" to avoid innocent victims in the murderous undertaking about to begin—the assassination of Orlando Cavalcanti da Silva, a major in the Espíritu Santo police force.[3]

Two of the hired assassins, Fausto Ferreira dos Santos and Antônio Gregório da Silva, who went by Toninho, were skinny and short.[4] The third, Alvaristo Vicente, was slender and a head taller than the others. All three had worked for Josélio previously, according to the media.[5]

The three pistoleiros positioned themselves near the entrance to the bar. Fausto and Toninho intently scrutinized the inside of it. Each had two revolvers, one in either hand, to anticipate the moves of Cavalcanti, a thirty-six-year-old officer and father of five.[6] Alvaristo also held two revolvers, but he had remained a bit behind and wasn't expected to use the weapons unless it was strictly necessary. Having worked on Josélio's fazenda for some time, he was trusted by the rancher, who was monitoring the operation from a pharmacy near the bar. Some days prior, Josélio had bribed the druggist to lace bullets with cyanide, said to be a common practice in Espírito Santo, where old revolvers didn't always deliver the precision and power expected.[7] Josélio had also ordered Fausto to shadow Cavalcanti for some days in order to prepare the attack. Initially they were going to ambush him in a small town where Cavalcanti frequently met a lover, catching him while he caroused with the woman, but the officer had later changed his plans, and Nova Almeida offered the best chance to kill him.[8]

Inside the bar, Cavalcanti was dressed casually and chatting with friends. Trained as a law enforcement agent in Rio de Janeiro's police academy, he carried a .32-caliber revolver on his hip.[9] He was widely known in the rural northwest of Espírito Santo, an area bordering Minas Gerais dotted with small, family-run coffee plantations nestled in the wooded hills. Press reports described this as one of the most lawless areas of the country, where factions of strongmen, or coronéis, used corruption and political violence to impose their will.[10] "The law was that of the gun," wrote the media.[11] According to historians and journalists, each faction was composed of fazendeiros, politicians, police, and bands of hired assassins.[12] Extreme polarization and deep-seated clientelism prevailed, and in the previous two decades, hundreds, if not thousands, had died or fled that region, known as the Contestado, meaning "disputed."

Cavalcanti was talking with the bar's cashier that morning when someone screamed, "Sai do meio!"—"Get out of sight!"—followed by a furious firefight that caused chaos among the customers.[13] According to the conflicting versions provided by the press, Fausto and Toninho shot Cavalcanti with twelve to fifteen bullets.[14] Alvaristo never fired his gun because the tall man brought up the rear and, as previously planned, would shoot only if something went wrong during the lightning assault. But nothing really did. One of the bullets, apparently the first shot by Fausto, penetrated the victim's mouth, and all the others hit Cavalcanti above the waist.[15] Still, the major did manage to exit the bar and grab Fausto, who was pulled into Toninho's line of fire and was hit in the arm.[16]

When Cavalcanti finally dropped, the three hitmen—"calm, shoving people aside," according to media reports—walked out toward a dark-blue jeep with a black top and no license plate.[17] They reloaded their revolvers during their short walk to the vehicle, where Josélio was at the wheel, and then they departed toward Josélio's hometown, the hamlet of Baixo Guandu.[18]

Located in the far west of central Espírito Santo, Baixo Guandu sits adjacent to the small city of Aimorés, which is just a few miles away but is located on the Minas Gerais side of the state borderline. The towns border each other and are both on the shores of the Rio Doce (Sweet River), a major waterway in the region. The Sweet River headwaters flow north from the hilly, mineral-rich region of southeastern Minas Gerais before its course follows a half-circle bend and continues east, toward Espírito Santo, where its lower path zigzags through a deep valley of rolling green hills—the Vale do Rio Doce (Sweet River Valley), a hot spot of social and political conflicts in the Contestado.

At around 3:00 p.m., the party arrived at the ranch belonging to Josélio's father, who produced livestock and coffee with the aid of sharecropper families.[19] According to the press, he had "resolved for his son . . . to be the organizer of the ambush."[20] An employee of the fazenda cleaned off the bloodstains left on the jeep by Fausto's wounds, and the injured pistoleiro received rudimentary medical treatment and changed his blood-soaked clothes before they began the second phase of their escape toward Minas Gerais.[21]

With Josélio in command, they switched jeeps and drove toward Belo Horizonte, the capital city of Minas Gerais, located some 260 miles west. As soon as they reached Aimorés, only a few miles west of Josélio's ranch, the criminals might well have relaxed, because they had just crossed state lines and entered a new jurisdiction, thus lessening the possibility of a quick arrest by Espírito Santo's police, who were already investigating the homicide. After several hours of driving, Josélio and the three killers reached Itabira, a hamlet where an acquaintance removed the bullet in Fausto's arm.[22] They hid there for four days and left in a different car. Driven by a sergeant of the Minas Gerais army police who was part of the plot, they covered the remaining 70 miles to Belo Horizonte smoothly.[23]

In the state capital, the group disbanded. Josélio remained for a few days while the three hitmen continued toward Uberlândia, a town some 300 miles farther west. According to the press, Josélio and his men had agreed that in the following weeks they might meet in one of the southern states of Brazil, but that at this point it was better to split up to reduce the risks of being caught.[24] Before they went their separate ways, Josélio gave Fausto, Toninho, and Alvaristo their wages for the murder.[25]

As the perpetrators rested in relative safety, all of Espírito Santo was shaken by the crime. The media picked up the story, and state legislators engaged in speculation and heated debates about the implications of the assassination. At that point Brazil was under the control of the military, which had taken power in 1964 and promised to restore law and order throughout the country in addition to crushing any remnant of communism. The state police forces of Espírito Santo and Minas Gerais, often at odds over territorial disputes in the Contestado, were instructed to cooperate to hunt down the killers.

Law enforcement agents suspected the Barros family immediately. "Twenty-four hours after the crime, I already had the names of the major's assassins," Espírito Santo's chief of police, José Dias Lopes, would later say.[26] Officers swiftly raided the Barroses' fazenda in Baixo Guandu, finding a small arsenal of two muskets, four revolvers, four carbines, and a rifle.[27] They also interrogated the farm's workforce for information. Pressured, the man who had cleaned Fausto's blood from the jeep broke down and revealed all he knew—the participation of

Josélio and the identity of the three hitmen, as well as where they'd headed after leaving the region.[28]

The heat was quickly mounting on the killers. Still, the three pistoleiros managed to evade the authorities for almost a month, until the third week of March 1967, when Fausto, Toninho, and Alvaristo, all three hiding in Uberlândia, were caught by surprise. A theft had taken place in the city, and a neighbor had reported the presence of three suspicious strangers who had arrived so recently. The group hadn't committed the crime, but, unexpectedly, they found themselves encircled by a cordon of police officers. Fausto and Toninho surrendered and confessed, but Alvaristo, the bravest of the trio, fled through a back door and managed to reach a nearby forested area. For six days, he hid in the wilderness and occasionally exchanged gunfire with the cops, who had been ordered not to kill him. At the end of March, with "feet and legs bleeding and raw from his nonstop flight through the bushes for almost 150 hours," according to the media, the resilient Alvaristo was finally arrested.[29] He was also shot in the arm.[30]

On March 28, 1967, the three pistoleiros, visibly distressed, were photographed by the press. "Only Josélio misses [out on explaining] the story of the killing," stated a headline in the newspaper *A Gazeta*.[31] A large archival picture of Josélio, arguably the most wanted man in the state at that time, was printed on the front page. Clean-shaven, wearing a dark suit and tie over a spotless white shirt, he looked handsome and debonair.

"Turn yourself over to the police," said chief officer Lopes, conveying a message to Josélio, who was still in hiding.[32] "Otherwise," he warned, "it won't be a surprise to learn today or tomorrow about the death [of Josélio] . . . by the very vipers he created and fed."

BY THE TIME of Cavalcanti's assassination, Josélio was already the father of several children and an admitted killer.[33] Some years earlier, he had murdered two men near a train station he'd suspected were hired assassins about to take the life of his father, José, who was planning to travel that day.[34] For that crime, Josélio had been absolved in a jury trial held in Baixo Guandu. Josélio argued that the jurors had been lenient because the double homicide was a desperate action to save his father's life, but prosecutors attributed such "impunity" to Josélio's "status as a wealthy cattleman."[35] Law enforcement officers quoted by the press stated that by the time of Cavalcanti's killing, Josélio had been involved in the slaying of "some ten people."[36] A confidential report issued by the Brazilian army would later say that Josélio was a pistoleiro and had been "a gang leader" whose criminal offenses dated back as far as 1958, when he was in his early twenties.[37]

However, not everyone in Baixo Guandu considered him or the Barroses to be outlaws. The family's story was in fact one of upward mobility through hard work and investments in education. Josélio's father, a descendant of Portuguese migrants who'd come to Brazil in the nineteenth century, was originally from the interior of the state of Rio de Janeiro, where he had studied to become a pharmacist before moving, at "seventeen or eighteen years old," according to the later account of a family member, to the Sweet River Valley.[38] There José, whose face was characterized by bushy eyebrows and prominent ears, had set up a pharmacy next to the farm of Cândido Afonso de Alcântara, one of the first elected council members of Baixo Guandu and also a descendant of Portuguese migrants. José fell in love with Mr. Alcântara's daughter, named Olga, and later took her as his spouse.

Eventually José left the pharmacy business to devote his time to the farm, which his wife had inherited along with a large wooden house with twenty-eight rooms but no electricity, according to family recollections.[39] The fazenda wasn't immense, but still, agriculture was a demanding task. A significant portion of the plot was covered with Mata Atlântica or Atlantic Forest vegetation, a biome less dense and diverse than the Amazon rainforest but nonetheless containing a "remarkable number of endemic species," in the words of an American historian who researched its subsequent destruction throughout the second half of the twentieth century.[40] Lacking tractors or power tools, the Barroses were forced to work the land manually.

The couple had one daughter, Josélia, and four sons: Josil, Jorio, Josélio, and José Francisco, who went by Chico. Large families like theirs were common in the interior of Espírito Santo, whose soils and climate were suitable for producing coffee and for logging. Chico, Josélio's younger brother, would later recall that parents in Baixo Guandu "didn't care if their children studied; they used them as free labor."[41] But José and Olga did want their offspring to be educated and worked hard to finance their teaching. Josélia, the eldest, became a professor, while Josil and Jorio grew to be doctors, the latter achieving the milestone of being elected first to the state and then to the federal legislature.[42] Josélio and Chico attended private schools, with Josélio enrolling at the Baptist "American School" of Vitória in his early adulthood. He was described by his teachers as a well-behaved and dedicated student.[43]

At some point, despite cutting corners, José and Olga could no longer support the education of all the children simultaneously, so the two youngest—Josélio and Chico—were taken out of school to lend a hand on the fazenda.[44] In his twenties, a young Josélio accepted his fate and devoted himself to the family business, occasionally traveling throughout the region to trade wood and cattle. His brother Chico would describe him as always "dirty and sweating, because he was always working."[45] That, according to Chico, earned Josélio respect and a reputation as a determined, hardworking person—traits that would mark him throughout his life.

Things began to be increasingly dodgy for the Barros family in 1950, when José successfully ran for councilman in Baixo Guandu for the National Democratic Union (União Democrática Nacional, UDN) party.[46] The UDN was opposed to the rule of President Getúlio Vargas, one of the most influential and controversial Brazilian leaders of the first half of the twentieth century. Vargas, who had just returned to power and was promoting a series of nationalistic economic policies dividing the society, was supported by a center-left alliance of the Brazilian Labor Party (Partido Trabalhista Brasileiro, PTB) and the Social Democratic Party (Partido Social Democrático, PSD).

At that time, Brazil was experiencing a rapid socioeconomic transformation due to, among other things, a demographic explosion and severe economic crises. The PSD-PTB and the UDN clashed over labor relations, wages, and the creation of national monopolies. The atmosphere became toxic, inflamed by cold war rhetoric. The pro-Vargas factions suspected imminent coups, while the UDN, backed by the army and supporting liberal and pro-wealth-holders' stances, engaged in defamation campaigns against the administration.[47] In Espírito Santo, newspapers carried accounts of legislators threatening their political opponents and carrying revolvers into the state legislature.[48]

Amid this extreme polarization and volatility, Vargas committed suicide. He shot himself on August 24, 1954, hours after he had agreed to a temporary "leave of absence" in an apparent attempt to appease the opposition, who were pushing him to resign. The right had routinely criticized the corruption and populism of Vargas's government, but what had precipitated the suicide was a wave of indignation following an armed attack against a prominent newspaper editor and fierce UDN supporter named Carlos Lacerda near the Copacabana Beach three weeks prior. Lacerda was only injured, but an air force officer protecting him was killed, triggering a political outcry. Vargas would deny any involvement in the scheme, but the pressure mounted when it became known that one of his bodyguards was behind it. A coup d'état by the army seemed imminent. Vargas, who appeared to have no relationship whatsoever with the failed attack, refused to step down until he was finally persuaded to accept a "leave of absence."[49] Hours later, he shot himself and left a dramatic suicide note accusing foreign forces of conspiring against him and the interests of Brazil, thereby becoming a national martyr. Protesters marched and threw stones at multinationals and at the embassy of the United States in Rio.[50]

Vargas's death failed to calm the political unrest of a country struggling to navigate the geopolitics of blocs and to cope with public dissatisfaction stemming from endemic corruption and the widening gulf between social classes. In 1956, a pair of reactionary pro-UDN army men "stole a combat plane laden with arms and explosives" in Rio de Janeiro, intending to "ignite a rebellion in central Brazil and to start a civil war."[51] They failed, but three years later, in 1959, air force officers hijacked an airliner flying to Belém and stole three military planes in a rebellion aiming to depose President Juscelino Kubitschek because "the Government was corrupt and Communist infiltration was widespread."[52] The coup would actually take place in 1964.[53] Washington backed it in order to quell the "socialist threat" posed by President João Goulart, a former minister in the Vargas administration who had planned to seize farmlands controlled by wealthy families and redistribute them among the poor.

In this atmosphere of turmoil, the election of Josélio's father—who opposed Vargas and the PSD—had serious implications for his family. By entering politics, José became an enemy of powerful rivals—local power brokers in the orbit of the PSD. The situation for the Barroses was made even trickier because Baixo Guandu was in the Contestado, a region where political violence was ongoing. Journalists would describe the Contestado as a place of "feudal lords holding the power of life and death over the unhappy people" who inhabit the area.[54] This misery in the face of lawlessness eventually led Josélio to take the life of Major Cavalcanti.

Encompassing some 3,800 square miles, the Contestado was a backwoods, hilly region of scattered outposts (today's Ecoporanga, Cotaxé, and Mantena, among other towns) where peasants made a living by raising livestock, harvesting timber, and growing fruit and crops, notably coffee.[55] At first glance, life here didn't look much different from life in other parts of the conservative rural interior of southeastern Brazil, save for one thing: a decades-long border feud between the states of Minas Gerais and Espírito Santo, causing a virtual state of anarchy in disputed territories along the state line and giving the Contestado its name.

For years, both Minas Gerais and Espírito Santo had been claiming jurisdiction over the same areas, and it wasn't a mere issue of formalities. On the ground, overlapping bureaucracies jockeyed for power. "A chaotic environment, marked by extreme violence and by the absence of the state," one scholar maintained, "shaped the society of the Contestado area, where the law of the most powerful prevailed, victimizing smallholders and peasants."[56]

Some villages, a researcher later noted, had three chiefs of police, three state judges, and three mayors.[57] Communities were required to pay the same taxes twice, the greedy officials of both states ignoring the plight of the inhabitants.[58] As one reporter put it, "Cattle rustling and horse theft, the invasion of lands, ranches, ran wild."[59] Hired killers took down political rivals and subdued voters, utilizing terror to get their candidates elected. (Eyewitness accounts describe pistoleiros stationed inside the voting booths to ensure that the *right* box was checked.[60]) The situation was so bad that some peasants had even attempted to rise up in arms to create a new, independent federal state in which they could be free of the chaos. In 1953, about a year after the rebellion, those peasants' hopes were crushed by officers from both Espírito Santo and Minas Gerais.[61]

The historic roots of the territorial dispute went back to the colonial era and originated in the greed for gold. During the late seventeenth and early eighteenth centuries, extraordinary gold deposits were found and tapped in the southeast of modern-day Minas Gerais.[62] The region was about to become pivotal in making colonial Brazil one of the world's top regions for gold production, the rush for the mineral prompting wealthy and stylish cities to flourish. The most notorious case would be that of Ouro Preto (Black Gold), at that time one of the most populated towns in Latin America due to the influx of fortune hunters.[63] However, the Portuguese rulers of the colony viewed that economic opportunity warily, worrying that gold could be smuggled out without paying the duties imposed in official ports.[64] They also feared a potential invasion of the resource-rich area by a foreign country that they anticipated would land troops on the barely patrolled coast near the deposits—today's Espírito Santo.[65] Therefore, to ensure that the wooded landscape of the Sweet River Valley would remain as a natural barrier to both smugglers and invaders, the colonial rulers purportedly limited the creation of donkey and horse trails connecting the mine to the Atlantic.[66] One of the consequences of this restriction would be that vast interior areas of central and north Espírito Santo would remain virtually inaccessible by pioneers and therefore largely untamed and unsettled for decades, their territories inhabited by Indigenous peoples and scattered communities of rural dwellers.[67]

During the second half of the nineteenth century, the region began to experience radical changes.[68] By then, the gold deposits of Minas Gerais were exhausted, Brazil had achieved independence, and authorities were promoting the settlement of hinterlands, turning the former no-man's-land into an economic frontier.[69] The colonization programs were open to foreigners as the nation attempted to transition from a slave-based agricultural system to a new one sustained by smallholding. Some American pioneers fleeing the United States after the Civil War ended up taking their chances not far from Baixo Guandu. According to one historian, between 130 and 400 Confederates, escaping the "lawlessness, violence, fear, hunger, poverty, ruin, and social disorder that were to mark the [American] South for many years" after the Civil War, settled in 1866–1867 on the banks of Lake Juparanã, a green paradise some ninety miles east of Josélio's hometown.

Because slavery remained legal in Brazil, although the importation of new African slaves was by then banned, the American southerners, who reached the Sweet River Valley on steamers after a land deal was cut with Brazil's then emperor, Pedro II, found in the South American nation not only a chance to escape the yoke of the victorious Yankees but also the opportunity to replicate their slave-based plantations.[70] The country, in fact, became a hot spot of immigration for Confederate families. It is estimated that "perhaps as many as 10,000 [southerners] went into exile in foreign lands" after the war, and Brazil was the most popular destination. The precise number of migrants from the US is unclear, but it ranges from two to four thousand.[71]

At Lake Juparanã, Americans would try to "recreate the large, successful riverfront plantations of the South" in fertile plots along the lake's shores, wrote a descendant of one of these settlers.[72] It was not long, though, before the families realized that their venture was more quixotic than pleasantly exotic, as one American researcher would put it:

In the first months, the families . . . found life to be satisfactory, if not extremely comfortable. . . . Crops were planted, homes were built, and families were raised. It appeared that the settlement would flourish. In 1868, however, all of the potential evils of the region beset the former Southerners. Mosquitoes became epidemic. Rain poured through thatched roofs, spoiling much-needed supplies, followed by a drought that destroyed patiently planted crops. Illness followed, with chills and fever affecting entire families. As food supplies became more and more scarce, talk turned to leaving for more hospitable areas such as Rio de Janeiro. The exodus that soon began was the beginning of the end.[73]

The fate of the Sweet River Valley, which seemed destined to remain isolated and unsettled forever, would definitively change three decades later, during the early twentieth century, when a railroad was built connecting Vitória to Minas Gerais. Paralleling the Sweet River, the Estrada de Ferro Vitória a Minas railroad would spark a rapid colonization of the interior by peasants eyeing plots for the growing coffee business. By the middle of the century, Espírito Santo became one of the top state producers of coffee in Brazil, the country dominating the world market.[74]

Economic development intensified the territorial disputes. Although no large-scale armed conflict ever occurred between Espírito Santo and Minas Gerais, skirmishes among the states' police were frequent in the 1940s and 1950s, as were belligerent speeches by politicians from both states, each routinely decrying the other as an invader of its jurisdiction, internal affairs, and tax claims. This chaos and misrule provided fertile ground for strongmen and bandits.

One whose status approximated a local potentate more than a fazendeiro was Secundino Cipriano da Silva, known to all as Colonel Bimbim. A tall, portly, balding man with thick lips and a chevron mustache, Colonel Bimbim, who lived in Almorés, had never been an army officer, but people addressed him as Colonel, the military rank frequently used in backwoods areas of Brazil as a moniker to express respect to strongmen.[75] He was a boorish but shrewd man, and before becoming perhaps the greatest villain of the Contestado, he had won the favor of the local people by killing horse thieves, the bane of smallholders who relied upon the animals to work the land and transport their produce.[76]

Over time, Bimbim grew in power and wealth, becoming a coffee trader, moneylender, and business owner, and, more crucially, a reliable—although often illegitimate—authority maintaining order (or at least what he viewed as order). Over the course of his life, Bimbim would be chief of police twice, vice-mayor of Aimorés for the PSD party, and, some years later, mayor under the umbrella of the UDN.[77] He was a much-coveted *cabo eleitoral*—a kind of canvasser or local political broker—of governors, lawmakers, and senators jockeying for the crucial rural votes he could deliver in the vast interior.[78] One of Bimbim's nephews, also a man of dubious integrity, was elected a state legislator in Espírito Santo for three consecutive terms in the mid-1950s and early 1960s before being named judge.[79]

Tales of Bimbim's wickedness spread far and wide through the Contestado, reaching the status of myth. During his forty years in power, Bimbim was alleged to be responsible for some eight thousand deaths, many committed by a putative army of five hundred gunmen acting on his behalf.[80] These numbers are almost certainly exaggerated, but Bimbim was unquestionably the epitome of an old political culture steeped in quid pro quo, collusion, and the subjugation of local populations.

Bimbim apparently came up with the name of the Criminal Syndicate to refer to the combination of bandits and the men in the higher political spheres who protected them. According to the press, anyone who tried to stop the Criminal Syndicate was silenced. "If a police detective had the courage to arrest a pistoleiro, open an inquiry, or hand over [a hitman] to the law to be processed," wrote a reporter, that detective "would be removed or assassinated."[81]

This overarching lawlessness was the environment in which Josélio had been raised, and it had profoundly shaped him. Chico, Josélio's brother, would recall having witnessed teenagers in Baixo Guandu brandishing *garruchas*—small double-barreled pistols—"in front of schools."[82] "In public parks, children gossiped about who had killed whom," said Chico. "It was horrible."[83]

In April 1964, at the age of sixty-nine, Colonel Bimbim died on his fazenda of a heart attack. According to a journalist, over five thousand people attended the strongman's funeral, and a legislator paid him a tribute in the state legislature, recalling Bimbim's "fair and good-natured character."[84] His death left a power vacuum, and many put themselves forward as potential successors.[85] Josélio's father—who, according to *Jornal do Brasil*, had joined a faction with party ties—was said to be among these. His political ambitions made him and Josélio not only feared men but also potential targets of a rival group led by Major Cavalcanti.[86]

Before Josélio orchestrated Cavalcanti's killing, José had survived not only the murder plot at the train station, but also another serious assassination attempt by a group of pistoleiros who invaded his ranch. Fortunately for him, he happened to be mounting a horse at that moment, and the animal, as soon as it heard the shots, broke into a furious gallop and saved his life.[87]

When Josélio himself became a target, he was forced to contemplate murdering Cavalcanti. Josélio would later describe several chilling episodes in which he'd faced possible death. Once mobsters disguised as cops had attempted to kidnap him while he was hiding in Minas Gerais. Another time he'd been attacked while he was driving his pickup along a dirt road: he miraculously survived that hit, which left the vehicle peppered with bullet holes.[88] But the crucial incident that prompted Josélio to attack Cavalcanti happened on February 4, 1967, just three weeks before the major was killed, when the son of a landholder who resembled Josélio and drove the same model of jeep was slain in Baixo Guandu.[89] No one could explain why he had been murdered until the hired gunmen—and Josélio himself—realized that it had been a case of mistaken identity.

Sometime later, as rumors spread about the botched assassination, Cavalcanti, who probably feared a retaliation, approached Josélio's father, who was accompanied that day by Chico.[90] "Hey, Barros, people are saying that I ordered gunmen to kill your boy [Josélio]," he declared, referring to the failed attempt. "But it wasn't me. I'll order the murder of those who committed the crime to prove to you that it truly wasn't me." According to Chico, though, instead of keeping his word, the major looked for new ways to kill Josélio.[91]

"There was nothing left for Josélio to do but to murder Cavalcanti," Chico bluntly admitted years later.[92] And that's exactly what Josélio did.

IN THE FIRST week of April 1967, about a week after Fausto, Toninho, and Alvaristo were arrested, Josélio turned himself in to the police in Nova Iguaçu, in the suburbs of Rio de Janeiro. He was immediately taken into custody and transported to Espírito Santo.

"[He was] well-groomed and spoke eloquently," wrote a reporter, describing Josélio the day he was presented by the cops to the media. Josélio confessed to the homicide and admitted that he had "ordered many crimes" prior to the murder of Cavalcanti, justifying his behavior by arguing that he was only doing what needed to be done to defend himself and his family against the "impunity of feral humans who commit all kinds of crimes without punishment."[93]

"I don't deny having ordered the killing of Major Orlando [Cavalcanti]," Josélio declared, displaying a certain arrogance and bluntness noted by reporters. "I knew that he was plotting my death through third parties, and before he killed me, I opted to order his death."[94]

Josélio claimed that his original plan was to turn himself in after the crime, but later, having second thoughts about the treatment he would receive from the police, he'd changed his mind. Afterward, Josélio continued, sources told him that chief officer Lopes was a "fair man," and so he finally decided to turn himself in. He admitted that if he had thought he would be denied the right to a fair trial, he would have fought "against everything and everyone" to survive.[95]

"I'm not a fearful man," Josélio would write of himself during those years.[96]

Officer Lopes also talked to the press the day of the arrest. He defined Josélio as a "polite boy . . . who was forced to commit crimes only because he lived among outlaws who were never sought by the authorities. . . . Josélio is a product of the context in which he lived, where impunity and protected bandits prevailed."[97]

In June 1968, at the end of a long investigation involving politicians, police, ranchers, and gunmen, Josélio stood before a jury in Vitória. Josélio insisted that he acted in self-defense; he was not an outlaw lacking ethical principles, he said in court. "I killed in order not to die; any other citizen in my position would have done the same, because there was just one choice: KILL OR DIE. I chose to kill in order not to die. I chose TO EXERCISE A RIGHT," Josélio would later write.[98]

Initially, his defense strategy didn't help him much. Josélio was convicted and sentenced to thirteen years in prison for masterminding the murder of Cavalcanti, and each of his three accomplices received a sentence of more than twenty-five years.[99] But he didn't stop there. In September 1970, on appeal, he got his sentence reduced to eight years.[100] A year later, his defense attorneys appealed again to request that due to his good conduct, he be put on parole. State prosecutors fought his release, but on March 7, 1972, an appeals court accepted the request and let him out.[101]

Because the crime had been committed only five years earlier, Josélio's father feared that his son might still be murdered. The border dispute between Espírito Santo and Minas Gerais had been solved by the official establishment of the state lines in January 1966, when a boundary marker in the form of a granite block was placed in Baixo Guandu.[102] (A local described the political deal thus: "Now we know who to pay taxes to, whose authorities to obey, and whose judges to appeal to."[103]) But old grudges die hard, and the murder of one of Josélio's close allies in 1973, less than a year after Josélio had left prison, renewed the Barroses' worries about a potential attack.[104] In response, José decided to sell a piece of land and give Josélio money to settle somewhere else.[105]

Paragominas, bordering modern-day Rondon do Pará, was the site of Josélio's chosen exile. In December 1973, while still on parole, he requested permission from the court to relocate with his family to a ranch he'd bought there.[106] The petition was accepted under the condition that he avoid committing any felonies and show up regularly before a notary public to report his status.

By moving to the Amazon, Josélio was seeking to turn the page on his controversial past and leave behind forever the overarching law-lessness that had colored his life and compelled him to become a criminal. He wanted a fresh start, the chance to be just "a happy man," as he said during that period to the court.[107] "I dreamed what any young man dreams. I wanted to be happy. I wanted others to live happily. I wanted a rich world both materially and spiritually."

Pará seemed like a proper place for a wealthy young rancher to pursue those dreams. In 1966, the military government had launched a broad campaign to develop and settle the jungle they named Operation Amazonia. It consisted of attracting agricultural and livestock companies to the region by offering generous economic incentives—tax breaks, special lines of credit, and cheap land. As Susanna Hecht and Alexander Cockburn would later write in *The Fate of the Forest*, investing in the frontier became "the surest of sure things. . . . And so, the Amazon's new boom began."[108]

The state-driven policies also proposed the construction of a grid of roads to bring settlers into a Brazilian region more than half the size of the contiguous United States but with a population density of only about one person per square kilometer, a demographic imbalance that worried the government.[109] "In this vast territory, humans are largely lacking," declared a Brazilian minister at the time, noting that the Amazon represented "the greatest demographic void in the world, apart from the polar regions."[110]

Expanding into that enormous territory wasn't done for strictly economic reasons. Under Operation Amazonia—later redesigned and combined with other programs that shared the objective of pushing the frontier north—the generals aimed to take full control of an uncharted area that shared thousands of miles of borders with seven countries, some of them with a long record of territorial disputes with Brazil.[111] The cold war and the rise of Fidel Castro in Cuba fed into suspicions that the tangled, secluded jungle could be a place for seditious groups to establish guerrilla forces. In short, Operation Amazonia was a response to a rainforest perceived by the regime as "empty and easily annexed economically, ideologically, and perhaps even territorially."[112]

The slogan "Integrate so as not to surrender" summarized the multiple potential dangers to national security.

But of course the Amazon *wasn't* empty, and the frontier area where Josélio chose to start a new life would soon feel the shock waves of the grandiose colonization plans. The scramble for jungle land was about to begin.

3

Terror on the Nut Road

A T the end of July 1969, while Josélio was serving his prison term for the murder of Cavalcanti, a wave of unrest originating in the wilderness competed for space on the front pages of Pará's newspapers with the greatest scientific milestone of that era—the first moon landing.

On July 25, only five days after the American astronaut Neil Armstrong planted "the first human foot on another world," it was learned that the Gavião-Kyikatêjê, an Indigenous group living near the Tocantins River, had used arrows and clubs to slaughter a peasant family.[1] The information available was scarce, and the cause of the attack wasn't totally clear, but the dispute seemed to stem from the control of a vast area of forest that had harbored the isolated tribe for decades and was now being increasingly encroached on by migrants coming from all over Brazil.[2]

Reacting to the news of the slaughter, about two hundred peasants and their children, fearing murder by the Gavião-Kyikatêjê, fled the ramshackle adobe and wood-planked hovels they had erected in scattered jungle clearings, leaving their animals and possessions behind.[3] They were headed for the Nut Road, a sloping road of saffron-colored soil that—following the path of a former trail used by hunters and nut gatherers—was being built to provide pioneers with a way to penetrate the frontier. The anxious crowd walked for hours to find refuge in Vila Rondon, a hamlet that had sprung up along the 137-mile road, which would be renamed PA-70 (and later BR-222) once it became fully accessible to cars, buses, and trucks.

The initial account of the murders was provided by Artur José de Oliveira, owner of the farm that was raided.[4] Artur, who had survived the rampage, reported at least two of his farmworkers had been killed by a rain of arrows shot by the Gavião-Kyikatêjê, popularly known in the region as Gaviões (sparrow hawks) due to the ethnic group's historical belligerence against trespassers. "The Gaviões drink water from skulls and roast meat on Christians' shinbones," an early colonist had written years prior, claiming to have witnessed ambushes against nut gatherers.[5] "Their bows are very powerful, and their arrows, when shot, have such force that they go through [riverboat] hulls made out of hardwood," a reporter had written, describing the techniques of warfare deployed by these "athletic warriors" in the following chilling way: "Every time they commit a killing, they make of the [victim's] body a real toothpick holder."[6]

Artur's unnerving story caused panic in Vila Rondon, as "it was said that the Gaviões had pierced one of the farmworkers with twenty arrows," explained Rita Belem, one of the residents.[7] The outpost—nothing but a clearing in the thick jungle—had been founded between the end of 1966 and early 1967, when the construction company charged with building the PA-70 (named Departamento de Estradas de Rodagem or DER) had chosen the spot to set up a base camp and accommodate employees.[8]

Only hours after word of the attack spread, some six hundred people invaded the DER's base, a handful of white wooden bungalows scattered in a large yard that was cordoned off by a low fence and depended on generators for electricity.[9] They were all "looking desperately for a secure place," recalled Rita.

Fear was also fostered by uncertainty. The DER's radio operator had promptly contacted officials in Belém to report the deaths and request urgent backup and supplies, but it was unclear when this help would arrive.[10] Under normal conditions, it would take the authorities at least a day to send police to Vila Rondon, located halfway down the Nut

Road, due to the poor state of the road. Besides, according to some unconfirmed accounts, the Indigenous people had set fire to wooden bridges along the PA-70 to virtually cut off Vila Rondon from Marabá, located about 75 miles west.[11] Therefore, any support would have to come from the east, some 55 miles farther, in the state of Maranhão, where the Nut Road crossed the BR-153, a 1,240-mile-long highway popularly called the Belém-Brasília. Also referred to as the Road of the Jaguar for the abundance of big cats encountered during its construction, it became the first large-scale overland connection between the Amazon and central and southern Brazil, including the new federal capital, Brasília, which was inaugurated in 1960.[12]

With seemingly no further options but to remain in the fenced-in bungalows for days, some colonists in the DER camp speculated about the possibility of organizing an armed counteroffensive.[13] The idea was to retaliate against the Indigenous people with a punitive expedition planned for the next morning. The target would be the tribe's main village, located south of the Nut Road.

The tribe's dominion was vast, reaching some twenty-five miles south of the road before hitting the north shore of the Tocantins River, and the assault was undoubtably risky as well as morally dubious. But pioneers reasoned that, after months of scuffles with the Gavião-Kyikatêjê, this could be a decisive opportunity to demonstrate to both the Indigenous people and the Brazilian government that they themselves had moved to the frontier to stay.[14] "We're Christians, we have the right to land, and we want to live [here] in peace," one settler would blurt to a reporter.[15]

It was then that Antonio Cotrim Soares intervened to tame the rage of the colonists. Hired a year earlier by the government's National Indian Foundation (or FUNAI) to prevent outbreaks of violence between the tribe and a white civilization expanding its presence as work progressed to complete the PA-70, the then-twenty-eight-year-old Indigenous expert—or *sertanista*, in Portuguese—persuaded the colonists that an attack was illegal and, even if carried out, would likely fail. By then, he explained, the Gavião-Kyikatêjê had already scattered across their territory, a survival strategy developed after decades of fights with a more populous rival who also possessed superior weapons—mainstream society.

By that time, Cotrim was fully aware of the dangers of the ongoing escalation. Days prior to the raid against Artur's ranch, Cotrim had found a corpse decomposing near a stream.[16] It was pierced by over a dozen arrows. The slaughter was apparently a response by the Gavião-Kyikatêjê to an arson attack by settlers against their gardens. Cotrim had immediately reported the issue to the authorities in Belém, prompting officials to send some police to Vila Rondon. A few Brazilian journalists had joined them to report about the case.[17]

That night of July 25, as tensions grew in the DER camp, Cotrim felt an urgent need to do something. He successfully persuaded the colonists to drop the plan of retaliating against the tribe. Then, to calm the settlers down and buy some time for himself, Cotrim announced that he would attempt to access the raided farm—located some 8 miles deep in Indigenous territory—to find out what had really happened there. After that, Cotrim said, they would figure out what to do next.

Over the next few hours, Cotrim assembled a group of fifteen people ready to accompany him on the fact-finding mission, including a federal police officer, the journalists, and three Indigenous men described as "already civilized" who shared a history and language with the Gavião-Kyikatêjê and knew the region well.[18]

ONLY A FEW hours later, on the morning of July 26, three groups of men, following the same path but separated from one another by several hundred meters, left the DER camp and penetrated the forest.[19]

"I want the minimum number of arms required," Cotrim had said. The idea sounded foolish to some of the participants, but the sertanista anticipated that the Gavioes would station lookouts through the jungle, probably hidden in the treetops, to inform about—or if necessary, kill—hostile trespassers, and he betted on semiotics to send a crucial message to the Gaviões: that he and his men were not there on a retaliatory mission.

They hiked for about six hours through hilly terrain and dense brush.

By 4:00 p.m., they reached the site of the massacre.[20] Next to the farm's rice store was a small clapboard house on a slope surrounded by vegetation. Here the body of Clovis Souza, Artur's son-in-law, was found. He had been killed with "one shot in the front [thorax] and twenty-one arrows in the head, neck, and ribs," according to a mission report.[21] The arrows—much longer and heavier than those commonly used by members of the First Nations on the Great Plains of the United States—had entered the man's body with such force that they had not only pierced the skin and internal organs but passed entirely through to the other side. The tips of the arrows protruded from the man like skewers.[22]

The macabre scene unnerved the group. Cotrim knew what those multiple arrows—probably made out of monkey bones and dipped in poison—meant: "Stay away." Then, as some of the men dug a grave for the corpse, the Indigenous interpreters accompanying them noticed some vultures nearby. There was another dead body.

About 300 meters from the first corpse lay Vicente Batista, another farmworker, who had likely been massacred while fleeing.[23] He had two shotgun bullets in the thorax, which confirmed that the Gavião-Kyikatêjê, in addition to arrows, were in possession of firearms. Another handful of men, exhausted and seemingly having second thoughts about the previous decision to bring only a limited arsenal with them, dug a second grave.

Before long, more vultures were spotted soaring in wide circles about 800 meters away, suggesting the corpse of a third person. The farm's owner, Artur, who had joined the mission, said that it possibly could be José Francisco, another farmhand present at the time of the assault. At that point it was getting dark, so the group wondered what to do next. Should they try to dig a third grave before setting off for Vila Rondon?

As they brooded over the issue, they heard "screeches coming from the jungle," unnerving the guides.[24] "They said we should leave that place before nightfall," a civil servant later wrote.[25] "There was a risk that the Indians could be watching us and believe that we were a punitive expedition." Cotrim ordered the group to abandon both the area and the unburied body.[26]

The memory of the skewered corpse made the return an excruciating ordeal, a second eight-mile march through a maze of branches, climbers, and brambly vines. "It was dark, and we had only four flashlights for sixteen men. The column fell silent and walked north [back to the Nut Road]," a reporter wrote.[27]

The dangers had, in fact, grown exponentially for the party—and not only because the Gavião-Kyikatêjê might take this opportunity to inflict more damage. The surrounding area was filled with dangerous animals of all kinds: vampire bats, boa constrictors, five-meter-long black caimans, and jaguars, feared not only by the other animal species that composed their usual prey—nine-banded armadillos, southern tamanduas, or sloths—but also by even the most skilled Indigenous warriors.[28]

"The jungle's noise was unceasing. Birds, howler monkeys . . . , and the darkness already made it impossible to see anything. We walked stumbling into lianas," a journalist later wrote, describing the extremely loud yells—audible at a range of over a mile—of red-handed howler monkeys, a species that displays domination over food and females by launching powerful hoots from treetops.[29]

By two in the morning on July 27, the men, physically drained (one of them had even fainted during the expedition), reached the Nut Road. "It was one of the greatest adventures I ever lived, but it was a real hell," admitted a reporter who was there.[30] Back at the DER's camp, Cotrim and the others passed on the news of the murders, which caused more panic among the people."[31] A group of twenty police agents had already reached Vila Rondon with food supplies, but nobody wanted to return to their clapboard houses and gardens in the forest unless they were certain that the threat was unequivocally past. Every night for the following two weeks, according to eyewitnesses, people would remain in the DER's cabins.[32]

As a deep sense of vulnerability engulfed the colonists, many wondered how to get rid of the *índios* or "Indians." Some wanted the government to expel or even annihilate the tribe. But Cotrim had radically different ideas. The issue worrying the sertanista was how to help the Gavião-Kyikatêjê, because assistance, and not terror, was what he had promised them when he had succeeded, only some months prior, in making peaceful contact with these extraordinary people. As he saw it, they had already faced racism, prejudice, and the violation of their most fundamental rights.

COMPARED TO PEOPLE'S experiences with other Amazonian tribes, Cotrim's initial contact with the Gavião-Kyikatêjê group had been relatively easy. The sertanista was hired for the mission in 1968, when he was already skilled and experienced in dealing with hostile Indigenous tribes, though Cotrim had never really planned to devote his life to that work.

The son of a well-off merchant in Maceió, capital of the coastal state of Alagoas, in his early twenties Cotrim had participated in leftist organizations advocating for agrarian reform. By the 1964 coup, Cotrim, who, according to a reporter, was "linked to the guys of the [Brazilian] Communist Party," envisioned resisting the dictatorship by joining a guerrilla force at the Peru-Brazil boundary.[33] In the end, he ran out of funds before reaching the border and abandoned his revolutionary ideas.

In 1965, Cotrim enrolled as a volunteer in the Indian Protection Service (SPI), the predecessor of modern-day FUNAI. Soon thereafter he moved to the southeast of Pará, where he took part in a series of expeditions to contact other groups of Gaviões. (Ethnic groups in the Amazon often split when clans disagreed on strategic decisions or when their population grew too large and the game or farmland available could not support the group.) In mid-1968, with the construction of the PA-70 progressing and Operation Amazonia luring migrants into the region, the Gavião-Kyikatêjê attacked some settlers, and the federal authorities responded by hiring Cotrim to scout out the area before eventually establishing contact with the tribe.[34] (As the prominent sertanista Sydney Possuelo would put it, "the glory [of a sertanista those years] was in the contact." The government believed that contacting an isolated tribe was the way to prevent conflicts. Later, Possuelo reshaped this policy of "contact to save" into one in which Indigenous groups were left alone and the official effort was put into protecting their lands against invasions.)[35]

Cotrim wasn't alone when he investigated the origins of the conflict near the PA-70. Sertanistas rarely enter the lands of a tribe without Indigenous guides. In many respects, the knowledge accumulated by non-natives about life in the forest could hardly equal that acquired by Indigenous groups, who had been passing down vital knowledge from generation to generation. Social scientists are still researching the full extent of their mastery. For a long time, the dominant theory was that tribes had never thrived in the Amazon and had subsisted by living in small settlements scattered throughout the forest. The recent work of anthropologists, archaeologists, geographers, and even analysts of spatial data and remote sensing have refuted those theories.[36] In pre-Columbian times, there were "fairly large-scale complex societies," including groups who built villages that were home to several thousand people and featured squares, canals, causeways, ditches, and roads.[37] By 4000 BP, according to estimates, local societies had evolved from foraging to farming systems that involved the domestication of native species and fed large populations.[38]

Scientists wondered how that leap forward was possible, because despite the popular misconceptions about its extraordinary biodiversity, the Amazon has poor soil for agriculture overall. One of the momentous discoveries—to a point of prompting a reexamination of the history of human settlement in the Amazon before 1492—was what experts called dark earth, or *terra preta*. These are highly fertile soils supportive of the expansion of agriculture, and they were apparently obtained by a long-term land management system that involved the use of composting, burning, and mulching. Found dispersed throughout the Amazon Basin and not far from ceramic remains, dark earth sites "have the potential to feed millions of inhabitants."[39] Thanks to terra preta, it is now estimated that up to ten million people—rather than two million, as previously hypothesized—may have lived in the Amazon before epidemics and European violence depopulated the area.[40]

Cotrim hired tribesmen he knew from previous missions to help him settle in the jungle and approach the Gavião-Kyikatêjê.[41] One of them was called Itacaiúna and was described by Cotrim as a great tracker in thick forests.[42] He and other tribesmen also shared some common language with the Gavião-Kyikatêjê.

Their first move was to set up a base camp about 9 miles north of the Tocantins River, close to where the Gavião-Kyikatêjê hunted and farmed. It wasn't a fancy place, just a bivouac composed of tarpaulin shacks and hammocks. Anticipating that he might spend weeks researching, Cotrim laid in vital provisions. Later, he would write his own essential packing list for future sertanistas charged with contacting tribes. It included coffee, sugar, beans, rice, instant soup, corned beef, antimalarial drugs, and "equipment to fish and hunt."[43] Last but not least were the "gifts" for the tribe to be contacted: machetes, axes, serrated sawback stainless-steel daggers, rakes, scissors, fishhooks, hammocks, blankets, and mirrors.

Such "gifts" might be considered paltry peace offerings by people in the modern world, but for seminomadic hunter-gatherers with no permanent access to metal, like the Gavião-Kyikatêjê, these axes and machetes were extraordinarily precious indeed. According to historian John Hemming, those industrial items represented a "technological miracle in a forested world where men's most wearisome task was felling trees to create garden clearings."[44] Years prior, groups of Gaviões had launched raids to grab axes and machetes, assassinating officials of the Indian Protection Service because they hadn't offered them enough tools.[45]

Once settled in the forest, Cotrim and his Indigenous guides activated the standard protocol for making contact with an isolated tribe. This consisted of leaving the presents in jungle clearings to entice the Gaviões steadily closer. Unexpectedly, the first contact took place only a few weeks later and was highly dangerous, because it caught Cotrim and his colleague Itacaiúna off guard.

The pair had gone to hunt, and as Cotrim would admit later, they "got too deep into the jungle." "Suddenly," he wrote, "we were caught unaware by three warriors. Face-to-face with us, they brandished their bows in our direction. Instinctively, Itacaiúna also aimed his shotgun toward them. I was a bit behind and immediately put my weapon on the ground. Right away, I placed the machete I had on both my palms, symbolizing a peaceful act of no aggression, and screamed to Itacaiúna to behave similarly, which he did, despite reservations," wrote the sertanista in a report.[46]

Cotrim's insistence that the two lower their arms could have looked suicidal, but he was following a well-known—though by no means infallible—strategy to respond to an initial Indigenous aggression. It was Colonel Cândido Mariano da Silva Rondon—the tireless and supreme Brazilian explorer who had founded the SPI before accompanying Theodore Roosevelt on his 1913–14 expedition through the River of Doubt, in the western tract of the Brazilian Amazon—who had evolved the theory in the dictum "Die if you must, but never kill."[47]

The tactic of nonaggression worked for Cotrim. The warriors responded positively to his goodwill gesture, and they also put down their weapons, subsequently coming closer "to get the machetes," the sertanista wrote. "This is how, totally unexpectedly, we had established the first contact," said Cotrim, who managed to communicate "with gesture and signs," as well as with some common words of Itacaiúna's language, to set a date for a new meeting. "Three days later, accompanied by Indian interpreters, we came back to the site, where there was a large group of Indians, including women. Interactions were aided by the understanding of the dialect spoken by our interpreters. The distribution of gifts—machetes, axes, and other items, including some shotguns—reinforced their trust in our mission."[48]

Over time, Cotrim gained the tribe's full confidence, becoming entitled to a privilege rarely accorded to strangers: access to their main village, where the Indigenous people stocked their arms and food provisions and where the elders, women, and children remained in safety while the warriors hunted or patrolled the territory. He would spend several months—"eight, ten," he said in an interview years later—coming and going at will. Cotrim estimated that the Gavião-Kyikatêjê group was composed of ninety-two members.

During that time of mutual trust and discovery, the sertanista absorbed the way the tribe lived and survived. He witnessed their traditional rituals, including demanding relay log races through the hilly forest and archery games requiring mastery of bows two meters long that represented training for real-life combat and hunting. He understood why they farmed in gardens that anthropologists would later report were scattered in sites smaller than a hectare throughout the jungle: they wanted to avoid being overdependent on a single source of agricultural output that might be plundered by enemies.[49] "This strategy would allow them to survive if their main village was attacked," wrote Cotrim.[50]

Fascinated and now personally invested in the tribe's survival, when Cotrim traveled to the nearby towns to report that he was well and that the contact was progressing steadily, he bought new clothes to avoid spreading diseases against which the Gavião-Kyikatêjê had no immunity.[51] But because most of the game was being driven away by both the building of the PA-70 and the increasing number of intruders brought by Operation Amazonia, prospects were rapidly changing for the Gavião-Kyikatêjê, and there was little that Cotrim could do about it. "They lived in fear due to the devastating presence of white men, whose invasion of their territory was constant," he complained, angry that "the dissemination of the news about the early peaceful contact," rather than prompting the authorities to protect the tribe and their lands, had only lured in waves of homesteaders. The federal government would eventually pass an executive order on October 31, 1968, nine months before the attack against Artur's farm, banning trespassers from an area of about 300,000 acres where the Gavião-Kyikatêjê lived.[52] But the decree had little effect on the ground, as pioneers—being aware of the government's limited ability to enforce the law—continued to reach the area.

The push into these lands was far from spontaneous. In the wake of the changes brought by the national strategy to settle the Amazon, unscrupulous real estate speculators and businessmen, realizing that the colonization of the area was taking shape haphazardly, took their chance. *Jornal do Brasil,* quoting a police report, published that fazendeiros with powerful ties in other states had illegally "demarcated great areas, including in the Indian territory, and sold the land, without any deed, to colonists."[53]

As the newspaper article put it, "the objective of the land-grabbers and large landholders is to expel the Indians from the region, and they have even armed groups of men with the mission of exterminating them."

Cotrim witnessed firsthand the radical changes that those actions entailed. Vila Rondon, "initially a settlement of adventurers of fifty, one hundred houses," as he would say, was transformed into a magnet for immigration.[54] The Gavião-Kyikatêjê, increasingly "harassed by civilization," decided to defend their lands by attacking "colonists at the edges of the PA-70."[55] Feeling betrayed, the Gavião-Kyikatêjê also began to view Cotrim with suspicion. On July 23, 1969, just two days before panicked settlers fled their homes toward Vila Rondon and settled in the DER camp, the sertanista was expelled from their village.[56] It was then that Cotrim had found the dead body with the arrows and had informed officials in Belém that the situation was growing more dangerous.[57]

ABOUT A WEEK after he was thrown out of the Indigenous village and only seventy-two hours after he had returned from the expedition to Artur's farm, Cotrim penetrated the forest once again to attempt a meeting—peaceful, he hoped—with his former friends.[58] His aim was to reestablish a dialogue with the tribe and negotiate a peace deal. Two days later, while hiking through the jungle accompanied by his loyal interpreters, Cotrim found a tribesman harvesting palm cabbage.[59] Distrustful at first, the man finally agreed to take him to meet the other Gavião-Kyikatêjê.

"The initial reception wasn't very welcoming," recalled Cotrim, who, in the village, witnessed the Indians displaying clear signs of bellicosity. He reported having observed tribesmen "heavily armed, uttering war cries, painted [with black dye]."[60] He was informed that the death toll from the earlier raid was actually six people, not just the three Cotrim had found slaughtered. Warriors who had participated in those attacks were carrying out "purification rituals, a way to chase away the spirits of their dead enemies," Cotrim noted.

He devoted about a week to securing a peace deal. After long discussions, the Gavião-Kyikatêjê gave him an ultimatum: either an armistice sealed through the immediate eviction of all whites from their lands or total war.[61] The leaders said they simply wanted to be left alone and their lands declared off-limits to intruders. Before Cotrim returned to Vila Rondon, he was given a horn made out of a pumpkin so that at the next meeting, he could announce his arrival.[62] Last, as an indication that the sertanista had lost their trust, the Gavião-Kyikatêjê kept three of his interpreters as hostages—they would be freed only when he came back with a response, expected in about ten days.[63]

The Gavião-Kyikatêjê appeared more determined than ever to protect their territory and their people, but behind the scenes powerful political figures and their associates were preparing a fight of their own.[64] When Cotrim stayed longer than expected during the negotiations without reporting his status, some hawkish politicians disseminated the fake news that the explorer had been killed in order to create a casus belli against the Gavião-Kyikatêjê. The mayor of Imperatriz, a town on the shores of the Tocantins River where sawmills were being constructed, announced on the radio that one police officer and four soldiers would participate in a retaliatory operation.[65] He also called on colonists to join the raid. A decade prior, one of his predecessors had machine-gunned the Gaviões after enticing them with gifts, according to media reports.[66]

The possibility of a Custer-like mission charging into Gavião-Kyikatêjê territory set alarm bells ringing for Cotrim, who was informed of the plot as he emerged from the jungle. Many feared that the mayor, who was suspected to be in collusion with fazendeiros and logging companies targeting prized timberlands, would not bother to change his murderous plan even if the sertanista was in fact unharmed.

Federal authorities in Brasília also worried about the implications of the case, which could further harm the international image of the dictatorship just months after a landmark report had been released detailing a "20-year scandal of murder, rape and robbery of Brazil's Indians," as Paul Montgomery wrote in the New York Times.[67] It would be known as the Figueiredo Report because it had been written by Attorney General Jader de Figueiredo Correia, who had traveled over 10,000 miles to investigate years of abuses by corrupted officials.[68]

Among the human rights violations, meticulously logged in twenty volumes, were "charges of widespread corruption and sadism, of crimes ranging from the massacre of whole tribes by dynamite, machine guns, and sugar laced with arsenic to the removal of an eleven-year-old Indian girl from school to serve as an official's slave," according to a *New York Times* cover article published on March 21, 1968.[69] The gruesome revelations, echoed by the *Los Angeles Times*, *Le Monde*, and the British *Sunday Times*, among other global outlets, caused a political storm. Over two hundred officials from the SPI were either charged with crimes or dismissed, and the governmental body was replaced by the newly created FUNAI.[70]

However, the tribal rights organization Survival International, founded two years later as a response to the Figueiredo Report, would claim that "no one was ever jailed for the atrocities."[71] Adding to the suspicion of political collusion was the fact that months after the scandal broke, the 5,000-page report was mysteriously "destroyed" in a fire before its full content had been made public. More than forty years would pass before parts were accidentally uncovered by a researcher in the archives of the Indian Museum of Rio de Janeiro.[72] Experts believe that the "documents were archived and opportunely 'forgotten'" by the dictatorship.[73] Nonetheless, the partial release of the Figueiredo Report was enough to spark intense international criticism, causing profound embarrassment for a regime whose colonization plans for the Amazon involved little real consideration for the fate of native peoples.[74]

Therefore, when the federal authorities were informed that the mayor planned to murder the Gavião-Kyikatêjê, they took action to prevent that from happening. The politician finally retracted his words, arguing they had been misunderstood.[75] For Cotrim, though, the issue wasn't fully settled. Fearing that the volatile situation would lead to the slaughter of the hostages, he went back to the jungle.[76] It would be his last trip to the Gaviões' dominions.

When he met with the leaders, Cotrim shared both the news of the punitive expedition and the idea gaining momentum among his bosses to bring a final end to the conflict: transferring the tribe to the Mãe Maria, or Mother Maria, an Indigenous reservation located some 65 miles west and accessible via the PA-70.[77] Comprising over 153,000 acres, the Mother Maria Reservation was already home to other Indigenous tribes and other groups of Gaviões.[78] The Gavião-Kyikatêjê were promised a place to settle in peace and exclusive access to forests abundant in game and gardens.[79] Though deeply reluctant, the group had no choice but to accept the deal.

In the following months, the tribespeople were removed from the rainforest and brought to the reservation.[80] Rita Belem, the homesteader who had found refuge in the DER camp, would recall having seen almost naked tribesmen and tribeswomen being transported in military lorries. The motorcade, she said, was ordered to stop in Vila Rondon so locals "could see that the Indians were being removed." Eyewitnesses took pictures of the tragic scene, but army officers confiscated the rolls of film.[81] Decades later, FUNAI claimed there were no images of the expulsion in their archives.[82]

Cotrim refused to take part in the removal process and left Vila Rondon before it happened. In the following years, he would still participate in some "contacts" with Indigenous tribes across the Amazon, where conflicts with groups living in isolation mushroomed as a consequence of the policies to open up the frontier. In May 1972, he resigned from his job as sertanista. Cotrim said that he was tired of "being a gravedigger for Indians," returning to his hometown in Alagoas to work in the family business.[83] Many other explorer-activists working in the Brazilian jungle during the early years of Operation Amazonia followed in his footsteps, dispirited at finding "themselves navigating around recently toppled forest giants, across ground still smoldering from recent burns, ducking gunmen in the pay of ranchers and land sharks, racing to contact besieged tribes before the goons got to them first," wrote a *National Geographic* reporter.[84]

Months after the tribe had been evicted, a medical mission by the International Committee of the Red Cross visited the Mãe Maria Reservation and found the Gavião-Kyikatêjê "in a most deplorable state of health, the worst of all that we had seen" in their three months of working throughout the rainforest.[85] (The mission was authorized by the Brazilian government to tame criticism over the findings of the Figueiredo Report.) Doctors wrote that they lived in "provisional shelters" under "poor" hygiene conditions, and "apart from dried manioc, no other food seemed available, and no source of protein." Several members of the group, in addition to living in a state of semi-starvation, were also found to suffer from untreated malaria, bronchopneumonia, and scabies. "There did not seem to be any old people or young children, and the men and women we saw looked very sick and apathetic," read the report, published in July 1970.[86]

WHILE THE TRIBE was confronting ravaging epidemics and the trauma of dislocation, colonists were transforming Vila Rondon.

Houses and businesses shot up in the dusty urban center, located near the DER's camp and along the main avenue—the finally completed PA-70, which cut straight through the village. About a year after the eviction of the Indigenous group, twenty-eight groceries, three barbershops, two drugstores, and one Roman Catholic and two Protestant churches operated in the hamlet, where the urban population had grown tenfold to surpass 2,500.[87] "Every week, a new family comes," one newspaper crowed.[88]

Isolation made life expensive for the colonists, as coffee, sugar, oil, medicine, and all other goods had to be freighted in, which wasn't an easy task.[89] Access from Marabá, the regional trade center, was often impractical despite the PA-70. During the dry season, usually running in Pará from March to September, the dust that plumed in vehicles' wake reduced visibility to almost nothing, making the PA-70 a treacherous route subject to frequent head-on collisions. The trip was an equally endless agony during the six long months of rain, when vehicles were stranded for hours or days, mired in mud as sticky as peanut butter. Reaching the Belém-Brasília Highway—only some 55 miles east—could easily take up to forty-eight hours.

In the absence of a local authority, some of the DER's top staff filled the gap and acted as unofficial mayors of the hamlet. "When migrants came to the DER to ask where to place a garden, the chief engineer replied, 'Just pick an area and clear it.' A new house was built daily in town," recalled José Coutinho de Queiroz, a truck driver employed by the DER and one of the town's earliest residents.[90] "None of us could prevent those people from gaining access to land, because it was government-owned."

The assumption of the pioneers was that by clearing an area, one virtually became the owner of the land. Still, as enticing as the prospect might have been now that the Gavião-Kyikatêjê had been evicted, the average colonist family did not have an easy time of it. Setting gardens and pastures involved first penetrating the entangled brush with machetes to open pathways. Once a proper area was located, often not too far from other peasants or streams, the process of clearing began with selective logging, which often involved cutting large trunks. At that time, the job was arduous and labor-intensive, because chainsaws were rare and expensive. The region suffered a chronic shortage of workers, so wealthy pioneers looking to set up large fazendas resorted to middlemen or contractors to hire squads of laborers. It is not known exactly how many transient laborers worked during that period clearing lands, but some estimates indicate that "there were anything from 250,000 to 400,000" across the Brazilian Amazon.[91]

After selective logging, fire was used to raze the area. This wasn't simple either, because, as noted by experts, "fires are not a normal part of the ecosystem in the humid tropical forests" like the Amazon.[92] Therefore, chopped logs and shrubbery were left to dry for several weeks to make them flammable. Scholars have noted that the practice of using man-made fires for clearing in the region is long-standing as Indigenous peoples have always set "fire to grasslands, to savannahs, to downed forests destined to be agricultural plots."[93] But the scale of the environmental destruction of those activities bore no comparison to the settlers' massive slash-and-burn operations.[94]

Over time, the DER staff also became involved in dealing with tasks less pleasant than the orientation of settlers. The first cemetery, according to Queiroz, was established when a brawl between laborers degenerated and ended with "one man suddenly cutting another in half" with a sharp machete. Informed about the homicide but unsure about what to do, the chief engineer ordered his men to get a bulldozer, clear an area, and bury the dismembered body in a grave without ever knowing the victim's real name.[95] "Things here were this way, because there was no one to report to, and also because no one knew who those migrants were," recalled Queiroz.

Despite the seeming lack of civilization, which many reasoned was the consequence of pell-mell development, people remained optimistic about the future. The country was undergoing a rapid transformation as a result of state-led policies to industrialize southern Brazil—especially São Paulo—and the national economy was booming. The GDP was rising "at an annual average of 10.9 percent from 1968 through 1974."[96] The "economic miracle," as it would come to be known, was unfolding as the government presented a new project—a refined version of Operation Amazonia—to "inundate the Amazon with civilization."[97] Called Programa de Integração Nacional (PIN), the new strategy, which was financially supported by international institutions, combined the ambitious construction of a set of federal highways with massive plans for direct colonization.[98]

The flood of migrant peasants and colonists, until then largely a result of moves by families pursuing economic betterment, prompted the government to try to establish some order in the process. In October 1970, about a year after the eviction of the Gaviões, the Brazilian president Emílio Garrastazu Médici traveled to Altamira, a small town located some 370 miles west of Vila Rondon, to kick off the construction work of the Trans-Amazonian Highway, one of the pillars of the new campaign.

A 2,600-mile road crossing the whole basin from east to west and passing through Marabá, the Trans-Amazonian connected the densely populated and impoverished northeastern region of Brazil with the heart of the jungle, further expanding the frontier deep into the hinterland.[99] "A necessary effort," declared Médici in a public speech given at the construction site, "to solve two problems: one of people without land in the Northeast and another of land without people in the Amazon."[100]

The plan to build the Trans-Amazonian Highway was said to have been conceived after a trip Médici had made a few months prior, when he'd witnessed how rural dwellers in the northeast starved due to one of the droughts that periodically affected that part of Brazil.[101] "Nothing in my whole life has shocked and upset me so deeply," said the president.[102] In a country slightly smaller than the United States but whose paved roads totaled only 38,000 miles at the time, the PA-70, the Trans-Amazonian, and other highway projects in the offing would represent a game-changer in terms of land connectivity.[103] They were integral to the larger plan to distribute plots to some 100,000 migrant families who, according to the federal authorities, would be provided with agricultural services, healthcare, and education in the villages and towns established along the roads.[104]

"The government will occupy the entire Amazon before the end of the presidential term," the minister of transportation in those years, Mário David Andreazza, declared.[105] "Distribution of these lands will undoubtedly be the largest and most important agrarian reform undertaken in the Western world since the historic Homestead Act, enacted by the United States in 1862, whose mass distribution of federally owned land, in the opinion of international economists, marked the beginning of outstanding progress by that great northern nation," wrote one Brazilian scholar.[106]

Regardless of the utopian ideals behind the regime's policies and the glowing words they inspired, perhaps the most profound shift of the new campaign to settle the Amazon would occur in people's minds. The national imagery surrounding the jungle and its riches was reshaped through nationalistic propaganda. In his book *Conjuring Property*, Jeremy M. Campbell, an American anthropologist who later studied the colonization process in Pará, would summarize the changes thus:

> The arrival of development paradigms in the 1960s shifted national attitudes, and brought with them a vision of the region as a frontier for state and market expansion. . . . Amazonia became a development frontier to the exclusion of almost any other metaphor for political

geography. In the popular imagination—and in the programming devised by domestic and international elites—it was a site of expansion, accumulation, and contest. For both elites and peasant colonists, the region had been remade into a blank space on the map of the nation, a *terra nullius* that called to be occupied and developed. The frontier had triumphed.[107]

Historian Frederick Jackson Turner, the father of the concept of the frontier, once described it as "a field of opportunity."[108] And that was precisely the idea that the government promoted through radio, television, and newspaper and magazine ad campaigns "emphasizing Brazil's superpower potential through westward expansion."[109] "Amazon: Enough with the legends; let's take advantage of it" and "Amazon, the challenge we will together overcome" were some of the slogans that would be burned into people's minds.[110] The one that stuck was Medici's reformulated and shortened "land for people for people without land."

Soon, events in Vila Rondon would demonstrate that things were far different than the propaganda would suggest, especially for humble migrant families dreaming of owning a plot. Josélio de Barros was about to make his name known in that remote outpost. There the former Contestado man would be confronted with the ghosts of his past.

4

The Chainsaw Murder

PIONEERS would recall the Vila Rondon of the 1970s as wild. "It was a violent place, mostly at night, when people went to parties, drank too much, and got involved in shoot-outs," explained Rita Belem, who, after hiding in the DER during the Gavião-Kyikatêjê raid, slept with a shotgun under the bed to protect herself from bandits.[1] The Italian missionary Giuseppe Castelli, who in 1968 became the first Roman Catholic priest to preach in Vila Rondon, recalled that the sense of constant danger led people to carry guns at all times, even during his masses. "Initially, hidden under a loose shirt; in subsequent years, fully displayed," he said, implying that the practice had become standard in the hamlet.[2]

Besides barroom brawls, the true motivation for conflicts was the control of land. During those years, people were engaged in a frenzied scramble to secure plots, gardens, lots, and parcels near the PA-70. The continual arrival of caravans of pickup trucks and long-haul interstate buses bringing new waves of settlers only added pressure in virgin areas and fostered a sense of urgency to get a plot. Some families even rented small trucks back in their home states to move en masse with relatives—sons, cousins, aunts, uncles—to Vila Rondon. These wide-eyed peasant colonists frequently carried wads of cash hidden in sacks of flour or seeds to invest in land right away, often without previously consulting with officials about whether a particular plot was available. Posing as real estate agents or legitimate sellers, an underground net of forgers sold land that was owned by the state or the federal government and that, not infrequently, was already occupied by previous migrants or local populations with user rights.

"Deceitful sellers in Vila Rondon argued they had land available for buying," recalled one early pioneer, a woman whose family had bought large areas along the PA-70.[3] "They drew a map of the parcel on the dusty soil of Vila Rondon, and said to new colonists, 'On this side of the property, there's a river; on that one, it borders with Mr. X's property, on the other, with Mr. Y's. . . .' And the deal was sealed. People here bought jungle, pristine jungle. And everything worked on the basis of trust."

Handwritten receipts were handed to buyers, but they had no legal value whatsoever. Relentless demand led to a spike in land prices, and the jungle became not only a potential means of production but also a sought-after commodity, drawing in more speculators and outlaws who soon became infamous.[4] As in other areas that were big draws for immigration, official and illegal land markets existed side by side, triggering disputes.[5]

During this period, national newspapers like *O Estado de S. Paulo*, *Jornal do Brasil*, and *Folha de S. Paulo* published stories about violent land feuds, some involving wealthy American families who had settled in massive spreads near Vila Rondon. The *New York Times*, the *Washington Post*, and other American outlets would also report on such conflicts, which often had a common theme—the recruitment of gunmen by so-called fazendeiros in order to push out peasants and homesteaders.[6] Media stories also denounced the use of "slaves" in the process of denuding the rainforest, and a civil servant would claim to have found graveyards near Vila Rondon where workers had been "disappeared."[7]

Among the many conflicts, one case in particular caused tensions to deepen. It took place in a swath of forest some nine miles west of Vila Rondon. There, a rancher from Bahia named Pedro Alves dos Santos had set up a large cattle ranch called Fazenda São Pedro and had begun urging his neighbors—about three hundred families of homesteaders—to leave the area without delay.[8]

Alves said he had moved to the land as early as 1967, while the

Gavião-Kyikatêjê still lived there, and that he had employed seventy laborers to clear adjoining plots he claimed to have bought from former owners.[9] But the homesteaders accused Alves of being a land-grabber and claimed that they had set up their gardens when the region was unsettled, some recalling having faced "jaguars and snakes" and having survived on a diet of "roasted monkeys and turtles" in order to clear the jungle and earn their plots.[10] Now Alves was threatening to steal their land, they said.[11]

In March 1972, the peasants brought their case to Belém.[12] An Italian Catholic missionary named Giuseppe Fontanella accompanied them. Born to wealthy farmers in a tiny village near Piacenza, in the north of Italy, the thirty-nine-year-old, five-foot-eleven Fontanella—a man said to be outspoken and quick-tempered—had been named Vila Rondon's parish priest after being sent to the area along with Father Castelli, who finally settled in Paragominas.[13] Upon their arrival in the late 1960s, the priests watched with mistrust as a rising class of fazendeiros quickly established links with judges, policemen, mayors, and lawmakers.

Like many other grassroots members of the Catholic Church living in Pará, the Italian priests decided to side with homesteaders and peasants and help them organize. In Vila Rondon, Fontanella drove a car with a loudspeaker, urging *posseiros*, a term for peasants without a proper land title but having user rights, not to abandon their lands.[14] He was also key in bringing the story of the three hundred families confronting Alves to public awareness.[15] He approached the media to draw the attention of the federal authorities, and the case eventually gained traction in the Brazilian congress in August 1973.[16]

One of the newspapers that decided to investigate the case was *A Província do Pará*, which sent a reporter to Vila Rondon.[17] Colonists denounced abuses and even a murder by gunmen who were allegedly operating on Alves's orders.[18] Aware that the case was seriously damaging his reputation, Alves also offered his version of the facts to the journalist. Casually dressed and sporting a fresh haircut, Alves admitted that "two men who take care of [Fazenda São Pedro] are armed."[19] But he denied the claims that these men were hired assassins and assured the reporter that "they have gun licenses" and would use the weapons only within the boundaries of the fazenda—and only for a legitimate reason. Asked by the journalist how he had acquired the disputed lands, Alves declared that he had bought some 60,000 acres from a woman named Maria Ismael dos Santos (same surname but not related), who had handed over a deed that was later authenticated by the state authorities.

But the journalist himself witnessed how far reality diverged from Alves's narrative. In his full-page story titled "Vila Rondon, the Place Where Pioneers Fight for Land," the reporter gave a firsthand account of the blunt display of power exerted over anyone interfering with the rancher's businesses.[20]

"It was Saturday, 12:30 p.m., when the reporter arrived" at Vila Rondon, the journalist wrote, referring to himself in the third person. "He entered the first tavern he saw to have a soda. Right away, he was dumbstruck by a cowboy armed with two guns and a shotgun." The man, the chronicle went on, rudely placed the shotgun over the bar top and ordered a soft drink, then turned to the reporter and, "with an index finger poking his chest, said: 'Young man, are you the journalist who wrote those stories about the estate of Mr. Alves?'" The reporter answered that he was, and the defiant cowboy asked him when he was leaving town, to which, unintimidated, the reporter answered, "In four or five days, the time [necessary] to verify everything related to the incidents between the colonists and Mr. Alves. . . . The objective of my newspaper is to find the truth."

The cowboy, who was actually an employee of the fazenda, then said in a threatening manner, "Careful, young man, don't fall from your horse." After that, he paid for the sodas, "slapped the bar top," exited the premises, and jumped into a jeep transporting other wranglers that soon disappeared along the sandy streets of the outpost.[21]

Up until that point in time, no one had mentioned Josélio de Barros, who by then had already settled in nearby Paragominas.[22]

FOR YEARS, THE dispute over the Fazenda São Pedro would come sporadically to the attention of the media. But the case would hit the front pages of newspapers in Pará only in mid-1975, right after two peasants involved in the land feud went missing. Their names were Honório Vieira Ramos and Antônio dos Reis Silva, and they had migrated to the area along with their extended families.

According to witnesses, on August 7, 1975, the pair had vanished after they "were invited [for a meeting] by the overseer of the Fazenda São Pedro, who claims to be a policeman but is in fact a pistoleiro."[23] A source would later say that the two men, while working that day in their gardens, had been urged by employees of the ranch to go with them to Vila Rondon to meet a visiting civil servant of the colonization agency who was supposedly handling deeds for plots.[24] Likely thinking that the bureaucrat was there to mediate the conflict, Reis and Vieira believed the account and left with the men. Neither of the peasants was seen again. According to the newspapers, they had been "kidnapped."[25]

A sister of Antônio dos Reis named Conceição explained that her brother was a brave man and had taken measures to prevent his lot from being seized.[26] "My brother put cattle on his plot to show that it was his land and would not give it up," she declared. "There were a lot of years of struggle. My father begged my brother to let it go, to give up the land, as he himself had done in another rural area of Vila Rondon acquired for cattle raising. But Antônio refused."

Weeks passed, and the families became desperate. "Find my brother, or at least his corpse," pleaded Geraldo, a brother of Antônio dos Reis and Conceição who had taken the lead in campaigning for the peasant's return. "Dead or alive, I need to know where he is," Geraldo begged reporters in Belém, where he had traveled in early September 1975 to inform both the media and Pará's legislators about the case.[27] A few days later, a handwritten note was slipped through a window in the Reis family's house. The crude writing scrawled on crumpled lined paper finally revealed the fate of the two missing men.

"I want to let people know," said the note, riddled with spelling mistakes, "that two dead bodies were spotted in the Big Gurupi River that borders the states of Pará and Maranhão, and no one is missing in that area."[28] The note continued: "Take the old Belém-Brasília road; at the Santa Rosa sign, turn right, then 25 kilometers ahead," sketching meager but sufficient directions to locate a wooded area of a fazenda in Paragominas crossed by a stream. The letter ended: "Be careful about the can of worms around there. I don't say who's writing because I don't want to get into trouble."[29]

Geraldo himself went to verify the information in the note. At the site indicated in the letter, he found the dead bodies of Reis and Vieira, riddled with dozens of gunshot wounds and showing signs of having been partially burned.[30] Geraldo could identify his brother's remains only by his clothes, because the corpse had been beheaded. He decided not to take the bodies, probably to avoid altering the crime scene, but he had the fortitude to get the shoes and some of the ripped clothes off his brother as proof of the crime. Then he traveled to Belém to try to find justice. He was so determined to expose the facts and press the authorities that he posed for the media with the evidence he had collected, his hands grasping the rags of his executed brother's trousers and shirt.

"Savage Massacre in Vila Rondon," read the front page of A Província do Pará.[31] The article reported that Reis's daughter, a ten-year-old girl named Eni, had initially found the anonymous letter. The sensational story and the shocking photographs sparked an immediate reaction in the state's legislature. The leader of the party in power said that the governor "would take vigorous measures, because he wasn't willing to accept the massacre of our peasant brothers."[32] Another influential legislator denounced the "primeval and enraging events that should not occur in the twentieth century, when man walks on the moon."[33]

The forces of law and order were urged to act. Squads of Belém-based detectives and agents were deployed to investigate and recover the corpses.[34] When the police arrived at the crime scene, though, only Vieira's body remained, leaving the investigators to wonder whether the murderers had spirited away Reis's body or if it had decomposed and sunk or been swept away by the current of the nearby stream.[35] Crucial questions remained unanswered. Who had killed Reis and Vieira, how had they ended up in that remote area several dozen miles from their plots, and why had they been slaughtered in such a brutal way?

The last issue was anything but trivial. Land-related violence had become widespread in Pará, but, those days, the standard practice for gunmen was to shoot the victim and escape, leaving the body unaltered or, at most, hidden in the bushes. If the crime was to remain unknown for days or even weeks, a usual way for pistoleiros to prove to their bosses that they had taken care of their assignment was to cut off the victim's ear and bring it to the client.[36] (Years later, the police would find in the office of a Pará-based fazendeiro a jar in which the landholder kept grisly "trophies" floating in formaldehyde—the severed ears of people he had ordered his men to execute.[37]) This time, the case seemed different. The way the slaying of Reis and Vieira had been staged—the beheading, the abundance of bullets in their bodies, and even the awkward letter—seemed to indicate that the double murder had been perpetrated not only to kill the two peasants but also to instill fear in the rest of the families in the area.

Rancher Alves was the main suspect—and for good reason. Even though he claimed he employed only nonviolent means to defend his interests, police records showed that he had already been prosecuted for another homicide. On that occasion he had received backing from the vice-governor of Bahia and ended up getting the case dismissed.[38] Another suspect was one of Alves's alleged hitmen and the overseer of the Fazenda São Pedro, Sebastião Canuto Batista, who had already been sentenced for another murder.[39]

The investigation, however, took an unexpected turn when Alves, the same day that Geraldo reported he had found the bodies, showed up in a police station to inform the authorities that he was no longer the owner of the disputed fazenda. Months prior, Alves said, he had divided the Fazenda São Pedro and sold the land to several cattlemen. One of the buyers was Antônio Fernando Machado da Cunha, a wealthy fazendeiro from Minas Gerais who had been investigated on suspicion of making a fortune selling fake deeds in Paragominas and elsewhere.[40] Like Alves, Fernando also had a network of high-ranking officials backing him. (He was the protégé of a top army officer—a *marechal* or marshal, formerly the highest rank of the Brazilian army—who had participated in the coup.[41]) The other buyer of the land, Alves declared, was Josélio.[42]

Josélio had settled in Paragominas two years earlier, while he was still on parole.[43] Now the father of six children—five daughters and one son—he quickly mingled with the incipient local establishment, mostly wealthy sawmill owners and fazendeiros who, like both Fernando and Alves, had moved to the region lured by the cheap land and the government incentives to raise livestock.

Initially, Josélio had bought a fazenda called Mirabela from Fernando.[44] It is unclear when Josélio and Fernando first met—whether they had crossed paths in Paragominas or before both moved to the Amazon. Whatever the case, they had gained the attention of local authorities for their partnership. Before they were mentioned in the newspapers in relation to the massacre in Vila Rondon, the pair had been accused in confidential police reports of invading the property of another landholder in Paragominas to log the timber and sell the lumber to a group of sawmill operators who were also accused of a murder.[45] According to a police report, Josélio and Fernando had done so with the backing of the local chief of the police.[46]

Josélio also had the support of influential authorities to help him settle on the frontier. "I've been sent to the region of Vila Rondon by senators Dirceu Cardoso and Eurico Vieira de Rezende," Josélio would say in those years, referring to the support he claimed to have from two Espírito Santo senators at the time.[47]

For a reason never fully established, but probably for the great business opportunities the hamlet offered, Josélio decided to leave Paragominas and relocate to Vila Rondon. He bought the Fazenda São Pedro about a month before Reis and Vieira were executed, according to official documents.[48] After years spent trying to evict the peasants, Alves had decided to sell the property in order to start anew.[49] By July 1975, Josélio was already investing in his ranch, which he renamed Fazenda Serra Morena. As Josélio would later admit, it was also at that time that he started to develop a plan to get rid of the people settled in the area.[50]

According to Josélio's own account, one morning that July, he met at the Fazenda Serra Morena with two hired assassins.[51] The real names of the killers would never come out, the two being referred to in official reports by their nicknames, Zezinho and Juarez. Josélio would claim that he had never contacted the killers; instead, he said, Zezinho and Juarez had simply shown up one day at the ranch and had offered him their services. This version is highly unlikely, the most plausible explanation being that Josélio had hired the men somewhere and wanted their identities to remain unknown forever. Regardless, Josélio would admit that, after a few days of discussions with the assassins, a deal was sealed: he would pay them 10,000 cruzeiros, the equivalent of about $5,000 today, to kill Reis and Vieira. The date of the murder was set for August 7, 1975.

That day, the killers went to the fazenda and met Canuto, the overseer who had been brought to the region by Alves. (Josélio had decided to keep Canuto as an employee.) Together, the two pistoleiros were taken to the plots of Reis and Vieira, located some miles further into the jungle.

Reis was in his house when the two mobsters, who claimed to be army men and were probably dressed as such, showed up and urged him to accompany them to Vila Rondon to discuss land issues. Reis immediately took the bait and agreed without discussion. Vieira was working in his garden when the assassins reached his shack, but once he returned home, he had no problem accompanying them.[52]

Afterward, the two smallholders were taken into the wilderness and promptly tied up. Then Canuto went to the fazenda to inform Josélio. "The two posseiros have been captured," he said.[53] It was early afternoon, the sun shining brightly in Vila Rondon, so Canuto and Josélio agreed that it would be better to meet some hours later on a dirt road crossing the estate. Josélio said he would bring his pickup truck.

By six, with darkness descending, Josélio arrived with the vehicle. The killers, armed with revolvers, stood by their victims. All the men except Canuto were ordered to get in the pickup. One peasant was put on the bed of the vehicle and the other inside the cabin. Then they headed toward the fazenda mentioned in the anonymous note.

It is unclear what the killers' original plan may have been. But Josélio would declare that while approaching the ranch, they encountered a wooden bridge over a stream that was in bad condition and therefore couldn't be crossed. At that point, the two pistoleiros got out of the car and took Reis and Vieira with them. Josélio remained in the vehicle, from where he heard four shots. Afterward, Josélio said, the assassins climbed back into the pickup. They received payment, and he dropped them at a gas station on the Belém-Brasília Highway. The outlaws fled the region.

This version would be disputed by sources in Vila Rondon, where the double homicide would henceforward be referred to as "the chainsaw murder."[54] (Two decades later, when the man who had sold the Te-Chaga-U to Josélio was interrogated by Detective Moraes after the raid to unearth the human bones, he said that Josélio was known because "he had the custom of chainsawing in half the peasants who displeased him by not selling their plots."[55])

Conceição would say that contrary to Josélio's account, her "brother wasn't killed by gunshots; his head was struck from his body."[56] Her narrative matched the one provided by an eyewitness, a smallholder who was also being pressured to leave the land near the fazenda. The source would say that Reis and Vieira, on the day of the crime, were misled and brought "to the middle of the forest."[57]

There were about twenty [gunmen], and they caught these wretches, they shot and shot, but Antônio dos Reis just wouldn't die. . . . [H]Onório [Vieira] was killed by the first bullet; they emptied six revolvers into Antônio Reis. The more they fired, the more he fought back. . . . I don't know why, but they got a motorized saw and used it on him. He was very wicked and cut Antônio's neck to kill him. . . . He cut his head off, and Antônio still jumped like a chicken with a broken neck. So, what did he do then? He wrapped the body in some cloth, together with that of [H]Onório [Vieira], tied them up well, placed the bundle in the back of a Ford F-100, and made tracks toward Belem. In Paragominas, he threw the bodies into a river. . . . A fisherwoman saw them, but she would not speak out, afraid they might also kill her and dump her in the river.[58]

This witness mentioned Josélio as participating in the scheme but didn't provide the name of the person who actually did the beheading.

ONLY HOURS AFTER Reis vanished, and long before the crucial note appeared and grabbed the attention of the media, the family already suspected Josélio was involved in Reis's disappearance and reported the case to the police. In fact, on August 8, about twenty-four hours after the murder, officers visited the Fazenda Serra Morena to interrogate Josélio.[59]

In this first meeting, Josélio denied any involvement and advised law enforcement to contact Canuto to learn more about the issue. He "may be able to tell you something about it," Josélio said.[60] Later, it would be known that Josélio was lobbying Canuto to eventually take responsibility for the murder. Canuto would say that Josélio had promised him a vehicle and "all kinds of comforts" if he served time in prison.[61] The overseer had considered the potential deal for a while, but afterward he'd turned it down and resolved to reveal the involvement of both Josélio and the two pistoleiros. Canuto would also say that contrary to Josélio's account, the killers didn't flee the region right after the double homicide. He would say he'd met Zezinho and Juarez in Vila Rondon a day after the murder, and they confirmed what they had done—Reis and Vieira, they said, were "having a bath."[62]

There's little doubt that the early stages of the investigation unnerved Josélio. Judges in Espírito Santo had agreed to his relocation to Pará while still on parole on the condition that he show up every sixty days before a notary public to fulfill his legal obligations and, of course, that he stay out of trouble.[63] By the time of the "chainsaw murder," he had served his sentence for the murder of Cavalcanti in full, but Josélio was now a repeat offender, and he faced the possibility of years, if not decades, in prison for the crime. Therefore, as soon as the bodies were found by the family and his name appeared in the press, Josélio vanished, apparently going to Belém and then to Espírito Santo.[64]

Josélio's attorney would speak for him in his absence. Initially, the lawyer denied any involvement by Josélio and his partner Fernando. He also made an odd request: he wanted a "preventive habeas corpus" to ensure that, although Josélio was suspected of participating in a double homicide, he could "testify without fear of being arrested."[65] The cops, the lawyer complained to the media, were acting hastily. Six weeks after the murder, Josélio and Fernando, who was also unreachable, were guaranteed "safe conduct" from a judge to avoid their arrest.[66] It was the first clear signal that the crime might remain unpunished.

"It is possible that pressure is being applied to reduce the appearance of guilt or even to cover up the crime," an officer working on the case would warn in a confidential report.[67]

However, just days after Christmas, another judge ordered the immediate arrest of Josélio and Canuto for "ordering the murder of the two dead peasants."[68] By the end of 1975, Fernando was apparently no longer a suspect after eyewitnesses testified that he wasn't part of the scheme. Without delay, Canuto and Josélio confessed to the crime, and they were brought to a prison in Belém.

It is likely that by then Josélio had already contacted a crucial ally of his early years in the Amazon, the powerful superintendent of the federal police in Pará, Lincoln Gomes de Almeida. An intelligence report would state that Lincoln was suspected of providing Josélio with cover after the double homicide, although the high-ranking officer would deny it.[69] Josélio, however, wasn't shy about his ties to Lincoln. When he was arrested, he claimed "to have a great relationship with security organs, the federal police, and the army," declaring to the police that he was a "friend of Lincoln."[70]

Eleven years older than Josélio, Lincoln—with a bony face, prominent ears, wrinkled cheeks, and a receding hairline—had a stellar career by the time he was named the top federal cop of Pará. The son of a well-connected family in Goiás, a state of central Brazil dominated by rural elites, Lincoln would assume many top positions during his life, including the secretary of public security of at least two states.[71] He would also be a source of controversy, suspected of participating in a scheme to steal a set of fifteen machine guns from the army.[72]

By early 1976, with Josélio still jailed, Lincoln was relocated to Brasília, where he was named superintendent.[73] The new job, though, didn't make him forget about Josélio. According to the account of an inmate in Josélio's jail, "Lincoln used to call from Brasília to the prison to talk with Josélio to know about the case, and the lawyer [of Josélio] was hired by him [Lincoln]."[74] The source said that he had learned of the links between Josélio and Lincoln from Canuto, who was jailed in the same prison. He also provided context to explain the motive for the assassinations of Reis and Vieira. "The deal made by Josélio with his friend Lincoln, according to the overseer, was to expel all the colonists from the area."

The informant was a smuggler who accused agents of planting evidence to convict him. The extraordinary thing about the story was that he would communicate it to the president of Brazil, General Ernesto Geisel.[75] Complaining that he was being unfairly jailed and probably seeking help for his own legal troubles, the man had sent Geisel a letter exposing the alleged involvement of Lincoln in the case of the two murdered peasants. An official stamp on the letter confirmed the information about Josélio and Lincoln was received by the office of the president three weeks after it was sent from the prison.

That wouldn't be the last time the chief of the nation would hear about Josélio and Vila Rondon. Geraldo, the brother of Reis, also addressed a letter to the president cosigned by seventeen other homesteaders who desperately called for help. In the two-page document, they decried renewed attempts on their lives by criminals supposedly tied to Josélio and benefiting from the collusion of local police.[76] "Your Excellency, permit us to require the sending of a commission of administrative and police inquirers, composed by people you trust, to verify here the truth of this denunciation," they wrote.[77]

It is unclear whether the president ever sent someone to Vila Rondon, but by that time, the top authorities had acknowledged that the situation on the frontier—especially in the south of Pará, probably the most contested region of the Amazon—was starting to unravel.

The army and the police warned in reports that the "clash of interests among fazendeiros, homesteaders, and land-grabbers, each of them arguing to have rights over land, have culminated in the establishment of a 'modus vivendi' resembling that of the conquest of the American West, when each [individual] started to exert his own power of police, resulting in the prevalence of the law of the jungle, which generated a climate of deep social tension."[78] One of the main problems, according to the written records of the secret services, was widespread corruption endemic in the judiciary of Pará, where individuals suspected of ghastly crimes were freed even in cases that seemed to be airtight.

CASES LIKE THE chainsaw murder caused communities of settlers, peasants, and laborers living along the PA-70 to feel extremely vulnerable and suspect that nothing good would happen to them as long as the country was led by a regime favoring fazendeiros over poor peasant families.[79] A prominent Brazilian intellectual would later write that "never in the history of Brazil did the *latifundio* [a Brazilian term referring to the landed elite] make such unbridled use of private violence as during the military years."[80]

In truth, the dictatorship, despite the plight endured by victims, did not tolerate the flourishing of any organization that might potentially engender what they might consider revolutionary ideas. Brazilian intelligence archives reveal fears that the region could be infiltrated by communist elements, including foreign missionaries like Fontanella. The Italian priest was in fact investigated for his early defense of the peasants near the Fazenda São Pedro—now Josélio's Fazenda Serra Morena—and for his alleged involvement in other land feuds between homesteaders and large landholders.[81]

Cold war ideology certainly held some sway over the Brazilian government, but the cautious approach vis-à-vis class organizations was also sustained by a shift in the strategy to colonize the Amazon. Brasília had yet again revised the development model for the region, now prioritizing the establishment of massive farms and spreads— some as big as 1.7 million acres—by corporations and fazendeiros enjoying tax breaks and subsidies.[82] The policy shift meant that the generals no longer supported the tens of thousands of sharecroppers and laborers moving into the area to claim their lots. The "land for people for people without land" policy had been dropped, and the driving force of colonization had become commodity production and agribusiness. This policy reversal had real consequences for rural dwellers and migrant peasants, and the Catholic Church of Brazil, the largest Catholic community on the planet, emerged as a pivotal institution supporting these rural communities. Under the influence of liberation theology—a religious movement advocating for social justice founded in the late 1960s and popular throughout Latin America's clergy— the Brazilian church became engaged in fighting poverty, defending human rights, and, in general, denouncing excesses of this economic model, which was widening the gaps between social classes and caus- ing an environmental disaster. In June 1975, bishops of the Amazon funded the Pastoral Land Commission (CPT), an agency to help peas- ants fight for their rights, including their right to claim a plot on the frontier. The landless of Brazil, the CPT said, had to organize through the creation of rural unions.[83] Many small rural workers' unions were founded during those years. Unsurprisingly, the police regularly sent agents in to infiltrate and monitor them and their leaders.[84]

Father Fontanella followed the CPT line and encouraged down- trodden rural dwellers in Vila Rondon to think outside the box and to organize to improve their lives. His help was sought soon after the chainsaw murder, and the priest mobilized bishops to reach the top judicial and political authorities of Pará.[85] But, in the end, the case had no major legal consequences for Josélio, who in the second half of 1976 was released on bail.[86] The case would remain open for years, but according to Conceição, it "ended in nothing," a version confirmed by Brazilian researchers who maintain that Josélio was never tried.[87] (As would happen with other criminal cases involving the gentry across Pará, the file of the case—which would confirm the actual legal outcome—couldn't be found in Pará's judicial archives when a journal- ist later requested access to it, and the federal police refused to grant access to their archives to verify the nature of the relationship between Josélio and Lincoln.)[88]

The Reis family, fearing further retaliation, decided to move away from Vila Rondon. On November 25, 1977, Josélio formally concluded the purchase of 7,260 hectares from Alves to form his Serra Morena estate.[89] He paid some 300,000 cruzeiros in cash—today's equivalent would be $62,000—for it.[90] The original deeds used in the sale would spark great controversy. In the ensuing years, federal officials would accuse Josélio of illegally expelling more settlers from their lands.[91]

Fontanella's activism would ultimately backfire against the priest. On December 13, 1976, while Josélio was still under investigation, the Brazilian authorities deported Fontanella to Italy.[92] The official reason was that his visa had expired. The police even suggested that Fontanella might have forged his resident permit.[93] Hours after returning to Europe, the priest attributed his expulsion to "the pressure exerted by some large fazendeiros to get rid of me, because I have always supported the cause of the peasants."[94] He would stick to this story until his final days in 2013.[95]

Years later, people in Rondon do Pará would say that Josélio and his connections at the federal police—which, in Brazil, deals with foreigners' resident permits—might have played a role in Fontanella's deportation, although no official record supports this account.[96] A more likely version of facts is that Fontanella was deported for his suspected but never proven involvement in the assassination of three Americans near Vila Rondon on July 3, 1976, when dozens of squatters ambushed the landholder, former air force pilot and Presbyterian missionary John Weaver Davis and two of his sons.[97]

The brutal crime—yet another related to the control of frontier land—unnerved the regime because it had been committed on the eve of the two hundredth anniversary of the American Declaration of Independence. The Brazilian authorities hinted at the possibility that it could have some further political meaning or could even signal the revival of a far-left armed group that had been bloodily crushed in Pará in 1974—the so-called Guerrilha do Araguaia.[98] The army and the police would come down with an iron fist in an effort to capture the killers, but the conflict involving the Davises would last for decades, and justice would take years of struggle in courts. Fontanella, who never returned to Brazil, would deny any involvement in the assassinations, although the Davis family would claim that the priest had orchestrated the ambush in order to evict them from their 247,000-acre spread, named Fazenda Capaz, and to redistribute the land, mostly covered by forest, among poor homesteaders—a claim never officially confirmed.[99]

The murders, reported by the *New York Times* and the Associated Press, earned Vila Rondon a reputation nationwide as a dangerous and lawless place.[100] Soon, it would be renamed Rondon do Pará and transformed into a municipality with its own local government.

Almost a decade after these crucial events left a mark in the early history of the region, Rondon do Pará would be the place chosen by Dezinho and Maria Joel to begin a new life as settlers. The young peasant couple were unaware at that time that they, too, would leave their mark in that corner of the frontier.

Part II

Rise and Fall

5

The Boomtown

MARIA Joel was twenty-one years old when she first set foot in Rondon.[1] She arrived there from her home state of Maranhão after a long trip on a dilapidated bus. It was June 1984, about a decade after the chainsaw murder, and she was accompanied by her two children and her mother, who had already established a place in the town. Dezinho was expected to arrive in the coming weeks.

The introverted woman with black hair and a sweet voice was excited to reach the city that, according to her mother, offered so many opportunities for rural immigrants like them. "There's a lot of abundance—come with us," her mother had insisted. More than half of her large family—Maria Joel had eleven sisters and brothers—had already moved in. Previously, they were sharecroppers in São Felipe, a tiny village situated in the municipality of Urbano Santos, and they had moved to Pará with the aim of becoming smallholders.

By the time Maria Joel arrived, the former hamlet had evolved into a hive of activity and noise. The first thing she saw in the raucous and dusty boomtown was an avenue clogged with a caravan of rig trucks delivering logs to sawmills. The rumbling, mud-encrusted vehicles, driven by shirtless men, passed sidewalks on which peasants sold their produce, women used parasols to shade themselves from the hot sun, and children played soccer.

Maria Joel quickly realized that Rondon was a far cry from the town of tidy residential neighborhoods and lush urban forests that she had envisioned. About two dozen sawmills running on diesel generators and employing about a hundred workers each had become the engine that drove the local economy.[2] The sawmills were located here to take advantage of the nearby reserves of Spanish cedar, fifty-meter-tall *maçarandubas*, and mahogany—highly valuable tree species "so abundant in Rondon's surrounding forest that it seemed like someone had planted them centuries prior," recalled a pioneer.[3] The economic potential of those riches was so great that some wood investors, after surveying the area, had relocated to the frontier town en masse, dismantling their industries back home and moving in with blades, power tools, vehicles, and qualified personnel.[4] They often did business with fazendeiros, thus developing bonds of interests with the ranchers.

The thriving lumber industry, spurred by domestic and global demand for wood products, provided stable jobs and tax revenues for the newly created municipal government of Rondon. Other businesses like garages, restaurants, ice cream parlors, pharmacies, photography stores, and shops selling all kinds of goods also proliferated in response to the demands of a rising working class. There was even a movie theater with a loudspeaker outside blaring out the unforgettable Ennio Morricone theme song for *The Good, the Bad and the Ugly* to announce that the next showing was about to start.[5]

Scores of municipalities were founded during that period in a Brazil expanding and consolidating its colonization project in the Amazon. But in southern Pará, Rondon seemed somehow special due to its thriving economy, and the city became known informally as "the little princess of Pará." The flood of migrants, crucial to supply the workforce necessary to sustain the economic growth, had pushed the number of residents past 20,000 by 1984, about a tenfold increase from the days when the Gaviões had been evicted.[6] Many settlers were unaware that the area had been subject to profound conflicts over land, believing instead that they were relocating into a sort of El Dorado. For those aware of the problems, the sense of possibility outweighed any reticence they might have faced due to the killing of peasants and subsistence farmers by pistoleiros suspected of operating under the orders of land sharks and fazendeiros.

Initially, Maria Joel also believed Rondon would be a good place to settle and prosper. But as she took in her new home, her high expectations were dashed. Maria Joel's early impression of Rondon was awful. She disliked it immediately and felt it was a terrible place to settle with her husband, her two-year-old daughter (Joelma), and her newborn baby boy (Joelson), who was about fifteen days old when she made the trip.

Besides its ugliness, Rondon was not an ideal home for a family with young children. Many sawmills had been built only a few feet from cabins, churches, and businesses, and in some neighborhoods the clatter of blades quartering logs and the smell of gasoline were so invasive that residents were left with no option but to endure the noise and the stink several hours a day. Rondon also lacked a modern medical system, so infant mortality related to malaria, malnutrition, and dehydration was also high.[7]

Though she wasn't aware of the land-related murders that had marked the town's young history, Maria Joel soon realized that crime was an issue. In the fifteen months following Maria Joel's arrival, the Pará media reported nineteen homicides.[8] Among the victims were a rancher killed with a .38-caliber weapon in broad daylight; two peasants assassinated in a fight during a party, one of whose hands and feet were amputated with a machete; and a laborer mutilated and then battered to death for raping his employer's wife—a sexual assault he had committed at knifepoint.[9] No one was free from the risk of death: victims included Pentecostal pastors, cowboys, a councilman, and a thief, whose body was found decomposing not far from Josélio's Fazenda Serra Morena.[10] Even young people settled their scores in blood. In a case that gained a lot of play in the media, a fourteen-year-old boy killed his stepmother with a gun because "he didn't approve of the marriage with his father."[11]

Maria Joel heard some of these accounts from her sister Eva, who had moved to the area before her. Two years older than Maria Joel, Eva was more outgoing than her highly reserved sister. Despite their different personalities, the pair shared a special relationship. "When one was ill, the other also felt the pain," Eva would say to describe their bond and mutual understanding.[12]

Eva hadn't had an easy time settling in Rondon, either. One night, while walking home, she and another female friend had been followed by a group of drunk men who had attempted to get the pair into a car. They had run away and hidden in the yard of a cabin before a passerby spotted them and decided to help. "He was an old man with a chainsaw who said he would take us home safely," Eva recalled. "He said, 'If those mobsters try to do something to you, I'll cut them!'"

Stories like this made Maria Joel reconsider settling in Rondon. Her feelings were in sharp contrast to the great enthusiasm of sharecroppers, landless peasants, and pioneers of all walks, who continued to move to the town because they believed that "the little princess of Pará" offered them great opportunities to prosper. But Maria Joel saw things in a different way. The tiny village where she'd been born was much poorer and certainly underdeveloped, but Rondon provoked in her a sense of fear and rejection. Maybe, she reasoned, it would be better to return to the Atlantic city of São Luís, the state capital of Maranhão, where Dezinho had remained while she scouted out the boomtown and sorted out logistics—a home, a job, and schooling for Joelma, whom the pair affectionately called Joelminha, "little Joelma." She planned to write to Dezinho to explain her doubts and propose a plan B. (Letters were the primary form of communication, as the city had no private landlines and the only way to receive or place a phone call was through a single paywall, where businessmen queued for hours to place orders and negotiate over their produce.)[13]

"I don't like this place," Maria Joel told her mother.

"Don't worry about it now," her mother insisted. "It's a good city to start a life."

BEFORE SHE SENT that letter, Maria Joel realized she should think through her decision. One central issue was the family's difficult economic situation. Just three years prior, Maria Joel and Dezinho had migrated for the first time, leaving the interior of Maranhão to settle in the state capital, and if they had expected to get stable and decent paid jobs in São Luís, things hadn't gone as planned. Dezinho had worked for a while in construction, but he was now jobless, and the couple had been struggling to pay the bills.

Much had changed in Brazil since the economic bonanza of the early 1970s. In the first half of the 1980s, the economy had shrunk, and Brazil, like other Latin American countries that had borrowed heavily from foreign bankers to sustain a decade of growth, was now struggling to honor its loan engagements. Both its external debt—the largest in the developing world—and the mounting balance-of-payments deficits had put foreign investors and lenders on the defensive.[14] Brazil's economy was also crippled by runaway inflation—192 percent in 1984, 225 percent in 1985—which had a terrible impact on families with low consumer purchasing power, like Maria Joel's.[15] As a consequence of all this, Brazil was shaken by recession during those years.[16]

Some economists would later refer to Brazil's 1980s as "the Lost Decade."[17] However, while the crisis drove masses of unemployed industrial workers to pockets of poverty and exclusion (the favelas), opportunities were blossoming in southern Pará.[18] There, the regime was developing a colossal economic project to mine the most abundant high-grade iron ore deposits known on the planet. Called the Greater Carajás Program, it had the potential to transform Pará's economy and provide tens of thousands of jobs. A scholar from the London School of Economics would call it "the largest 'integrated' development scheme ever undertaken in an area of tropical rainforest, anywhere in the world," and a Brazilian minister would say that the Greater Carajás Program was "the most alluring project" of its time.[19]

The initial discovery of the deposits had happened almost by accident.[20] In 1967, Brazilian geologists surveying areas of forest for a local subsidiary of the United States Steel Corporation landed their helicopter to refuel in the Serra dos Carajás, a range of low mountains some 200 miles southwest of Rondon. To the surprise of the scientists, the surface of the hilltop contained scant vegetation and what seemed to be iron ore rocks. They took samples and analyzed them, promptly confirming the extraordinary potential of the area.

By the late 1970s, after much planning and back-and-forth, a program was launched to extract and export the reserves. The plans involved the construction of both a massive hydroelectric dam on the Tocantins River to provide energy and a deep-water exporting port near São Luis that would be connected to the mining area by a 550-mile railroad cutting through the wilderness. The dam and the railroad would have a deleterious impact on hundreds of communities of rural dwellers and Indigenous tribes. Initially valued at an astounding $61 billion, the Greater Carajás Program aimed to transform that forgotten swath of jungle into one of the most promising iron ore mining regions on Earth, a strategic asset to earn much-needed foreign dollars to pay down Brazil's debts.[21] With reserves projected to last several decades, the regime also promoted the establishment of a local steel industry. Unsurprisingly, it didn't take long before the Greater Carajás Program became "a giant population magnet."[22] Later estimates showed that the workforce employed in the program jumped from some 3,000 in the 1970s to about 50,000 over the ensuing two decades.[23]

But Carajás wasn't the sole opportunity in that region of Pará for families who, like Maria Joel's, lacked capital or the qualifications necessary for good jobs. Just a few dozen miles east of the iron ore deposits lay another mine, this one filled with gold. Its name was Serra Pelada, which can be translated as Bald Hill or Naked Mountain, drawn from the process of stripping away the vegetation from the knoll to pave the way for the excavation of gold.

The gold of Serra Pelada was said to have been discovered in 1979 by a cowhand who had chanced upon a nugget in a stream near a low hill.[24] The man had apparently shared his treasure trove with the landowner of the area, who sent the sample to Marabá for analysis and confirmed that the stone contained gold. "Word leaked out, and within a week, 1,000 prospectors had descended on the farm. Five weeks later, there were 10,000," a *Time* reporter wrote.[25] Eventually, about 100,000 men from all walks of life—including many of Rondon's inhabitants and even Percy Geraldo Bolsonaro, the father of Jair Bolsonaro—would gouge a crater 600 feet deep and half a mile wide.[26] In Rondon, located about half a day's drive from the deposit, "many men left their jobs in the sawmills to find a future there," causing a labor shortage, recalled one pioneer.[27]

By 1983, a year before Maria Joel's arrival in Rondon, annual gold production in Serra Pelada reached 13,000 kilograms, making it one of the world's largest open-pit gold mines.[28] According to one scholar, Serra Pelada was "the richest gold discovery in Amazonia in living memory."[29]

The Brazilian photographer Sebastião Salgado would masterfully capture the essence and antlike frenzy of the Serra Pelada goldfields with a series of astonishing images in which seminude wildcat prospectors known as *garimpeiros* jockeyed to dig the soil while others carrying weighty bags of ore climbed across towering wooden ladders called *"adeus, mamae"* or "bye, Mom" at the risk of immediate death for any who slipped and fell.[30] Salgado's photographs, some observers noted, brought to mind Dante's Hell, the mine benches suggesting the allegory's circles while the gaunt gold diggers, often working for other garimpeiros or for small investors, resembled desperate lost souls seeking in the dazzlingly rich netherworld a redemptive path for their errant lives.[31]

It is not totally wrong to say that the mine and the shantytown sprawling near it—the modern-day city of Curionópolis—were a sort of inferno. Mercury flowing from the sluices polluted the waters, while handfuls of diggers were buried alive in the deposits due to the recurrent landslides; roving gangs extorted and assaulted upstart garimpeiros, and underage prostitutes were smuggled through the forest to the area around the mine, where women were banned in an attempt to prevent bloody conflicts.[32]

"Everybody had a shotgun. There was firing in the air every time there was a discovery—maybe 200 times a day. No one was killed. But real violence was just a matter of time," a gold prospector wrote, describing an unpredictable, anarchic place not appropriate for all types.[33]

For most fortune hunters, though, opportunities outweighed the risks. The luckiest garimpeiros unearthed nuggets as large as 137 pounds, although record discoveries weren't really required to sustain the legend that Serra Pelada offered the chance of a lifetime—a 15-pound nugget would immediately yield some $100,000.[34]

Maria Joel's brothers fantasized about the place, but for the young woman, Serra Pelada was never a real option. She was a conservative, devout Roman Catholic who'd stayed close to her family and away from trouble throughout her life, and even if she disliked the chaotic hustle and bustle of Rondon, Maria Joel couldn't have felt more alienated anywhere in the world than in that mining area, where perdition and death lurked around every corner. "We knew there were a lot of opportunities. But we were too scared," Maria Joel would say.

Maria Joel's cautious approach to life stemmed from her childhood, when she had learned discipline, obedience, and order from her strict father, who involved his children in caring for the garden. She would reminisce about her father's admonition that the children in a landless peasant family had little time for fun, no matter how young they were. Maria Joel and her siblings worked to produce rice and manioc, and they had to walk over a mile to reach the nearest school. Only the weekends offered some respite and diversion, when the family put on their best clothes—shirts, dresses, and skirts made by a sister of Maria Joel from a single bolt of fabric—to go hear the Word of God and participate in community events.

Although this harsh childhood had had a sobering effect on Maria Joel, it also meant she was a highly determined person. She had loving parents who stayed together, and Maria Joel would recall that despite the hardships, her childhood did not lack joyful moments. The lessons of those early years—struggle, diligence, resolve—would become crucial when she later confronted the greatest adversities of her life.

But as a young mother, Maria Joel wanted a different life for her children—one in which food on the table, formal education, and leisure were everyday certainties. She also aspired to one day own her own house—one that, even if modest, had electricity, running water, and a bed for each of her children.

Despite its Wild West atmosphere, a trait common with many other frontier outposts of Pará, Rondon did seem to offer her the opportunity to accomplish that dream; besides, in Rondon she had the support of her extended family. So she weighed her options. Should she send that letter to Dezinho? Would she be doing the right thing for her children if she pushed them to move somewhere else just because she didn't have a good feeling about the place? After considerable thought, Maria Joel ultimately decided that she would say nothing to her husband and would give "the little princess of Pará" the chance it deserved. After all, she posited, they would manage to stay out of trouble.

6

Early Challenges

DEZINHO arrived about a month after Maria Joel had settled
with Joelminha and her newborn, Joelson. They stayed a few
weeks with relatives, and then the family moved to a small clapboard
cabin they rented in the suburbs, in a neighborhood of sandy streets
and scattered vegetation. Influenced by Maria Joel's lack of enthusiasm
for the town, Dezinho disliked Rondon at first blush. But the pudgy
and ebullient man had an outgoing personality that contrasted with
that of his wife, and his chattiness and folksy manners helped him to
socialize and adapt more quickly to the boomtown.

Six years older than Maria Joel, Dezinho had also had a tough childhood. He had no memories of his mother, who had died as a result
of the complications of pregnancy along with her unborn baby when
Dezinho was just two years old.[1] After the tragedy, not uncommon in
the Maranhão of those days, his father had remarried, and Dezinho
and his only full brother, Valdemir, had been raised by grandparents
and uncles. Eventually, his father would marry two more times before
finally resuming primary care of the boys. Still, he was an active phi-
landerer who couldn't devote his full attention to his eldest sons. "We
suspect," Valdemir admitted later, "that our father had some forty-five
children in his life."

Dezinho's family wasn't as poor as Maria Joel's. His father had inherited a plot of some 300 acres, a decent amount of land for the social
context in which they lived, and even though Dezinho and Valdemir
weren't fully spared from work, they were paid for it, allowing the two

brothers to earn a living and buy shoes, jeans, and even a bicycle. "We remained a poor family, but wealthier than others, because we owned our land," Valdemir recalled.

Their financial situation made it possible for the two brothers to move to the neighboring city of Urbano Santos when they reached adulthood. The little town in western Maranhão was anything but a bustling capital—in 1980, only 4 of the 19,000 people living in the municipality had a college degree.[2] Still, it provided Dezinho and Valdemir with an opportunity to meet different people and to lead a more stimulating and enriching life than they would have been able to in the backwoods interior, where family traditions and severe economic insecurity often determined a person's future.

Gifted with charisma, Dezinho was outspoken and had leadership qualities. "As an older brother," said Valdemir, who had two years on Dezinho, "I should have been looking after him, but he wanted to take care of me." In fact, Dezinho wanted to take care of many other people in his community, notably landless peasants seeking help from CPT priests and lawyers. He enrolled in a peasants' union and joined a local Catholic church with the aim of beginning a career as a grassroots activist. One day at church, he met Maria Joel, an arrestingly beautiful teenager with shiny black hair and a soft, delicate voice. Timid and pensive, traits that the mischievous Dezinho couldn't have liked more, she had a wide and honest smile. Before long, they were engaged.

Those years molded Dezinho's political ideas. Crisscrossing the interior of western Maranhão, he witnessed skinny children with stick-thin arms, their parents, despite recruiting everyone in the family to farm the land of others, struggling to get enough food to keep from dying of hunger. "For every hectare of the property farmed, sharecroppers had to deliver to the landholder ninety kilograms of corn, ninety kilograms of manioc flour, and a final ninety of rice," recalled Valdemir, noting that this wasn't illegal, but it was abusive. "Peasants worked two weeks per month for the profit of powerful landowners, and the other two to just be able to eat something every day."

Influenced by the progressive views of an Urbano Santos–based French missionary who, like Father Fontanella in Vila Rondon, was later arrested and interrogated by the authorities for "subversive" left-ist ideology, Dezinho considered the situation pure injustice.[3] In his view, extreme poverty in Brazil stemmed from unequal access to land, and agrarian reform became his personal cause, a crusade that would define his adult life.

In Rondon, Dezinho was keen to become a homesteader, but his expectations were crushed soon after he moved in. He was told by civil servants of the colonization agencies that the only plots available were located in the distant wilderness, over a hundred miles from the town. That area of jungle lacked electricity and was accessible only via unpaved roads unreachable by public transportation. Other Rondon pioneers had done it before—cleared the rainforest to relocate to ar-eas cut off from the world—and Maria Joel and Dezinho debated for a while whether to take that chance. In the end, they concluded that the project wasn't viable for a family like theirs. The children were too little, and they had no access to vehicles, tools, capital, or bank loans.[4]

Before long, the family's life would face sudden disruption when Jo-elminha and Joelson fell ill due to pollution. Their health, Maria Joel recalled, would be "compromised for years" due to difficulty in breath-ing and bouts of diarrhea that led to alarming weight loss. If the Ama-zon was often imagined as one of the most uncontaminated places on Earth, Rondon, in truth, was no such place. Nurses and social workers of that time reported that children suffered and some died from respi-ratory diseases resulting from air contamination.[5]

The source of the contamination was the relentless expansion of the frontier, as Rondon was located at the epicenter of the so-called Arc of Deforestation. This was a region heavily impacted by man-made environmental destruction that spanned from the northeast to the southwestern border of the jungle—that is, from the Atlantic coast of Maranhão almost to the base of the Andes in Bolivia and Peru. Cover-ing some 1,850 miles, this arc stretched along the southeastern edge of the Amazon, the devastation advancing toward the heart of the forest from south to north and from east to west.[6]

In Rondon, there were three specific drivers of deforestation. The first was logging to supply the timber industry. The pressure on the local forests was more intense than in other areas of Pará because Rondon had an advantageous location, close to major transportation routes like the Belém-Brasília Highway and the ports of São Luis and Belém. Also pivotal in the frantic rate of destruction were the now-ubiquitous chainsaws, an invention that would "prostrate at the feet of man" the once-indomitable jungle, as John Hemming wrote.[7]

A second driver was the use of all kinds of trees to produce charcoal, which was sold to the steel plants built around the Greater Carajás Program. The production of charcoal became an easy source of extra income, especially for landowners and loggers with access to timberlands, so thousands of igloo-shaped kilns consumed all sorts of trees. Eyewitnesses recalled that the town was often enveloped in plumes billowing from the brick ovens, contaminating the air.

But the true face of obliteration emerged during the dry season, when pioneers undertook slash-and-burn activities. The land-clearing process followed the same pattern as in years prior, but on a much larger scale. The actions were carried out simultaneously throughout the Arc of Deforestation, and satellites would log more than 5,000 fires on a single day throughout the basin. Each year between 1978 and 1988, the Amazon lost an area of forest bigger than the state of Connecticut.[8]

The population reacted to that destruction with mixed feelings. Residents in Rondon acknowledged the ecological disaster it constituted and the health consequences. (Later studies on Amazonian superfires confirmed that "inhaling the air polluted by wildfire smoke can cause short- and long-term diseases that range from wheezing and coughing to more serious diseases, such as asthma, pneumonia, chronic obstructive pulmonary disease and lung cancer.")[9] Environmental awareness was growing across the planet, and the Brazilian government was routinely criticized by the media for its policies, but the prevailing view on the frontier was drawn from the regime's early promotion of economic and demographic expansion. Thus, the harm to nature was somehow perceived by pioneers as a toll necessary for progress. Many argued that Brazil—a continent-sized country aspiring to become a global power but with millions still suffering from extreme poverty—had to follow in the footsteps of other first world societies that had caused great damage to the environment in the course of their own development.[10]

Although the issue was highly controversial, those favoring development over preservation found in historical precedents an argument to support their views. In the nineteenth century, for instance, America had drastically culled the bison from its frontier, reducing a population of 30 to 60 million bison to some 300 individuals in order "to destroy the basic economy base of the Plains Nations" and open the Western frontier to homesteaders.[11] In the contemporary world, Japan, the ascendant power of the 1980s, was experiencing an economic miracle while being routinely criticized for excessive logging in tropical areas and for the continued use of devastating drift nets for fishing, which caused dramatic destruction of marine life.[12] The ultimate example would be China, which was already undertaking a comprehensive state-driven strategy to make the country "the factory of the world," a historic push that would raise tens of millions from poverty but would devastate its natural resources.

The fate of the Amazon caused heated debates among policymakers, but for Maria Joel and Dezinho, the most urgent consequence of this destruction was the quick deterioration of Joelminha's and Joelson's health. "I worried I could lose Joelson, who was very little, or even both kids," said Maria Joel, recalling thick, stinking clouds of smoke produced by superfires that occasionally caused daytime blackouts during the dry season.

With the specter of death looming over the children, Dezinho had to take the first job he could find to pay for healthcare. He was hired to open skid trails for trucks to access new timberlands. Accompanied by gangs of chainsaw operators, he spent several weeks clearing paths, a job that drove him to the edge.

The workers doing this task lacked any technical expertise, and as they toiled on uneven terrain and in torrential rains, accidents were frequent. Dezinho himself had never done this type of work before. Sometimes a lumberjack would lose a hand or a forearm when, engulfed in a cloud of splinters caused by the friction of the blades against heartwood almost as sturdy as concrete, the person would stumble over bulky roots and lose control of the chainsaw. Death could come without notice for those men trembling with the fatigue of holding heavy tools for several hours every day. Every now and then an immense log would lean in the wrong direction and crush not only hundreds of smaller trees and plants but also a hapless lumberjack working nearby. Falling branches were also dangerous, and nurses recalled that it wasn't uncommon for doctors to perform emergency surgeries on broken bodies to try to save the lives of laborers who arrived in critical condition.[13]

At night, Dezinho would rest in his hammock at the loggers' bivouac, his dim lantern attracting clouds of mosquitoes. Covered with mud and sporting a sparse bushy beard, he wrote reassuring letters to Maria Joel, which were delivered by truck drivers transporting the logs to Rondon's sawmills. Dezinho chose not to share everything that he was witnessing, but when he returned home, he wasn't always able to hold his peace. Sister-in-law Eva recalled how Dezinho, after weeks in the depths of the forest, rested in a hammock at his cabin, his body rocked by the slight movement. Absently gazing at the sky, a cigarette dangling from his lips, he would say, "This life is not easy; this is not right." Eva and Maria Joel wondered whether he was talking to them or just thinking out loud. When they would ask, Dezinho would explain how the families of his colleagues were destroyed by accidents. "My dear," he crooned, referring to his wife with the affectionate Portuguese expression *meu bem*, "this isn't fair. Something needs to be done."

Those who endured the worst working conditions were the day laborers hired informally for seasonal jobs. "They lost a foot or a leg and were simply brought back to town and received no compensation. Others died in the forest, and they were put in a car, driven to Rondon, buried in the cemetery, and that was it; the person was simply gone. Dezinho couldn't endure the injustice," Eva would recall. Another issue was payment, which all too often failed to fulfill the expectations of the workers. "When they returned from the jungle, the boss argued that the laborers in fact owed him money because of the food they had consumed while working, which was charged to them at outrageous prices," said Eva.

LUCKILY, AFTER A long period of struggle, things improved for the family. Joelminha and Joelson overcame their health problems and recovered their strength. Likewise, the couple's financial situation began to stabilize, and Dezinho and Maria Joel befriended neighbors and joined social events at the Catholic church. They also succeeded in buying a two-room mud-walled house with a large backyard and an outdoor toilet that they paid for in installments, accomplishing one of Maria Joel's early dreams. In 1986, Joélima, their second daughter, was born; two years later, Joelina arrived, their fourth and last child.

Dezinho had by then left his jungle job for a less dangerous position loading trucks in a sawmill where Maria Joel's brothers also worked. It allowed him to be at home more, taking the little girls to school every day and picking them up on his bicycle. Sometimes when he neared home after a day of work, Dezinho announced himself by humming a popular tune. He would then enter the cabin and head straight to the kitchen to grab Maria Joel's waist and pull her into a *forró* or samba. Drawn by the rhythm and the sound of laughter, the children would gradually join the party, marveling at their parents' passion. Happiness filled their lives.

Over a decade had passed since Maria Joel and Dezinho had first met, and since then, no matter the hardships they faced, they had remained a solid, affectionate couple. Friends in Rondon, recognizing the couple's chemistry, would start to call the two soulmates Casal 20 (Couple 20), referring to the Portuguese translation of the American TV series *Hart to Hart*. A great success in Brazil, the series presented the story of a romantic married couple who were amateur detectives and devoted their days to solving crimes and mysteries.

Before long, Maria Joel would realize that the comparison, much to her regret, would fit them even better than expected—Dezinho was about to resume his career as a grassroots activist.

7

Crickets and Cattle

ARLY in November 1988, Brazil was about to hold mayoral elections. For the first time in more than two decades, over 75 million electors would freely choose their mayors and council members from candidates belonging to some thirty political parties.[1]

Three years earlier, Brazil had returned to civilian rule, and the country was now under the command of President José Sarney—"the perfect oligarch," according to one commentator.[2] A former senator and governor of Maranhão and a large landholder who had joined forces with the dictatorship after initially opposing the coup, Sarney had ascended to the presidency under tragic and shocking circumstances. His running mate, the seventy-four-year-old Tancredo Neves, a former minister for Vargas who had been chosen by an electoral college to lead the first civilian government in over a quarter of a century, had died from illness on the eve of being sworn in.[3]

The nationwide mayoral elections were an important political test for a country transitioning to democracy. It would be the first vote after a new constitution had been instated in October 1988.[4] The draft of the constitution, a pair of scholars later wrote, had been "the outcome of the most democratic constitutional debate in the history of Brazil," and the final text was indeed modern and progressive.[5] One of the greatest achievements was article 231, which granted the Indigenous populations "their original rights to the lands they traditionally occupy, it being incumbent upon the Union to demarcate them, protect them, and ensure respect for all of their property."[6] In the ensuing decades, some 690 Indigenous reservations would be created, their territories spanning over 13 percent of Brazil's total landmass, almost all of it in the Amazon.[7] In addition to attempting to address the pain and suffering inflicted on previously displaced and dislocated tribes, the creation of reservations would prove to have an extraordinary impact on taming deforestation, with studies showing that Indigenous groups "with recognized tenure and forest management rights are some of the world's best forest protectors."[8]

For millions of Brazilians, including Dezinho, the mayoral elections were also an opportunity to gauge the strength of the parties that would compete for the presidency a year later, in 1989. Dezinho was enthusiastic about the political prospects and the rapid ascension of a husky-voiced man named Luiz Inácio Lula da Silva. Valdemir would say that Lula was his brother's "most important political reference," and Maria Joel would recall her husband saying that Lula was "the only man who can change the country." Polls indicated that his Workers' Party had a chance of pulling off a historic *sorpasso* in the mayoral elections, including in São Paulo, the largest, wealthiest, and most entrepreneurial city of Brazil.[9]

The seventh in a poor family of eight children, Lula was seven when his family boarded a flatbed truck and moved to Santos, on the coast of São Paulo, where his mother anticipated a better life than in the drought-ridden interior of the northeastern coastal state of Pernambuco, where Lula had been born in 1945.[10] Lula worked during his childhood and adolescence as a shoeshine and delivery boy, street seller of oranges and peanuts, and helper in a laundry shop. At seventeen, while employed in a factory, he lost the little finger of his left hand during an accident.[11] At the age of twenty-five, the future leader of Brazil would be struck by the most overwhelming drama of his life when both his first wife—the twenty-two-year-old Maria de Lourdes Ribeiro da Silva—and their unborn son died during childbirth. According to later reports, both deaths were the consequence of medical negligence.[12] Her doctors had ignored Maria de Lourdes when she complained about an enduring malaise that was in fact a symptom of hepatitis. This caused pregnancy complications and ultimately the death of her baby and herself during an emergency C-section.

Later, Lula remarried and began to make a name for himself in the ABC region—a suburban industrial area of São Paulo referred to as "the Latin American Detroit" due to the large number of car companies that had been established there.[13] He became known for his abundant charisma and for the strikes he organized in massive vehicle assembly plants employing tens of thousands of workers. Due to his activities as a trade union leader, Lula was spied on by Volkswagen and jailed by the dictatorship, although only for four weeks.[14] The *Times* compared him to Lech Wałęsa, and in 1986, Lula was "elected to Brazil's Parliament with the most votes for such a contest in Brazil's history."[15]

Many Brazilians like Dezinho saw in Lula the right man to spark the profound changes the country required. At the time of Lula's ascension, the "economic miracle" had not only evaporated but was being reinterpreted by economists as an important cause of the widening gulf between the country's social classes and its regions—the traditionally wealthy South versus the impoverished and "backward" North.[16] With inflation at staggering annual rates surpassing 600 percent, the working class struggled to stay afloat.[17] Corruption also marred the reputation of a political establishment that continued to be dominated by barons exchanging favors for benefits and using all possible means to remain in power. "In Brazil, a poor man goes to jail when he steals. When a rich man steals, he becomes a minister," Lula would say in 1988, an aphorism that would come back to haunt him later in his career.[18]

The wind of change led Dezinho to become more engaged in politics than ever before. On the eve of the mayoral elections, he was already a member of the Sindicato, the farmworkers' union that he would lead and transform, and it's very possible that he was beginning to think about becoming a full-time activist as his anger grew by the day over the appalling abuses he witnessed around him.

The end of the dictatorship had enabled civil society to organize and begin to grasp the true magnitude of rural conflicts, with the number of reported land-related crimes increasing sharply.[19] According to estimates, debt bondage caused dozens of deaths, although the real number was more likely in the hundreds, and the brutality endured by the victims was unthinkable.[20] On a fazenda near Rondon, employees were found chained to their beds at night to keep them from escaping.[21] On another ranch in southern Pará, the federal police discovered that overseers—emulating the savage techniques of the Italian Mafia to deal with corpses—fed pigs with the flesh of rebellious laborers they had tortured and killed.[22]

The barbarous treatment was not limited to the rural estates, nor was there any consideration for the victims' age or gender. A later investigation revealed that children were picked to feed the charcoal kilns of Rondon, thus exposing little boys and girls to the risks of irreversible or even lethal burns when they entered the suffocating ovens.[23] Girls under fifteen were also submitted to child prostitution.[24]

Back in Urbano Santos, Dezinho had been aware of the favored treatment of the landed gentry and of the sporadic episodes of violence endured by sharecroppers, but the staggering abuses reported in southern Pará, often by CPT agents and priests working on the front lines of social conflict, went far beyond what he had previously seen. "The issue of slave labor shifted Dezinho's perception of the whole rural problem," Maria Joel explained. Rondon's farmworkers' union, despite providing some assistance to victims, did little to change these realities.

Months before the elections, Dezinho met a man who shared his aim to tackle those issues now that Brazil was marching toward democracy and freedom of speech. His name was José Soares de Brito, known as Brito, and he would be crucial in helping Dezinho understand the roots of another offense with serious environmental and social implications—one that would develop into an obsession for both of them: land-grabbing.

Six years older than Dezinho, slightly taller but equally paunchy and charming, and also from Maranhão, Brito was the Workers' Party candidate for mayor when Dezinho met him. Brito had landed in Rondon in 1977, after being hired to work at the Instituto Nacional de Colonização e Reforma Agrária (INCRA) office. This bureau of the federal colonization agency, Brito would say, was opened in the aftermath of the murders of the three Americans (the Davises) that had preceded the deportation of Father Fontanella; its mission was to "regularize land issues" in order to prevent further violence. Then in his mid-twenties,

Brito would become an eyewitness of how *grilagem* (land-grabbing) worked, receiving multiple complaints from desperate peasants about fazendeiros and loggers threatening them with expulsion. The land-grabbers were called *grileiros*.

The words *grileiro* and *grilagem* derive from the Portuguese word for "cricket" (*grilo*) and allude to the action of fraudulently claiming possession of someone else's lands. This form of fraud is named for crickets because these insects allegedly played a historic role in forging deeds and land titles.

Some experts date the origin of grilagem to as far back as 1850, when the imperial government reformulated the land policy, but the fraud reached an unparalleled level after the launch of Operation Amazonia. According to geographer Susanna Hecht, the regime, with its series of policies to settle the frontier of the 1960s and 1970s, had transformed the Brazilian jungle into "the site of one of the most rapid and large-scale enclosure movements in history as more than 100 million acres pass[ed] from public to private ownership."[25] The problem was that, as demonstrated by the eviction of the Gavião-Kyikatêjê, this colonization process was haphazard—or at least not as organized and systematic as its scale required—and land-grabbers took advantage of the chaos to commit massive fraud.

The classic scheme, as Brito revealed to Dezinho in their early meetings before the pair became fast friends, worked in the following way. Initially, grileiros falsified property deeds, copying the format, date, and language of old authentic deeds. The foundations of the fraud lay in the claim that the governor or some other high authority decades or even a century back had transferred or sold a specific area—now claimed by grileiros—to a certain person. In order to give a timeworn look to the bogus deed, grileiros placed it in a box with crickets; the insects' droppings and bites would artificially age the document, an effect pursued to offer some appearance of veracity.

In order to back up the fraud, land-grabbers and their associates—often savvy lawyers and corrupt civil servants working in federal and state colonization agencies—then produced a paper trail of receipts and contracts connecting the fabricated, seemingly aged deed to the present, and to the grileiro in question. One common method, according to anthropologist Jeremy M. Campbell, who researched the roots of grilagem in a frontier area of Pará, was to buy and sell the land, at least on paper, to relatives, associates, or henchmen in order to "further legitimize the claim, since the land sale would produce receipts, legal notices, tax documents. . . . The accumulated paper was proof of sanction, or at least an impressive legal bulwark that might dissuade any challenger eyeing the claimed parcels."[26] Professor Campbell would call grilagem an act of "conjuring property."[27]

Though widespread in Rondon when Brito was at INCRA, these practices weren't exclusive to the Amazon. In São Paulo, according to one historian, "the fakers had diabolical imagination and skill":[28]

> They obtained crested paper with the Imperial arms, imitated old handwriting, affixed old stamps, deliberately yellowed their document, tore out pages from registry office books. They transplanted 20- or 30-year-old coffee trees to the forest clearings they were claiming. They transferred parts of old buildings to their houses, filling them with antique furniture to create the right atmosphere and make it seem as if the land had been occupied for many years.

But the scale of the fraud committed on the frontier, especially in southern Pará, was unparalleled. A police investigation would find that, in the region near Rondon, "an area larger than 2,000,000 hectares"—that is, some 5 million acres, or over twelve times the size of the city of London—had been the object of grilagem.[29] In some parts of the state, the bogus claims were so extensive that the land claimed by private individuals equaled five or six times the actual size of the area, a consequence of overlapping claims and counterclaims—of grileiros jockeying for tracts of the frontier that were actually owned by the state and federal governments. Predictably, conflicts often erupted.

In this lawless universe, the idea that prevailed was that whoever cleared a plot became its immediate owner, no matter the legislation. So grileiros invested heavily in leveling large areas with chainsaws and fire. Afterward, they resorted to the "magic" of Nelore cows, another

beast being used as cover for the fraud.[30] Because the ranching prac-
ticed in the Amazon required only one cowhand to move some 270
head over a pasture extending as far as the eye could see, humpbacked
cattle were frequently used to turn public lands into sham private
property.[31] Studies on productivity showed that many fazendas in Pará
had less than one cow per hectare. Often, this mattered little to grileiros, because in the context of the hyperinflation affecting the Brazilian
economy, the business they pursued wasn't beef or milk production.
Grileiros wanted land, a hedge against an inflation spiral devaluing
fortunes that sat stagnant in bank accounts.[32] Besides, ownership of
a ranch or a farm also provided the grileiro with access to subsidized
state funding and fiscal incentives.[33]

But why didn't the authorities simply discount the false claims and
press charges against manipulators and lawbreakers? Why were grileiros rarely faced with forfeitures of land and criminal prosecutions
for fraud and environmental devastation? Why instead did grilagem
become, as Campbell wrote, a "mechanism whereby the public [land]
would become private and the illegal would be made legal"?

Professor Campbell recalls that, in the era of westward expansion,
the United States had made cadastral surveys of the frontier to map and
document the boundaries of land ownership.[34] These maps, logs, and
registries would be used as the basis for the subsequent colonization
and homesteading process that would radically transform the country.
But Brazil lacked such an instrument; there was no central, comprehensive, and government-controlled land registry that civil servants
could easily check to verify whether a claim was legitimate or fraud-
ulent. Much of the data was scattered in privately owned registration
offices run by notary publics, called *cartórios* in Portuguese. There are
thousands of cartórios across the country, and they are a fundamental
part of the bureaucratic life of any citizen in Brazil: cartórios register
and handle certificates on vital issues like marriages, deaths, and births,
among many other things. They keep these registries in their books.

In frontier areas of Pará with a limited state presence, cartórios
charged with the task of recording land transactions faced many problems.[35] Their clerks did not always use a systematic and trustworthy

protocol to describe the physical location of the land being traded. They compiled registries and sale contracts in complicated, arcane, and frequently inaccurate ways. The use of degrees of longitude and latitude, for instance, wasn't always the norm. Rather, the location and boundaries of a property were described by the cartório staff with reference to its position with respect to local rivers, state lines, adjacent fazendas, or neighbors, who were frequently referred to only by their first names. It was an imprecise, flawed system that left room for multiple interpretations—and deceit.

To make the bureaucratic chaos even greater, it wasn't uncommon for cartórios of different districts—out of collusion, corruption, or simply incompetence—to log several land transactions involving the same deed. It was as if several people in several regions were simultaneously buying, selling, or dividing the same piece of Amazonian land. The result was that, when litigation arose over a given area, often between a fazendeiro and rural dwellers, the authorities could take months to establish as a matter of fact who was the legal owner of the disputed plot.

Eventually, INCRA or the Land Institute of Pará state (ITERPA) began checking the original registry books to verify whether the governor had actually sold that land a century back, as stated in the deed presented by the grileiro, or whether the claim was a fraud. They also resorted to forensics, tasking officials with examining the ink and the writing of the forged deed to establish whether it was genuine or fake. Still, each bureaucratic process could take several months, and enforcing the law—for instance, seizing the land of the grileiro—was not the norm. Consequently, disputes were frequently settled on the ground, and related parties resorted to extralegal means to resolve them, causing these conflicts to escalate. Between 1964 and 1989, some 556 peasants, lawyers, and Catholic priests defending poor people's rights to land and denouncing grilagem were murdered in Pará.[36] Campbell also noted that despite this violence, "civil authorities often moved to fully legalize *grileiros'* claims via legislative act or judicial decision, meaning 'fraud finds an accomplice in law.'"[37]

❧

GRILAGEM WASN'T LIMITED to the upper class; some small-owner peasants and subsistence farmers did buy forged deeds or fabricate their own fake documents to claim ownership of acres they illegally logged. But during this period, what captured the attention of the Brazilian media were incredible stories about brazen land sharks operating in Pará.

One example of such a scheme was the case of a mysterious man named Carlos Medeiros. Never seen in public and acting always through legal representatives, Medeiros claimed to be the owner of no less than 29 million acres—an area larger than Cuba or Honduras—located mostly in Pará's national parks and Indigenous reservations.[38] He used forged deeds dating to the early nineteenth century, "and there was not a municipality in Pará where he did not have at least one piece of land," wrote the press.[39] Rondon was no exception; this shady figure, who was suspected to be a front for a group of wealthy and well-connected people hiding behind fake identities, claimed some 8,800 acres in the municipality.[40]

Another notable grileiro, according to police records, was Antônio Fernando Machado da Cunha, the business partner of Josélio in Paragominas and an early suspect in the chainsaw murder.[41] Fernando, who had sold Josélio a ranch, claimed ownership of hundreds of thousands of acres via dubious documents.[42] "He makes illegal land transactions," a confidential police file on Fernando reported.[43] Journalists wrote that Fernando used notary publics in the state of Goiás and then resorted to "a circle of friends that buy and sell plots to each other to give the transactions some appearance of legality."[44]

Grilagem would also blotch the image of Josélio, a man who "was in the apogee of his fame" in the 1980s, Brito would say. Only months after he'd left the jail in Belém for the chainsaw murder, Josélio was again investigated for evicting peasants near his Fazenda Serra Morena, although those families actually had rights to their plots, according to a confidential report from INCRA.[45] Civil servants accused him of intimidating peasants and legitimizing the forced ouster of the smallholders with fake receipts. "[Josélio] even forces them to sign invoices stating that he has bought their lands for 200,000 or

300,000 cruzeiros, but he doesn't pay them a cent," read an INCRA file.

The rancher was also accused by law enforcement of using other strategies, like writing letters and giving misleading statements claiming that his fazenda was being invaded by alleged communists. The federal police would suspect that Josélio, referred to in reports as the "pseudo-owner" of the land, was lying to "give a subversive connotation to the issue with the intention of enlisting the sympathy of the security departments to his cause."[46]

These actions likely stemmed from Josélio's anxiety over the fate of the Fazenda Serra Morena. Only months after he had bought the land from Alves, the authorities had in fact informed him that the deed was a fake.[47] Dated 1913, the deed stated that the governor of the time had transferred the area to a woman named Maria Ismael dos Santos. But there had never been a Maria Ismael, or if she had ever existed, she had never received this land from the government.[48] The deed had been forged, and according to official records, the area where the Fazenda Serra Morena had been established was actually owned by the government.[49] Police reports never say whether it was Josélio or Alves who had faked the document.[50] But Josélio was aware that the deed wasn't real, because just months after formally concluding the deal with Alves in a cartório, he had asked ITERPA to issue a certificate validating the authenticity of the deed. ITERPA examined the document before concluding it was a fake, and then published a note in Pará's official gazette informing the general public of this.[51]

Still, the Serra Morena was never seized, and Josélio continued to acquire land from dubious sellers. In the mid-1980s, he bought the Te-Chaga-U, whose land had originally been sold by the state of Pará to a man named Hugo Muniz de Queiroz.[52] According to police records, Muniz was "widely known as a grileiro."[53] Still, Josélio's ownership of the Te-Chaga-U was never disputed by the authorities, and he accessed special lines of credit from the government to develop the ranch.[54] Some of the funds would be transferred to him at the same time as Gil found the human remains, according to official documents.[55]

Those land deals made Josélio one of the largest landholders of Rondon and unfolded at the same time as new accusations were made against him for alleged participation in homicides. Scholars in the region in those years wrote that Josélio was responsible for the death of a rancher who was fighting with him over some land.[56] The CPT would also link Josélio to the murder of a candidate for vice-mayor who had criticized Josélio's power.[57] Later, Brito and advocacy groups accused him of masterminding the abduction and later disappearance of another settler near the Serra Morena.[58] Still, none of these crimes were conclusively tied to Josélio by the authorities.

Some cases gained wider attention and were reported in the media. Paulo Fonteles de Lima, a progressive legislator and a lawyer supporting the cause of peasant families in Pará, declared to the state legislature in 1984 that there was an active conspiracy to kill him and other activists fighting grilagem. The murderous scheme, which Fonteles had been alerted to by confidants, was being orchestrated by a group of landowners and politicians from Paragominas that he claimed included Josélio, although the legislator didn't provide evidence to support the accusations.[59]

Three years later, Fonteles was murdered near Belém, hit with three bullets in the head.[60] The pistoleiro who shot him was later arrested, but the masterminds were never identified, and Josélio's name was never mentioned again.[61] Fonteles's family claimed that behind the plot was a consortium of fazendeiros and grileiros with bonds of mutual economic interests. These murky alliances, secretly established by members of a landed gentry who successfully dodged prosecution by using multiple middlemen to farm out the crime to pistoleiros, were routinely denounced by the CPT.

Exposing the roots of violence was just one part of the CPT strategy to change the status quo. With the arrival of democracy and the emergence of the left, social movements in the Amazon believed it was time to organize and oppose the injustices. Dezinho and Brito, supported by the CPT, crafted their own plan to fight those injustices in Rondon do Pará. The farmworkers' union would be their platform to build an opposition movement against the most influential people in town—fazendeiros and sawmill operators.

8

No Longer Meek

JUST as Dezinho had hoped, the Workers' Party achieved historical results in the mayoral elections. By electing some 1,500
council members across the country and three mayors in state capitals, including São Paulo, the party confirmed its "ascension as a real
force on the political scene," according to *Folha de S. Paulo*.[1] The eu-
phoria among its supporters was such that, hours after the vote, an
exultant Lula declared he would win the forthcoming presidential race
"by a great margin."[2] The left was gaining power just four years after
the military rule had ended.

For Brito, however, things went sour. A well-known fazendeiro
named Olávio Silva Rocha was elected mayor of Rondon. Despite his
election, the wider debate over inequalities and grilagem was gaining
a foothold, and Brito and Dezinho launched their struggle. Standing
on the bed of an old truck and speaking through loudspeakers, the pair
spoke at union events in public squares, declaring that Rondon was
a lawless place due to corruption, a lack of consequences for illegal
actions, and the frightening combination of hired killers and powerful
grileiros. Some of these events, supported by a CPT working to build
a network of grassroots activists in frontier areas of Pará and the Amazon, attracted hundreds of people in the early 1990s. The number of
attendees might have been small in absolute terms, but the intent was
direct. Since Father Fontanella's campaigns, no one had really confronted the landed gentry so openly.[3]

In 1990, Dezinho was elected director of the board of the Sindicato, and three years later, his gift for networking and his stated determination to undertake radical changes propelled him to the presidency. On April 18, 1993, the union held a meeting to elect a new leader, and the forty-one union delegates voted unanimously for Dezinho.[4] One of his campaign promises was to break with the policy of previous administrations, which had regularly compromised with fazendeiros, if not totally submitted to them. Dezinho told his colleagues that he planned to build an organization with "a combative style," one whose priority was "the fight for land."

Dezinho named Brito his special advisor and devoted his time to expanding his power base. The Sindicato had several hundred members but lacked cohesion, and most rural dwellers had no class consciousness. Peasants were of course aware of the injustices they endured, but they didn't realize that their situation could be changed by organizing and fighting. Events like the chainsaw murder remained etched in people's minds, and the inaction of the government regarding grilagem made them doubt that wrongdoers would be treated any differently now, under democracy, than during the dictatorship.

Aware of this, Dezinho saw his immediate mission as getting people to stop feeling small and to believe in the possibility of change. In his speeches, he insisted that they were not alone in their struggle and that they could count on an expanding network of lawyers, activists, and members of the CPT and FETAGRI with connections to the Workers' Party. The law was on their side and should be used as the main tool to fight injustices and abuses. The Sindicato, he said, would lead the charge against the outlaws.

Dezinho was a man of action, good at following his instincts. Years later, union members would admit that Dezinho's confidence in himself and in the power of a unified movement prompted them to be more assertive in defending their rights.[5] "Dezinho hadn't studied much in his life, but he was very intelligent," Brito would recall. "He understood the reality of rural dwellers' lives and could instill a feeling of trust and the possibility of change in them."

Although the union offices were located downtown, Dezinho often visited the vast interior of Rondon do Pará to meet victims and get to know their stories. The trips were risky: Dezinho had to travel one or two days along dirt roads cutting into a landscape increasingly dominated by pastures. Neither the Sindicato nor Dezinho had access to a vehicle, so he had to wrangle free passage from a truck driver transporting logs or mount a peasant's horse.[6] Sometimes Dezinho traveled by foot from one settlement to another, the walk often taking several hours. It was the perfect scenario for a gunman to end his life.

"*E aí, meu companheiro!*"—"What's up, comrade!"—he would cry on his arrival to hamlets, a wide smile on his face, his stout frame surrounded by weather-beaten peasants in shabby clothes. Most families of the area would then gather in a shack, and a long debate would ensue over their struggles and the strategies to confront them. It was difficult to organize large assemblies because of the relative isolation and poverty of these groups. Dezinho believed that cementing the sense of community was fundamental, so he and the Sindicato devoted several months to preparing meetings that gathered hundreds of peasants living in villages scattered throughout the region. The power of a movement on the rise and the ambitions of a leader who seemed "almost crazy," in Brito's words, in his willingness to confront the establishment, were on display during those forums.

Brito invested long hours planning the assemblies with Dezinho, often staying overnight in his house. Because they didn't trust the local authorities, they traveled to Marabá and Belém to report to FETAGRI and the CPT the accounts they gathered from peasants and laborers. In this process of building a network of collaborators, Brito became "like a brother" for Dezinho. The children of Maria Joel met him so frequently that he was referred to as Uncle.

The young wife observed her husband's actions and pondered what all this exposure would mean for her family. Maria Joel knew Dezinho was an activist at heart, and she didn't oppose his working at the union. Occasionally she attended meetings herself and even brought the children so they could hear Dezinho's speeches. But she worried about the potential consequences of his actions, because it was commonplace for the leaders of small unions to be attacked, sometimes murdered. "They had a calm life, but suddenly . . . she felt very distressed over his work," recalled Eva, who had also joined the Sindicato and helped with paperwork.

The life of the family also changed from a material perspective. They had never been in a comfortable financial position, though Dezinho's job at the sawmill had allowed them to get a small loan and enjoy an average working-class life. But the Sindicato's finances were in tatters because many members couldn't afford to pay their monthly dues, so Dezinho now earned only a few hundred reais per month. To make ends meet, Maria Joel, in addition to taking care of the children, also became the household's breadwinner. She bought clothes in larger cities that she and Eva later sold to laborers at the entrance of sawmills. Maria Joel would admit to her sister that if it hadn't been "for the great love I feel for Dezinho and because I believe he defends a fair cause, I would never do this."

WHILE DEZINHO WAS rising as leader of the Sindicato, an organization called the Movimento dos Trabalhadores Rurais Sem Terra (Landless Workers Movement) was transforming the way peasants fought for their rights, especially regarding access to public lands for homesteading purposes. The MST, as it was known, was founded in the south of Brazil in 1984 with the support of Roman Catholic priests and sectors of the CPT as well as left-leaning intellectuals and political parties, although its most well-known thinker was the economist João Pedro Stédile. Over time, the MST would become, in the words of a prominent intellectual, the "most effective movement for the modernization and re-socialization of rural populations in the history of Brazil."[7]

The essence of the MST's distinctive strategy rested in forceful campaigns and acts of civil disobedience organized by its members to win their objectives. Their broad range of tactics, described by some as radical, included days- and weeks-long marches, protest camps, road and highway blockades, and the occupation of buildings.[8] The core of their master plan to push for a comprehensive agrarian reform in the country was the occupation of land—land of dubious provenance and often connected to elites.

MST occupations, later emulated by farmworkers' unions and organizations like FETAGRI, were not intended to immediately take over a specific ranch or farm, even if it was confirmed to be the product of grilagem. Occupations were instead designed to pressure the authorities to enforce the law and expropriate the area to distribute plots among landless peasants. Planned and meticulously carried out by MST militants but fundamentally relying on the courage and resilience of underclass citizens to succeed, the occupations were carried out in the following way. First, MST militants and well-educated sympathizers investigated a specific area through the relevant public records—cartórios' logs, official files, judicial sentences, police reports—to uncover frauds. Then, they researched the business nature of the ranch or farm to be occupied to find out whether it was productive or simply speculative.

If reliable evidence was found that the area was the object of grilagem or could potentially be seized because it was unproductive, a secret plan of occupation was created. On a specified day, often at dawn, a caravan of trucks and buses transporting hundreds or even thousands of families—depending on the size of the land targeted—entered the area and set up huts. Participants often came from marginalized urban areas or poor rural communities, but even some former gold prospectors of Serra Pelada joined the movement. The MST called these initial camps *acampamentos* and described the act of settling in a fazenda as "cutting the wire," referring to the fences surrounding the properties.

These temporary settlements weren't the ultimate goal of MST. The organization instead aimed to create a situation in which the authorities, impelled by the potential conflict between the occupants and the landowner, were forced to investigate the land and eventually seize it and redistribute it among the landless families. If the MST succeeded in this process, an acampamento became a settlement, an *assentamento*, a permanent rural community that produces and sells food in a similar fashion to cooperative farms. In the best-case scenario, an assentamento can take years to materialize, and many occupying families quit due to the numerous hardships, like armed backlash by the "security personnel" of the occupied fazenda or by the police as well as poor sanitation in the precarious shelters, which often led to disease or death. Long struggles have always led people to give up and move on, but history has demonstrated that despite the great challenges and many setbacks, the strategy can work; about 100,000 families were eventually given a plot in Brazil through such occupations.[9]

The MST has often been framed by the conservative media as an extremist organization, and its members have been demonized for plundering fazendas, destroying public offices, and retaliating against fazendeiros. The MST argues that it doesn't engage in violence and maintains that despite being a radical act, occupations are legal, because, even in the cases when they target legally established farms, the Brazilian constitution states in its article 184: "It is the task of the Federation to expropriate, on social grounds, for the purpose of agrarian reform, rural property which is not fulfilling its social function."[10] The "social function" of property is a rather abstract legal concept, but the overall idea is that in a country with millions of hungry people, agricultural lands should be devoted to raising families from poverty rather than serving members of the elite as a financial investment.

Until the emergence of the MST, farmworkers' unions and even the CPT, which was quite revolutionary for its time and context, had expressed their demands in a less disruptive and abrasive way. Initially, occupations were not on the agenda due to their risky nature. But the MST's success in its goal "to rebalance the nation's social order and strengthen capabilities among its underprivileged population," as one scholar wrote, led other organizations to emulate and incorporate some of their tactics. This resulted in the number of occupations exploding by the time Dezinho became president of the Sindicato.[11] In 1988, about 10,000 landless families had taken part in occupations across Brazil; by 1993, that number had doubled.[12]

Landowners responded to the rise of the MST by creating their own network of organizations defending "the right to private property." The most reactionary had no trouble in publicly declaring that they wouldn't just sit and wait while their lands were "invaded" by MST, threatening armed retaliation.[13] But some fazendeiros agreed that both hunger and forged deeds were widespread problems in the rural world, and it is unfair to depict all cattlemen who joined these movements as a homogeneous class resorting exclusively to violence. Besides, while some ranchers were aware of holding dubious titles, they nevertheless opposed occupations because they had bought their lands during a period when chaos reigned and then had invested millions to set up their businesses on the frontier. Well-connected and politically active, they disputed the narrative that they were greedy land-grabbers or speculators, arguing that fazendeiros toiled in the fields to expand the agricultural production of Brazil and contribute to the economy.

In Rondon, one of the most active movements defending landowners' demands was the Associação Agropecuária Rondonense (Agribusiness Association of Rondon), which had been created in 1977 by Josélio and other pioneers.[14] In 1993, the same year Dezinho became leader of the Sindicato, Josélio was elected president of the Associação.[15] It was his second term.

This proved that Josélio, no matter what people in the Sindicato or in the peasant movement thought about him, enjoyed respect among his peers. Families of wealthy fazendeiros and sawmill owners in fact had an ambivalent view of him. "We heard many stories about Josélio, because everyone spoke about it, but he helped anyone who was a fazendeiro," explained Odeb Moreira, a wood investor from Espírito Santo who had bought land in Rondon and was a neighbor of Josélio's.[16] Moreira described him as "polite" and "calm," and he and other Rondon-based businessmen would recall that, despite his frightful reputation, Josélio was always willing to lend his trucks or contribute economically to undertake some community project, like improving a road used to transport production. Others said that Josélio often used his contacts in high places to get state funds for agribusiness projects like ExpoRondon, a livestock convention and exposition to promote locally produced beef and milk that would grow over the years and would contribute significantly to the local economy.[17]

His image as a determined, hardworking person helped Josélio occupy a prominent social position, but he wasn't the only businessman with substantial clout among his peers at the Associação. Another influential member was Décio José Barroso Nunes, who went by Delsão (the suffix *ão* is used in Portuguese to denote something or someone big). A tall, burly man with ice-blue eyes and an aquiline nose, Nunes was twenty-two years younger than Josélio and therefore part of another generation of pioneers.

Born in Minas Gerais, Nunes moved to Rondon in the late 1970s, when he was in his twenties. He came from a family of fazendeiros and timber businessmen, but when he settled in the Amazon, Nunes had little capital and no high-ranking contacts backing him. "He didn't even have a chainsaw," a relative would say. The formula that had changed his life was work and canny business decisions. "He's the most hardworking person I've ever met," recalled an employee who worked closely with him.[18] "He wakes up at four a.m. and goes to sleep at eleven p.m." A friend of Nunes would tell a reporter that Nunes was "a true human machine."[19]

Stubborn and brimming with self-confidence, Nunes had developed his own ideas about how to maximize profits in the timber sector. Contrary to sawmill operators who relied on subcontractors for the supply of logs, Nunes pursued vertical integration. He aimed to control the whole supply chain, from the harvesting process in the depths of his own jungle to the delivery of planks through his fleet of rigs. This strategy made him independent and extremely reliable, and he earned the nickname "the King of Wood."

Unlike other entrepreneurs in Rondon—haughty white pioneers who often dealt with their workforce as if they were their servants—Nunes won the hearts of the working class. Even after carving out a successful career, he continued to roll up his sleeves to toil alongside his laborers. "Once Nunes came home to ask my husband, who was one of his truck drivers, to go to São Paulo to get a rig he'd bought that was just off the assembly line," recollected a woman. "My husband was on holidays and replied that he couldn't do it, because we were doing some home improvements. Nunes replied, 'Go get the truck; I'll do the job for you.' The next day, here he was, one of the largest businessmen of Rondon, changing the doors of my home himself."

Nunes had already opened his first sawmill when Dezinho became a member of the Sindicato. Wood was the core of his business, so Nunes bought large swaths of the frontier with the goal of securing untamed, lucrative timberlands. But concurrent with Dezinho's rise to the presidency of the Sindicato, Nunes began to diversify his investments. He became a cattle rancher and a large landowner with the acquisition of over 247,000 acres in a remote area north of Rondon called Fazenda Lacy.[20] The area, where timber reserves remained abundant, had its own landing strip, and Nunes envisioned creating a large ranching/sawmill complex where hundreds of employees would work in two-week shifts. He wanted the Lacy business model to last for decades, so he invested in reforestation to guarantee the availability of future raw material (trees), a rarity in a region where the pillage of natural resources and boom-and-bust models prevailed.[21]

Dezinho heard rumors that the Fazenda Lacy deal was murky, and soon he would pay attention to it—and to the King of Wood. But before that, Dezinho had more urgent issues to deal with.

IN RONDON, THE first land occupation had taken place in 1988, when a group of peasants had settled in a ranch called São Jorge. Expanding over 11,800 acres, the estate was claimed by a man named José Hilário Ávila.[22] Born in Espírito Santo, Hilário had relocated with his family to Rondon, where he'd invested in land, timber, and cattle.[23] For years, he and the occupant families had managed to coexist without outbursts of violence, as Hilário's primary business had been harvesting timber. But by the time Dezinho became leader of the union, conflict in that interior area was escalating. Hilário sought now to recover control of the entire land, which meant the peasants had to leave.

Neither the Sindicato nor the MST had apparently participated in the earlier occupation, which had been spontaneously organized by families squatting in an area they called Vila Gavião (Village of the Sparrow Hawks). However, when the conflicts between Hilário and the peasants developed, Dezinho roused the Sindicato to their defense and, more importantly, personally involved himself in bringing their case to the top authorities.

On May 28, 1993, only weeks after being named president, Dezinho made it clear that Vila Gavião had become a priority for him. He participated in a meeting with the president of INCRA in Marabá along with other leaders of farmworkers' unions and requested that the high-ranking official ensure that Hilário's land claims would be investigated, because he suspected the Fazenda São Jorge was the product of grilagem. Dezinho petitioned the president of the federal colonization agency that, even should Hilário be found to be the legal owner, INCRA would redistribute the land among the two hundred occupant families who had now been living there for years, duly compensating the owner financially for the expropriation. However, if Hilário was found to be a grileiro, the authorities should "begin the process of regularization of the rural workers currently occupying the area."[24] In other words, the Fazenda São Jorge should be seized without payment. This might seem a rather obvious request, but never before had land been forfeited by a fazendeiro in Rondon as a consequence of grilagem.

Before long, a murder brought Vila Gavião to the attention of the media. On June 13, Dezinho reported the assassination of Alfim Alves Fagundes, an occupant of Vila Gavião, who was shot in the head and heart.[25] "He was a father of four," wrote Dezinho, accusing Hilário of the crime.[26] In a report he sent to FETAGRI and members of the Workers' Party, he criticized the police of Rondon do Pará for not investigating the crime and for threatening the occupants with a violent eviction. Apparently, law enforcement was coercing the peasant families to sign receipts stating that they had already been compensated to leave the area. His denunciations were echoed in Belém, and state legislators urged the authorities to take measures.[27]

From that time on, suspected pistoleiros began to stake out Dezinho's house.[28] "Everyone knows he's the next to fall," warned a Brasília-based organization affiliated with the Sindicato in a communication addressed to the top authorities of Pará requesting protection for Dezinho.[29] Protection was provided, but about a month later, on August 3, when Dezinho was already under police escort, he reported to colleagues that he didn't feel safe. He said that Hilário was planning to take further lives, including his, as a response to the conflict over Vila Gavião.[30]

Maria Joel would recall a chilling episode during those days in which her husband, while walking on the street, crossed paths with a well-known criminal.[31] The police officer escorting Dezinho, seemingly recognizing the outlaw's identity and assuming he might be about to ambush the union leader, signaled to the criminal that he had no bullets in his weapon, and thus the mobster might proceed without opposition. No attack was carried out, but Dezinho, who realized what had just happened, requested his protection be removed only a few months after it had been granted. Never again would he trust the law enforcement of Pará for his safety, instead taking his own protective measures, like leaving the city for short periods when he sensed that he was in imminent peril or being escorted in and out of hamlets by peasants. His family also moved to another house, because the cabin where they lived had a large backyard and an outdoor bathroom, which had proven dangerous for activists. In 1988, the famous rubber tappers' union leader and environmental activist Chico Mendes had been killed by a gunman hidden in the bushes near the backyard of his home in the Amazonian state of Acre, near the border with Peru and Bolivia, when he was about to take a shower.[32]

The family struggled to adapt to a harsh reality that unfolded just weeks after Dezinho had become president. The possibility that Dezinho could be murdered now became real for Maria Joel and the four children—Joelminha, Joelson, Joélima, and the little Joelina. Every so often, when he was departing for Vila Gavião, Maria Joel asked them to say goodbye to their father one by one. "Now, wish your father well, because we don't know if he'll be back," she said, unable to conceal her fright. The children, in tears, were left with no alternative but to follow her orders, and they began to blame Dezinho for the hardships they were experiencing. Joelson, now a preteen, didn't understand Dezinho's long absences and joked with his sisters that they "didn't have a father, because he's never home." They also complained about the restrictions imposed on them to prevent attacks.

Dezinho cared about his family, but he felt that he had a mission to accomplish—a mission that had just begun and for which he had been elected. With the occupations gaining momentum across the country, he perceived Vila Gavião as a litmus test for the Sindicato, so he made it clear that he would not be intimidated. "The powerful can kill one, two, three roses, but they will never be able to stop the arrival of spring," he wrote in an official communication, quoting a sentence that he attributed to Lula.[33] His brother Valdemir recalled a visit during those years when, sensing that Dezinho was in danger "in that town where ranchers walked accompanied by gunmen in the streets," he knelt before him and asked him to return to Maranhão. Dezinho looked at him and replied, "I'll shed my own blood for the good of this people if necessary."

Like the MST, with whom he sympathized (though he was not a member), Dezinho envisioned occupations as a tactic to spark bottom-up agrarian reform in Rondon. In his reasoning, the fight around Vila Gavião could deliver tangible gains to the movement. If peasants resisted and Hilário was eventually found to be a land-grabber and the ranch was finally seized by the government, this would represent a momentous victory. The case could set a legal precedent that the Sindicato—with the help of FETAGRI and the CPT—could follow with other fazendas suspected of grilagem. The seizure of Vila Gavião could also serve Dezinho personally, showing union members and influential people in Belém that he could deliver as a leader of the movement.

Many in the Associação of big landowners anticipated the move, immediately making Dezinho an enemy to watch carefully. "Almost the whole area of Rondon do Pará had been land-grabbed, so whoever confronted a grileiro would confront the whole community of landowners. . . . Fazendeiros knew that a threat to one of them represented a threat to all of them," attorney Carlos Guedes, a CPT member and a close partner of Dezinho during those years, would explain.[34]

Josélio himself sensed that the issue could have implications for him. In October 1993, at the time of the intensification of the violence in Vila Gavião and while Dezinho was pressuring INCRA to investigate the deeds of Hilário, he once more requested that the authorities give him a certificate of ownership of some of the land he'd been controlling for decades, as if presuming that some of his ranches could now be at risk.[35]

In March of 1994, a new attack was launched at Vila Gavião. Dezinho reported that Hilário, accompanied by six men, had shown up at the garden of one of the occupants to plunder the plot, "spreading terror" and stealing tools.³⁶ Dezinho went to the area and was told by settlers that the fazendeiro had warned them he would soon be back with some thirty men. In the Sindicato, some believed this was a desperate action by Hilário because the authorities had begun to signal that the fazenda could be seized with no compensation.³⁷

As the conflict escalated, the Sindicato suspected that other fazendeiros from the Associação were helping Hilário, hoping to keep him from losing his ranch. Although Hilário had done some business with Nunes, the King of Wood, Dezinho was especially wary of Josélio, whom he considered "the mentor of a group of fazendeiros with common interests," according to the account of a close friend of the activist.³⁸ Members of the Sindicato also suspected that Souza, apparently an overseer at the Te-Chaga-U, had threatened settlers in Vila Gavião.

Then Sueny Feitosa Cavalcante showed up at the union hall and told the story of the bonfire Gil discovered on Josélio's ranch. This unsettling account of slave labor and murders of migrant workers on the ranch was understood by Dezinho as a real opportunity to advance his confrontational strategy. Dezinho believed that with eyewitnesses Luiz and Gil committing to testify against Josélio, he could bring the rancher to justice in a coup that, coupled with the eventual forfeiture of Hilário's ranch, could flip the balance of power between fazendeiros and farmworkers.

Ultimately, the Te-Chaga-U raid didn't end in Josélio's prosecution, although the case would have some real consequences for the rancher and his family. Before that, however, one important issue had remained unsolved: why Luiz and Gil, who claimed to have been threatened by death in the Te-Chaga-U, had been able to leave the ranch unmolested. And also, why wasn't Josélio at the ranch the day it was raided?

The answers to those questions rested on the fate of Souza. Souza was, according to laborers, the overseer who had admitted to taking pleasure in seeing the grimaces on farmhands' faces before they were executed. In Gil's account, Souza had departed in the car with Josélio and Rai, another overseer, the day that Ceará had disappeared from the Te-Chaga-U.

Who was Souza?

Josélio had known him for over three decades, and by the time Souza was found on his fazenda, the former Contestado man had begun to feel that his invulnerability might be in serious jeopardy.[39]

9

Hunting Souza

IN June of 1994, almost the same time Dezinho crossed paths with the escaped laborers of the Te-Chaga-U, Detective Antônio Carlos Corrêa de Faria, a bearded, well-built officer of the state police of Minas Gerais, confirmed that a tip he'd received two months prior from Pará might finally result in the capture of his most wanted fugitive.[1] "We have him!" Faria, then in his forties, exclaimed to colleagues in Belo Horizonte, referring to Adélcio Nunes Leite, an outlaw with a horseshoe mustache who belonged to one of the most notorious families of the state.

The Leite (meaning "milk" in Portuguese) family had been operating for decades in the Vale do Mucuri (Mucuri River Valley). This was a hilly, violent, rural region in northeastern Minas Gerais not far from the Sweet River and in the area of influence of the former Contestado.

One of eleven siblings, Adélcio was a suspect in no fewer than forty-six homicides. He would be described by the media as "one of the worst pistoleiros in the history of Minas Gerais."[2] Since the 1970s, Adélcio and several members of his family had been accused of killing rival families, ranchers, politicians, cowboys, and anyone who allegedly opposed their goals, which, according to reporters, were mainly two: political power and land.[3] Strictly observing the oaths of omertà, said officers who investigated them, the Leite never confessed their crimes.[4]

Detective Faria, who had interrogated some of the family's hired guns, had heard multiple accounts of their hair-raising crimes. "They often ordered their pistoleiros to kidnap someone and then tortured the person, skinning him alive or slitting his penis," explained Faria, arguing that the Leite had what he called "a psychopathy." "If they focused on a specific target, they had to kill the entire family, not just one of its members." The police had found clandestine cemeteries on their fazendas where bodies of victims had been dumped, according to *O Globo*.[5]

Yet when any of the family members were arrested, rumors indicated that investigations were impeded or cases were thrown out. One family member, Alírio, a former policeman accused of murdering a rancher, had been absolved in three consecutive jury trials.[6] Prosecutors struggled to get witnesses to testify and jurors to sentence them.[7]

Journalists who posed inconvenient questions were also threatened. "This is good for a knife," one Leite answered a reporter interviewing him while he touched his belly.[8] In the little town of Malacacheta, the stronghold of the Leite, a reporter of the Globo television network said of the clan, "They are the police; they are the justice; they make the law."[9]

But on February 15, 1990, a crime brought the Leite to the attention of the nation and changed the family's fate. At around six in the morning, a group of six men wearing police vests entered the ranch of the Cordeiro family, thought to be a rival of the Leite. Claiming to be investigating a recent case, the group requested that all the Cordeiros gather in a single room in the fazenda, located in Malacacheta. Ten people, including some children, were having breakfast in the dining room. Adélcio waited outside in a jeep, to make sure nobody escaped.[10] The sun had barely risen, and the air held the chill typical of hilly Minas Gerais.

After a long and apparently cordial discussion in the large dining room, one of the Cordeiros asked the officers, who were holding forged warrants, to show his credentials. The mobsters promptly opened fire, killing seven people.

One of the three survivors of this massacre, which would come to be known as the Malacacheta Slaughter, was a ten-year-old girl named Maria Luiza. She'd escaped only because she'd set off for school before the rampage began. Four months later, fearing that the slaughter would once again go unpunished, the young girl sent a desperate telegram to Brazil's president, Fernando Collor de Mello, who had defeated Lula in the 1989 presidential election. The telegram read:

> Dear President, I paint, and I don't see the colors. I cry, yearning for Dad and Mom. Mr. President, arrest the killers of my family. Make justice for me.[11]

President Collor, a burly, debonair former karate champion belonging to a wealthy family, was said to be touched by the orphan's plea and ordered his minister of justice to do everything possible to arrest the culprits.[12] A task force of over 150 heavily armed police officers combed the region. They inspected houses, interrogated suspects, and seized weaponry. It was like a "war operation," wrote the press, who published photographs of officers closing access to villages and frisking people in public squares.[13] Commanded by Detective Faria and some of his colleagues, the operation yielded fruit in October, when Adélcio was arrested. He was accused of having orchestrated the killing through hired assassins. According to the police, the crime was the culmination of a long, bloody dispute over land between the Leite and the Cordeiros.[14]

However, in June 1993, as he awaited trial incarcerated at a police station in Belo Horizonte, Adélcio persuaded five officers to open his cell and join him "for a barbecue washed down with beer" in a home the family owned in the state capital.[15] Two wardens and Adélcio left the jail together, and a few hours later, the forty-two-year-old outlaw was on the run. According to the media, Adélcio had manipulated his wardens by offering them paltry gifts, like cheeses and meat cuts from his ranches.[16]

It was a public humiliation for the police, and from that moment on, recapturing Adélcio became a "point of honor" for Faria. For weeks, he deployed agents across Brazil—Minas Gerais, Espírito Santo, Bahia, Rio de Janeiro—to track the fugitive, trying to anticipate Adélcio's next moves. Initially, he thought the killer might hide in São Paulo, Brazil's most populated state, where he could easily disappear in the crowd while waiting for the uproar to subside. Then Faria speculated he might go to Rio de Janeiro, a state where organized crime was rampant due to the burgeoning drug business, with all kinds of à la carte services available, including protection and fake IDs. The last tip Faria got before finally finding out where his quarry had really landed was that Adélcio was intending to flee to Mexico, where he had allegedly bought land and was planning to start a new life. That plan never materialized because a source told Faria that Adélcio was in a little Amazonian boomtown called Rondon do Pará.

Rondon was then unknown to Detective Faria, but not to the Cordeiros. José Dário Godinho Rodrigues, a distant relation of the victims of the Malacacheta Slaughter, had bought land near Paragominas and traded wood and cattle. One day Dário noticed a chubby woman with long black hair who looked like Adélcio's wife, Eva Nilma, on a fazenda. After confirming her identity, Dário began to suspect that Adélcio was also hiding on the ranch and might have been planning to kill him.[17] He indeed had good reason to suspect so, because, after the killings in Malacacheta, more Cordeiros had been murdered. Helvécio Augusto Cordeiro, for example, had been executed by a pistoleiro in the state of Rondônia, also in the Brazilian Amazon, where the victim had relocated.[18] Apparently, the only reason for the murder had been the old feud between the Leite and Cordeiro families.[19]

Through a mutual acquaintance, Dário passed the information on Adélcio's suspected location to Faria, offering his fazenda as a base for a lightning-quick operation to recapture the killer. Faria, born in Aimorés, the town bordering Josélio's Baixo Guandu, agreed. (When he was six, Faria had relocated with his family to Belo Horizonte to escape the oppression of the Contestado.[20])

At midnight on June 13, 1994, three days after reaching the region, Faria and five officers armed with shotguns, revolvers, and an Uzi submachine gun entered the rainforest to carry out their mission. Divided into two cars and wearing bulletproof vests, the agents of the police of Minas Gerais advanced along a rutted dirt road, the headlights of the vehicles being swallowed up by the darkness of the forest and melding with the cloud of dust thrown up by the motorcade. Not knowing the area, Faria checked the odometer for guidance, as he'd been told by local sources the ranch was some fifty miles from the PA-70 highway. After a couple of hours of driving, they thought they were close to their target. Still, Faria needed a source to confirm where exactly Adélcio might be staying. When they spotted a house along the road, the detective jumped out of the car and pounded on the door. "Police! Open!"

Two men opened the door, panic on their faces. The cops, carrying flashlights, showed them a photograph of Adélcio. "Do you know this man? Do you know where he is?" Faria barked.

One of the two men looked at the other before replying, "He is Mr. Souza. He lives in the ranch of Mr. Josélio. Two leagues from here." The name of the estate was Te-Chaga-U, said the man, who was asked to guide the squad there. Faria ordered his men to complete the last part of the journey on foot so they could take Souza by surprise.

Faria had a long record of confronting frightening criminals, and he'd attended courses at the New York State Police Academy and in Italy, the home of the Mafia, but as the group walked the last miles through the jungle toward the fazenda, Faria realized that he was afraid. "Everyone who works with me needs to know fear," the detective used to say to his agents to prod them to think twice before engaging.

"There he is," said the civilian, pointing at the Te-Chaga-U.

The man was dismissed, and Faria gave the last instructions to his officers before hitting the ranch. "I don't want any of you hurt," he said solemnly. "Any armed retaliation from them must be contained immediately."

His officers then split into two groups. Some dogs barked, but they weren't aggressive, so the officers didn't have to resort to shooting them. Everything seemed eerily quiet when the task force reached the perimeter of the fazenda's headquarters. Faria's heart raced as he wondered how many men might be inside and whether the hush was actually an indication that Adélcio was expecting them.

One of his men saw a window ajar and peeked in. He spotted Adélcio standing in the kitchen, brandishing a revolver.

Suddenly four or five shots were heard.

"Freeze! Police of the Minas Gerais state!" yelled the detective.

The task force quickly entered the building, a "simple home with new furniture," according to Faria's recollection. The detective realized that one of his officers had been shot, but the bulletproof vest had saved him. Adélcio himself had been hit by three bullets. With him inside the house was his wife Eva Nilma.

There were more people inside the fazenda. A young woman wearing a large white cotton T-shirt and jeans glared spitefully at the cops. Next to her was a shirtless fat man who claimed to be a livestock buyer, but who Faria suspected was a pistoleiro and the woman's companion. Standing hands up next to the alleged hitman was Josélio. Bare-chested, he wore loose jeans with his belt loosened and his belly bulging over the waistband. He looked like he'd just woken up.[21] The rancher's blue eyes, appearing slightly darker in the dim light, "blazed with anger," Faria would describe.

"You don't know who I am," Josélio said in a deep voice, according to Faria's account. "But you won't get away with this. My daughter is a district attorney."

One of the officers, unintimidated, slapped Josélio in the face. "Let's take them," bellowed another agent, referring to all the people in the fazenda.

If they had been in Minas Gerais, Faria probably would have agreed. But they weren't federal police; they were officers of Minas Gerais already on shaky ground here in the jurisdiction of Pará, and Faria had to consider the actions he took carefully. There were at least two crucial things to consider. First, one of Josélio's daughters, Josélia Leontina, was the district attorney, which Faria believed could jeopardize the recapture of Adélcio from a legal perspective. Besides, he and his men still had to get away from the ranch and from Rondon safely—and with Adélcio and his wife in custody—in order to fly back to Minas Gerais. "It was too dangerous," Faria would later say to justify why he had decided that day to arrest only Adélcio and Eva Nilma.

Before placing the wounded fugitive and his wife in the bed of a pickup truck they confiscated from the ranch, the officers seized fifteen firearms, including shotguns and revolvers, according to Faria's account. They also destroyed both the local radio and the phone systems installed in the estate. The tires of all the vehicles on the fazenda were shot out to avoid pursuit. Under the watchful eye of Josélio, the team left the Te-Chaga-U at daybreak.

The operation inside the house had lasted some fifteen to twenty minutes, but the completion of the mission would take much longer. Faria and his men drove several hours before reaching the airport of Imperatriz. The detective, Eva Nilma, and Adélcio boarded a private plane that was forced to make an emergency landing soon after taking off from Maranhão because Adélcio had lost a lot of blood and was at risk of dying mid-flight. Ultimately, he was hospitalized in Bahia to be stabilized and then transported to Minas Gerais to receive further medical care and face justice.

While still recovering in a hospital bed, Adélcio agreed to be interviewed by a reporter. Faria believed that Adélcio and Josélio knew each other from the old disputes in the Contestado and had even committed a murder together back in 1984 in Maranhão, when Josélio was already living in the Amazon. Adélcio denied any of this and said he had never committed a crime. In a statement to the media, his distinctive mustache now shaved, he declared, "I'm a hard worker. In this country, [a man] who produces is [considered] a criminal."[22]

Adélcio admitted, though, that he'd planned to leave for Mexico before finding refuge "on the ranch of my friend Josélio."[23] Adélcio never implicated Josélio in any crime, and he would deny having participated in attacks against laborers at the Te-Chaga-U or against settlers in Vila Gavião.[24] He simply said that he and Josélio were pals and that he was on his ranch "just on holidays."[25] Asked by the reporter why Detective Faria had found so many weapons on the Te-Chaga-U, including one under his pillow, on the day of his recapture, Adélcio said that they were a precaution to defend against jaguars.[26]

Afterward, Faria, whose picture was printed in newspapers next to stories praising him as a hero for Adélcio's recapture, would retrace the outlaw's escape route. Before calling Josélio for help, the detective explained, Adélcio had traveled through the states of São Paulo, Espírito Santo, Rio de Janeiro, and Santa Catarina, where he'd apparently bought a fake ID under the name Roberto Souza. He'd traveled to Rondon over land. The date of this trip would remain unclear, but Adélcio said he had been in Josélio's fazendas (he also claimed to have visited the Serra Morena) for about a month.[27]

On September 30, 1994, over a year after escaping from jail and about three months after being recaptured by Faria, Adélcio was sentenced to 57 years for murder.[28] This would be the first in a series of trials that would increase his prison term to 229 years.[29] Some of his brothers and sisters were also tried and sentenced.[30] Still, although Adélcio was imprisoned for a time, he would never serve all those years in jail.

THE CAPTURE OF one of the most infamous criminals of Minas Gerais on the Te-Chaga-U was a real warning for Josélio. A fugitive was found on his fazenda and that could cause him serious legal troubles.

Faria's raid also took place in the wider context of the Sindicato's growing scrutiny of fazendeiros, especially after the conflict over Vila Gavião escalated and the media, again, covered land-related violence in Rondon. It is very possible that the new scenario, a result, in part, of Dezinho's campaigning, influenced Josélio's decision to have Luiz leave the Te-Chaga-U. After his role in helping Adélcio was exposed in the media, Josélio might have sensed that it was better not to draw further attention to himself.

But in this tangled story, there was yet another issue that influenced Josélio's actions: the threat of Dário, the relative of the Cordeiros who'd tipped Faria off about the presence of Adélcio at the Te-Chaga-U. According to officers of Minas Gerais, after the recapture of Adélcio, Dário had tried to kill Josélio in an ambush.[31] The attack was apparently to prevent Josélio from retaliating against him for helping Detective Faria. If it ever took place, however, Dário failed.

On February 2, 1995, Dário was murdered on his fazenda; three months later, on May 23, another rancher and close relative of Dário, Renato Tadeu Dalla Sily, was killed along with his bodyguard near Paragominas. The suspected mastermind of both murders, according to police records, was Josélio, who denied any wrongdoing.[32] Ultimately,

Josélio was investigated but never prosecuted for the homicide of Dário and Tadeu. Still, his name appeared printed in Pará's newspapers in association with both crimes by the time Secretary Câmara was pondering a field mission to the Te-Chaga-U to verify the denunciations brought to him by state legislator Zé Geraldo. Then human bones were uncovered on the fazenda.

The unfolding of events was as follows:

• April 1993: Dezinho is elected president.
• July 1993: Josélia Leontina becomes district attorney of Rondon.
• April 1994: Ceará disappears from the Te-Chaga-U.
• June 1994: Adélcio Nunes Leite (Souza) is recaptured.
• June–July 1994: Sueny meets Dezinho and then she, Luiz, and Gil depart the Te-Chaga-U.
• February 1995: José Dário Godinho is murdered.
• May 1995: Renato Tadeu is murdered.
• June 1995: The police find human remains on the Te-Chaga-U.
• June 1996: Josélia Leontina is transferred to Marabá.

Although Josélio wasn't prosecuted after the human remains were found on his property, the raid did have consequences for him, according to Zé Geraldo. Geraldo maintains that the rancher was arrested after the police unearthed human bones. "He sold about five hundred cows, paid a lot of money, and was freed," the state legislator claimed, suggesting that the fazendeiro had bribed members of the judiciary to dodge the consequences of a case initially built by Dezinho and Brito. No documentary evidence indicates that Josélio was arrested or had bribed anyone, although, oddly, a judge named Paulo Cesar Pedreira Amorim would later issue an arrest order against Josélio and Rai, the second overseer mentioned by laborers at the Te-Chaga-U, only to reverse his own decision less than twenty-four hours later.[33] (This judge would later be involved in great controversy and would be accused by Dezinho of siding with the fazendeiros.)

Whatever the case, the police operation ordered by Secretary Câmara in mid-1995 seems to have convinced Josélio that he could face prosecution, and he resorted to his inner circle—his children—to make his defense. He had faced many difficult moments in the past, both in Espírito Santo and the Amazon, but this was the first time his family was publicly involved in his defense.

Dezinho believed that Josélia Leontina was playing a crucial role in helping her father avoid prosecution, and he said so to Maria Joel and Zé Geraldo.[34] The Workers' Party state legislator believed Dezinho, and at the same time as he pressured Secretary Câmara to send the cops to investigate the Te-Chaga-U, he also reached out to the attorney general of Pará requesting that Josélia Leontina be removed from her post as district attorney in Rondon.[35] She had passed an aptitude test to become a state prosecutor, Zé Geraldo was told, so she couldn't be removed without tangible proof that she'd broken the law, no matter her father's suspected felonies.[36] On June 4, 1996, about a year after the raid, the authorities transferred her to Marabá.[37] Zé Geraldo said that the move was the result of his lobbying "to evict her from Rondon."[38]

Josélia Leontina denies this, arguing instead that her new assignment simply resulted from regular job rotation among civil servants.[39] She also resorted to the courts to defend her integrity. Years later, she would sue the *Jornal do Brasil* for suggesting that while she was a prosecutor in Rondon, she was involved in irregular activities in trying to protect her father, a defamation case that she won in Rondon's court of justice. (The judge ordered the paper to pay her some $12,000 as reparation.[40]) Afterward, she also sued a reporter for writing in a book that she had been "prevented" (*impedida*) from working in Rondon do Pará because human bones had been found on the Te-Chaga-U. Josélia Leontina claimed she had never committed any wrongdoings as district attorney and accused the reporter of defamation. Courts in Pará dismissed the case, finding that the reporter hadn't injured her reputation and had simply reported issues described in police investigations, but Josélia Leontina appealed. The case was finally dismissed in Brazil's Supreme Court after a series of appeals in the lower courts.[41] According to the records of the internal affairs branch of Pará's prosecution office, Josélia Leontina, throughout her career, has been investigated for suspected misconduct four different times, but she has never been found guilty, and the cases were not related to her years in Rondon.[42]

At the time Dezinho was building the case that ultimately led to the raid on Te-Chaga-U, all suspicions in the Sindicato were against Josélia Leontina. The truth, though, is that for a long time Josélio had been counting on another daughter to help him deal with his troubles: the thin, pale-skinned, short-haired Shirley Cristina.

Four years older than her sister, Shirley Cristina was a wise and ambitious woman who had been trusted by her father to be his legal representative even before the Te-Chaga-U case exploded. At least as far back as 1992, according to documentary evidence, Josélio had given Shirley Cristina "full powers to resolve all kinds of issues" related to him, including land, tax, judicial, or financial.[43] Married to João Malcher Dias Neto, the son of a high-ranking public servant, Shirley Cristina would throughout her life be a staunch supporter of her father, in addition to being a shareholder in many of his land investments. Townspeople would hear Josélio saying that his real successor—"my real man," in his words—was her and not his sole son, José Eloy, who also worked in the cattle sector.[44]

Already a lawyer when the Te-Chaga-U case came to the fore, Shirley Cristina played an important role in both the legal and the public defense of her father. When newspapers published the stories of the "enslaved" Gil and Luiz, she refuted the allegations and accused members of the Workers' Party of defamation.[45] One of her partners—the attorney Eliana Vilaça de Lima—would represent Josélio in the case. Shirley Cristina and Eliana had a Belém-based law firm called Barros & Vilaça Lawyers whose address was the same as Josélio's office in the state capital.[46]

The firm would also defend Rai—whose real name was José Antônio da Silva—when he was arrested for his suspected participation in Ceará's disappearance.[47] A member of the Associação, Rai would deny any involvement in violence, claiming he didn't have any links with Josélio at all, which was untrue.[48] Aside from the fact that, coincidentally, he was represented by the firm of Josélio's daughter, the rancher had previously admitted in a signed statement that Rai had worked for him.[49] Also, the day Rai was arrested, both a handwritten letter signed by Josélio and his business card were found on him.[50]

Rai spent only four months in prison. Judge Amorim would free him under the argument that no evidence indicated his involvement in any murder at the Te-Chaga-U.[51] Rai would henceforth become one of Dezinho's foes—only one on a rapidly lengthening list of increasingly powerful names. Brito would be also targeted.

10

Nowhere to Hide

O N February 11, 1995, in the period following Adélcio's recapture but before Secretary Câmara finally ordered the raid at the Te-Chaga-U, Rondon was shocked by news that the rancher José Hilário Ávila had been murdered in cold blood. At around 5:00 p.m. that day, Luiz Pacheco Neto, one of the occupants of Vila Gavião, had shot Hilário at a garage where the fazendeiro was dealing with an issue regarding a truck. After that, he sped off on a bicycle.[1]

Seriously wounded, Hilário died two days later in a hospital. But before passing away, he revealed the murderer's identity to his wife, Josefa, and linked the attack to the "invaders" of his fazenda. Hilário's widow explained to the police that less than a month before the homicide, Hilário had been negotiating with Dezinho and Brito an amicable agreement on the years-long feud, but the five-hour meeting had ended without reaching a deal. "About ten days after the meeting, we started to receive anonymous phone calls asking if it was the residence of José Hilário," Josefa said, hinting at the possibility that the threats originated with the Sindicato.[2]

Josefa also reported having witnessed Dezinho speak menacingly in addressing peasants in Vila Gavião. She described a situation in which the police, apparently enforcing a temporary injunction favorable to Hilário, had gone to the area to expel some occupants, and the union leader had said to the worked-up peasants, "Don't give up the land; the land is yours. Only hand over the land if blood is spilled!"[3] Prior to the murder, Josefa's statement went on, the squatters had warned her husband that he might win in the courts but would not get the land because they would kill him.[4]

The murder of Hilário left the landed class unnerved, wondering whether the Sindicato was raising the stakes and orchestrating deaths.[5] The threats against Hilário apparently weren't unique; the wives of other businessmen like the King of Wood would also report threats against their families by landless peasants. Reached by the media, Rondon's agribusiness association depicted the murder of Hilário as proof that contrary to the dominant narrative that the landowners were the source of the violence, they were actually the victims.[6]

Without delay, law enforcement looked for leads. Officers interrogated eyewitnesses and tried to hunt down Neto, the killer, who had left the region. At first the cops thought that Dezinho could be an important source of information in the case—or perhaps something more. Still, he would give his statement only on March 24, about a month and a half after the assassination, and under rather strange circumstances. Dezinho and Brito showed up in Rondon's police station accompanied by two lawyers working with FETAGRI and three state legislators, including Zé Geraldo, in an apparent display of political muscle to prevent them from being arrested. Both activists denied any involvement in the crime or in any violent acts. Hilário had been murdered for reasons unrelated to the land conflict, Dezinho explained. He also declared that he feared for his life.[7]

Rondon's chief of police didn't believe him and stuck to the theory that the Sindicato was linked to the killing. "Dezinho was from the beginning of the probe a natural suspect due to his persuasive ability, which he exerts over those humble people," the lead officer wrote.[8] In Vila Gavião, he reasoned, occupants would do "what the chief orders."

Agents who had interrogated the peasant families, apparently using violence, were nonetheless told a story that didn't accord with these assumptions.[9] Sources said the motive was personal, not related to land, or at least not a direct response to orders from the Sindicato. One of the lawyers accompanying Dezinho later recalled that by the time the murder occurred, the government had already signaled that it would seize Hilário's farm, so why would the Sindicato plot to kill the fazendeiro if they were about to win the legal battle?[10] Also, neighbors of Neto explained that the cause of the homicide was an issue of honor involving Neto's fiancée—a woman named Neuza who had apparently been threatened and assaulted by Hilário.[11] "Hilário smacked her after a discussion," declared a witness.[12] The same source also said that Neto wasn't a member of the Sindicato.[13]

Weeks passed, and Neto remained on the run, likely pouring salt into the wound for the bereaved family of Hilário. On May 10, a seemingly retaliatory attack was carried out when an occupant of Vila Gavião was hit with two shotgun rounds "that spread fourteen pellets in his body."[14] He survived, and members of FETAGRI and the Workers' Party published a release accusing "a group of fazendeiros" of the violence. More families were also at risk, so a few weeks later, the Sindicato got them out of the area.[15]

If the murder of Hilário made some wonder about Dezinho's real commitment to nonviolence, the issue didn't really impact his leadership, now reinforced by the Te-Chaga-U raid, which unfolded only weeks after Hilário was assassinated. In early 1996, Dezinho was unanimously elected for a second term. Brito was now his secretary general. They had run unopposed.[16]

The support of the unions emboldened the two leaders, leading them to plan more forceful actions. Whispers began to circulate in town that a first Sindicato-led occupation was being set into motion. The rumors sparked a backlash. On August 2, 1996, "three hitmen heavily armed with shotguns, revolvers, and a submachine gun" entered Vila Gavião looking for a community leader who luckily wasn't at home that day.[17] A week later, armed strangers came into the settlement and set fire to a shack and fifty sacks of rice.[18] Dezinho warned that if the authorities didn't take action to stop the pistoleiros, there would be a confrontation.[19]

Brito also saw that the situation was becoming more fraught with danger, and he responded by temporarily moving his family out of Rondon, although he decided to remain in town himself.[20] On August 23, he was leaving the Sindicato when he was approached by a man who introduced himself as a policeman. Dezhino was away on a trip.

"Brito, you need to come with me," the man said after flashing a badge. He was dressed in civilian clothing and argued that he wanted to interrogate him "for an issue linked to drug trade." Rondon remained a small town, and Brito often dealt with the police to report incidents involving members of the Sindicato, so he realized he had never seen that purported officer before. The timing—around 7:30 p.m.—also raised his suspicions. Why was he being summoned in the evening if he had been in the office the whole day?

"No," Brito replied, "I won't go."

The so-called cop became indignant. "What?! *You won't come?*" Brito realized that this man wasn't an officer but a crook, and he started to shout for help.

Right away, a car moved toward him. A second man jumped out of the car and tried to put Brito into a stranglehold. Brito was only five-foot-five, but he was stocky, with a robust neck and meaty hands, so it was hard to subdue him, although the thugs eventually got a grip on him and tried to shove him into the car's back seat. Brito's mind raced, and he regretted the lack of a machete or a gun.

The men managed to stuff his head into the car. In the rearview mirror Brito could see the upper part of a face that looked familiar. Running out of options, Brito pretended to give in, offering no resistance when his two captors, emulating police officers when they put a suspect into the back seat of a patrol car, pushed almost his entire body into the vehicle. They responded to Brito's sudden compliance by exerting less pressure on his neck and head. Brito then slammed a powerful elbow into the rib cage of one of the men, which allowed him to escape and head toward a nearby street.

The three men rushed after him. Until then the gang had tried to take him alive, but Brito believed that because he had recognized the driver, the mobsters now wouldn't hesitate to murder him if he came into their line of fire. A local who had watched the attempted kidnapping called out to Brito, offering his home as refuge. Brito didn't think twice and dashed toward the cabin, finally escaping the gangsters.

"The plan was to murder both Dezinho and me," Brito later declared, recalling that Dezinho had told him that "the scheme, hatched by Josélio, was to kidnap Brito and take him to his fazenda, where he would be tortured to death and his body dumped in a ranch somewhere in the middle of nowhere, so the corpse would never be found."[21] Dezinho, Brito said, had known about Josélio's alleged involvement from an informant, one of the sources who had access to the rancher's entourage and occasionally provided the Sindicato with intelligence.

Brito reported the attack to Rondon's police but anticipated nothing would be done about it, so he traveled to Belém with Dezinho to inform law enforcement at the state capital. In a long statement given on August 28, five days after the assault, the two union leaders declared that Josélio had joined forces with the family of the dead Hilário to recover the land of Vila Gavião. "Josélio told me that someone would kill me one day as a consequence of my involvement in the Sindicato," Brito said.[22] Dezinho accused the rancher of hiring "over thirty pistoleiros to go to Vila Gavião with the aim of executing eighteen peasants and their wives and children." Both claimed to have once witnessed Josélio being supplied with "three machine guns" by state officers.[23] That same day, a lawmaker denounced the attack against Brito in the Brazilian congress.[24] Josélio denied any wrongdoing.

Reporting the case didn't ease their situation. Only weeks later, a peasant heard from an alleged hitman in a bar that Dezinho would be killed during the 1996 mayoral elections, planned for October. On October 2, a man was found hidden in the offices of the Sindicato by the secretary of the union. The burglar told her that "he was looking for the president."[25] The incident, Dezinho wrote in an official communication to the authorities, "was immediately reported to Rondon's chief of police, who refused to investigate, arguing he had no patrols available."[26]

Then Hilário's murderer was found in a fazenda near Marabá. Neto, who had dodged the police for over twenty months, admitted to the crime and declared that Dezinho had planned the homicide in early 1995, during a meeting in Vila Gavião. The original idea was to use Neto's wife to provoke Hilário and prompt him to physically assault the woman, which would provide an alibi seemingly unrelated to the land conflict. "This is the first time I have had troubles with the law,"

Neto said, stating that "the 'heads of the movement' [Dezinho and Brito] told me that if I didn't help in the plan, I wouldn't be allowed to remain on the land." [27]

Immediately after hearing Neto, the police requested authorization to arrest Dezinho, Brito, and other peasants, but the judge in charge of the case refused to issue warrants.[28] She argued that the confession of the killer, which the Sindicato found biased and untrue, wasn't enough evidence per se to prove their involvement.[29] The judge instead urged the police to investigate further to verify Neto's version and to interrogate other alleged participants in the meeting during which the murder of Hilário was supposedly planned.

Soon afterward, Neto reversed his testimony. The assassin said that when he had been arrested, he had been coerced by police in Rondon to implicate Dezinho. "They forced me to say, under threat of death, that a plot was schemed for Hilario's murder," he declared.[30] Neto now claimed that he'd murdered Hilário in retaliation for the aggression against his sweetheart, framing the crime as an issue of honor. He also said that a member of his family had been killed in retaliation, and one of Hilário's sons had told him that he personally preferred Neto not go to jail, so he could kill him by "putting out his eyes and amputating his hands."[31] Despite the statement and the fact that Dezinho wasn't prosecuted, fazendeiros and some townspeople continued to believe he might have played a role in the crime.

According to Brito, the man driving the car the day he was almost abducted was a son of Hilário named Olivandro, who later denied any involvement in the issue. Decades later, Olivandro maintained that Brito and Dezinho had masterminded the murder of his father.

THE ATTEMPTS TO implicate him in Hilário's murder led Dezinho to conclude that law enforcement and ranchers in Rondon had joined forces to form a united front against the Sindicato. In his account of things, the coalition was now resorting to the courts to destroy him and squash the movement.

Dezinho's family endured a particularly rocky time in those days. In addition to what Maria Joel believed was a campaign to defame her husband, the threats were never-ending. Menacing notes were left at Dezinho's home or at those of his relatives, and well-known local outlaws were spotted by friends and neighbors near his house. Dezinho's name even appeared on a list of "men marked for death" that was publicized by the CPT in order to prevent the murders from being carried out. (Many activists and priests cited on these hit lists were executed, in spite of the publicity.) Fearing the worst, left-leaning organizations sent letters to the authorities to bring the dangers faced by Dezinho to their attention. They warned that if something happened to him, "Josélio and his group would be considered responsible."[32] They also requested that the authorities seize "immediately, and without compensation, the lands of those who use pistoleiros and slave labor." About a month later, another occupant of Vila Gavião was executed in cold blood.[33]

Prodded by Maria Joel, who argued that she preferred him to be "away but alive rather than close and dead," Dezinho began to leave the city for long periods. Their four children, now between seven and thirteen, also had to deal with the strict security measures. "Once you're out of school, come directly home!" Dezinho instructed Joelminha, Joelson, Joélima, and Joelina. Not only could they not choose how to spend their free time, but they also were seen as outcasts. The parents of many other children did not want their children to associate with Dezinho's kids, assuming that it was dangerous to hang out with the activist's daughters and son. "What are you doing?!" an anxious mother would shout. "Don't you realize that you could be hit by a stray bullet?" According to Joelson, teachers were uneasy when he was around, and he perceived school "as a hostile place" where he no longer wanted to go.

Dezinho struggled to deal with the tension in his household. "This is not the life I've chosen!" complained Joelminha, the eldest. Only two when they had moved to Rondon, Joelminha was now a teenager who had inherited her mother's beautiful dark brown eyes and her soft voice. She felt hemmed in by the security measures and the strict routines. "Every private matter had to be examined and discussed with my parents, including my dates," she would say with bitterness. "My father even interrogated my boyfriends to know who their families were and whether they had any involvement with the ranchers."

Outside of the family, Maria Joel was often seen as a woman struggling to cope with her husband's revolutionary ideas and the tumultuous events that surrounded him. But the truth is that Maria Joel was a tower of strength for Dezinho. She took care of the children, earned money essential to keep the family going, and advised her husband on how to run the union and dodge the dangers. She fretted about the possibility that Dezinho or another family member might be killed, and friends and family members worried that she also could be attacked. But even if she was terrified about the potential consequences, Maria Joel believed in Dezinho's cause and his integrity, and she showed him unconditional loyalty.

"Dezinho is not doing anything illegal," she said to neighbors, alluding to the issue of Hilário's murder, her calmness on display. "Those on the wrong side are the land-grabbers."

Without her support, Dezinho would never have been able to launch the Sindicato's first occupation, his most forceful action since taking control of the union. But before that, appalling events would take place in Pará and news of this would encircle the globe.

11

Nothing Shining in Eldorado

O N April 17, 1996, two months before the kidnapping attempt
against Brito, Brazil was staggered by a massacre that, if not
a game-changer for the landless movement, would represent an important inflection point in the struggle of organizations like the MST,
FETAGRI, and the Sindicato. The events occurred in the municipality
of Eldorado do Carajás, a rural area southwest of Rondon not far from
the Serra Pelada mine, now closed.[1]

That day, nineteen peasants and land activists were killed and several dozen wounded when members of the state police opened fire
against approximately 1,200 people, including children, participating
in a demonstration of the MST, whose influence and strength had only
grown over the years.[2] The protesters were blocking a tract of highway
linking the north and south of Pará to press the authorities to settle landless families in a series of fazendas suspected of grilagem that
spanned some 100,000 acres.

The killings happened after an initial skirmish between the police
and the demonstrators at around 4:30 p.m. The officers argued that
the peasants—armed with stones, sticks, and, allegedly, some guns—
had escalated the fight.[3] But the images captured by a local television
crew covering the demonstration, as well as evidence from autopsies of
the people killed, would paint a radically different picture. Some of the
victims had been killed at close range, while others who had escaped
the police bullets were chased and beaten to death by law enforcement.[4]

"The police were there to kill," declared a member of the MST to the press. "I saw a child who, after being shot, was kicked to death by a cop."[5]

The massacre appeared to be premeditated.[6] The police had initially hemmed in the protesters from both south and north so they could not escape. Also, prior to the shoot-out, agents were seen stripping their name tags from their uniforms so they could avoid identification.[7]

Referred to in the press as the Eldorado Massacre, the case monopolized the front pages of newspapers for days.[8] Images from the slaughter were broadcast nationwide and abroad, triggering a shift in the narrative about the nature of occupations, previously seen as unlawful acts. Even top figures of the state showed solidarity with the MST and its fight for social justice. President Cardoso said that the "landless [in Eldorado] were fighting for a fair cause," and he accused the police of excesses.[9] The president of the Supreme Court of Brazil declared that "all Brazilians committed to human rights . . . were ashamed by the brutality."[10] The Inter-American Commission on Human Rights urged Brazil "to ensure respect for the rights to freedom of expression and assembly, life, humane treatment, and liberty."[11]

Analysts emphasized that land-related murders like the Eldorado Massacre were the result of runaway impunity. The available data supported the idea that those perpetrating crimes against landless peasants and rural dwellers were almost certain to avoid prosecution. From 1985, when the country had initiated its transition toward democracy, until the events in Eldorado do Carajás, only fifty-six trials had been held for land crimes nationwide, most of them to prosecute hired killers or pistoleiros, not the instigators, called *mandante* in Portuguese.[12] These numbers were in stark contrast to the number of assassinations; during that same period, at least 976 people had been slain over land disputes, and another 891 had been attacked, according to the CPT.[13] Pará was Brazil's most violent state, with homicides at "near-epidemic proportions," as one American scholar put it.[14]

Experts noted that in addition to the near absence of rule of law, extreme inequity in land distribution was also a fundamental cause of the problem. By the time of the massacre, about 45 percent of the arable areas belonged to just 1 percent of the population.[15] Many factors explained the disparity, but a critical one exacerbating it was land-grabbing.[16]

The massacre, despite the dark reality it exposed, provided an opportunity for peasant organizations to advance their agendas and gain wider support. A poll published in April 1997, a year after the killings, showed that 94 percent of Brazilians considered the cause of the landless people to be just, and 85 percent expressed support for nonviolent occupations of fazendas as a way to achieve agrarian reform.[17] Already on the rise at the time of the massacre in Eldorado do Carajás, the number of families occupying estates, ranches, and rural areas would explode in the ensuing years, going from 20,000 in 1994 to 76,000 in 1998.[18] The government of President Cardoso, internationally embarrassed by the massacre, also responded by seizing some illegal farms and ranches to settle families.[19]

Unsurprisingly, the country's landed gentry worried about the potential implications, prompting agribusiness associations and conservative political parties to use their influence to prevent any major change in legislation. (Later, *Folha de S. Paulo* revealed that the army was spying on the MST in Pará, although a general denied it.)[20] One of the most vocal and reactionary was the Democratic Rural Union, or União Democrática Ruralista (UDR), which was created in 1985 as a kind of response to the MST. Some UDR members warned that, if the number of occupations continued to grow, landholders might retaliate violently. The media published reports in which fazendeiros in Pará admitted they were establishing armed militias to repel "invasions."[21]

A week after the Eldorado Massacre, the Associação also aired the worries of its members in a blunt statement. "The rural class of Rondon do Pará feels its private property rights threatened," wrote the fazendeiros.[22] They denied being "tyrants who enslave their underprivileged brothers" and claimed that instead they were rural entrepreneurs providing food and jobs. "We urge attention be paid to this note in order to avoid new victims," they warned in a message clearly directed at the Sindicato.

Dezinho and Brito followed the events in Eldorado do Carajás closely. The massacre had made the demands of the landless a national issue, and the pair of activists weighed the risks of launching their first occupation. "It was an opportunity for the landless people," a lawyer of the CPT named José Batista Gonçalves Afonso explained. Batista, a thin, savvy man of humble origins, arrived in Marabá in 1996 and had become a legal advisor to Dezinho and the Sindicato.

Many families in Rondon signaled to Dezinho that they would follow the Sindicato if it launched an occupation, but the union leader had second thoughts. The Sindicato lacked the resources and the near-military organization of the MST, which were crucial to achieve success. Also, the maximum number of occupant families they could deal with at one time was fifty or sixty, which meant some two hundred or three hundred people—a number much easier to intimidate and expel from a fazenda than the seven thousand the MST had managed to organize for some of its most representative occupations.[23] Dezinho knew his reputation had already been affected by the murder of Hilário, and now he feared that deaths would result if he behaved irresponsibly.

Ultimately, what made Dezinho feel it was time was Rondon's socioeconomic context.[24] Two decades of uninterrupted logging had depleted the accessible timberlands to the point that many sawmills were no longer profitable, and they had begun to lay off hundreds of workers.[25] The end of the logging boom was also a consequence of stronger federal policies to curb deforestation rates, another source of international criticism against Brazil. Pará remained Brazil's top producer of logs, and the profitability of the wood sector continued to be substantial despite the restrictions, so many entrepreneurs had moved to the black market.[26] But the years of prosperity, when several thousand people in Rondon were employed in the sawmills, were over, and although other businesses were flourishing, especially agribusiness, they couldn't generate enough jobs to replace those lost by the closing of the sawmills because they weren't labor intensive.[27] As a sign of the gloomy economic outlook, the municipality, which had some 30,000 inhabitants in the late 1990s, experienced a sharp drop in population, the first on record.[28]

"If we don't bring our people to plots to work and implement agrarian reform," Dezinho would say in a speech, worried about the consequences of the exodus of poor families from rural areas toward sprawling urban peripheries in Belém and Marabá, "we know that our children will die of hunger; we know that our children will become street kids; we know that our underage daughters will become prostitutes."[29]

In November 1997, Dezinho participated in a twelve-day protest camp in Marabá. Some 8,000 members of the MST and FETAGRI got together to urge the government to distribute land and devote more funds to agrarian reform. Only a few months after the slaughter in Eldorado do Carajás, men, children, and even pregnant women joined occupations, marches, and blockades. Two protesters even gave birth during those acts of civil disobedience.[30] These displays of courage and determination prompted Dezinho to launch a first occupation in Rondon.[31]

About sixty families crossed the fences of Fazenda Jerusalem, thought to be the product of land-grabbing. Families had registered with the Sindicato, which according to Brito had charged each person about a dollar in order to finance transportation in rented trucks. The occupants entered the 7,400-acre ranch, claimed by a fazendeiro named Anterino Pereira Rocha, and set up shacks.[32] The attempt to expel them came a few weeks later, when a group of armed men entered the camp and shot at the smallholders, who nevertheless refused to leave, according to the account of the CPT.[33] Dezinho reported the case to the police three days before Christmas and accused federal officers of siding with the gunmen; Josélio, he alleged, was preparing to take action to expel the families or even to take some lives, including his.[34]

Instead of backing down, the Sindicato upped the ante. "We want peace in Rondon," read a pamphlet distributed in town in early 1998.[35] "Rondon do Pará is known for the violence and the action of organized crime by latifundia that for years have been threatening farmworkers and union leaders." The document enumerated a series of killings, including the chainsaw murder and another execution committed, according to the pamphlet, "by gunmen of Josélio."

Dezinho, Brito, and Francisco de Assis, a member of FETAGRI stationed in Marabá who often visited Rondon's interior, reached out to the media to further expose the threats. For the first time, they referred in public to the existence of a criminal group formed by landowners working together to suppress any challenge to their domination. The list of alleged outlaws included a fazendeiro named Manoel "Duca" Lopes, another named Durval Ferreira, and a so-called Hugo "Bad." All of them, the Sindicato leaders said, were under the command of Josélio, who they claimed acted as a kind of godfather. "The fazendeiros were never satisfied with just killing the rural workers but also want the leader," Dezinho declared, accusing Josélio of contracting out his assassination to Rai.[36] The existence of the criminal group was also reported to Pará's top security authorities.[37]

Soon thereafter, as they increased the pressure with public denunciations, a second occupation was planned. It was launched in January 1999, when sixty-eight families settled on the ranch Primavera II (Spring Two).[38] It was owned by a cattleman suspected of grilagem named Jucelino Favoreto. He reported the "invasion" to the police and was given an injunction to expel the families.[39]

Despite—or perhaps because of—the pace of occupations, tensions developed between Dezinho and Brito. After having been joined at the hip for almost a decade, the two friends were now showing signs of estrangement. Brito felt that Dezinho was taking too many risks on his own. Sometimes, he would recall, Dezinho would meet with laborers and informants in his home, an endeavor Brito didn't agree with, because it was dangerous, but also because it kept him in the dark about important matters.

As Dezinho approached the end of his second term in early 1999, the strains in their relationship became more evident to union members. The issue of succession turned into a battle of wills. According to the union's rules, Dezinho couldn't seek a third consecutive mandate, and he'd previously told Brito that he would support him as the new president. But Dezinho wasn't totally sure that his friend would maintain the pace of occupations, which Dezinho believed couldn't stop now, so he thought about supporting other candidates.

Word spread that Dezinho might explore the possibility of changing the Sindicato's statutes so he could run for a third term. Brito didn't agree and thought it was undemocratic. But Dezinho found it difficult to accept stepping down from the presidency at that crucial point and being relegated to a minor position. He'd instilled a revolutionary soul in the Sindicato, and the control of the union had also served him personally, allowing him to establish connections with members of the Workers' Party, the media, and FETAGRI. These contacts were useful to air threats, gain a name, and look toward the future, perhaps serving as a stepping-stone to a career in the field of politics. Without a formal position at the Sindicato, Dezinho feared he would lose influence and also become more vulnerable. Eventually, they reached a compromise and Brito was elected president while Dezinho was named director.

Although he was no longer the leader, Dezinho continued to organize occupations. On March 4, 1999, the judge of Rondon summoned him.[40] In a ruling literally two lines long that didn't say why Dezinho was being called to testify, Amorim ordered him to show up at Rondon's courthouse and threatened him with contempt if he didn't obey. The ruling came after Dezinho had apparently said to the media covering conflicts at an occupied ranch, "We're armed and ready for everything."[41]

Dezinho decided to ignore the order, denying that he had said those words. On May 18, Judge Amorim responded by ordering that Dezinho be detained. The order was immediately enforced by the police.[42]

Dezinho was incarcerated at the city's police station, and Maria Joel, who had joined the union and dealt with issues related to women, called CPT lawyer Batista to try to get him out as soon as possible. Fearing he could be tortured or even killed, she organized a sit-in with some three hundred supporters who surrounded the police station.[43] Brito, in the nearby city of Dom Eliseu when Dezinho was arrested, rushed to the jail. Dezinho, frightened for his life, said, "Pal, don't leave me alone here. I fear that they may kill me."

Batista filed a motion, and Dezinho was freed on bail three days later. Dozens of rural workers celebrated his release with shouts and congratulatory hugs before marching through the town, led by Dezinho walking hand in hand with Maria Joel. In the ensuing weeks, some 3,000 people, according to the press, protested what they considered an attempt by Judge Amorim to intimidate Dezinho and by extension the movement.[44] Later, civil rights groups would argue that "Dezinho was put in jail despite the fact that no indictment had been filed and no procedural process had begun in the case. There existed no record of the proceedings, no eyewitness reports, and no requests for preventive detention by police officers."[45] The suspicion was that Judge Amorim had ordered Dezinho's arrest to stop him from launching new occupations.

A few days after Dezinho was freed, Josélio filed a complaint against the Sindicato for enticing people to "invade" the Te-Chaga-U and the nearby Nova Delhi, another ranch he also owned.[46] Nunes followed him, reporting to the police that his Fazenda Lacy was at risk of imminent takeover.[47] The King of Wood had continued developing his 247,000-acre spread, but he was increasingly suspicious of the Sindicato. Weeks after Nunes reported the case to the police, FETAGRI publicly accused Nunes of threatening employees, causing illegal deforestation, and practicing debt bondage on his Fazenda Lacy.[48]

The controversial arrest of Dezinho would be one of the last rulings by Judge Amorim. Less than a year later, the judge would be suspended for misconduct after the internal affairs office found that, in just a few months, he'd handed out favorable sentences of dubious legality to many companies and individuals.[49] No one ever investigated whether he'd received money to rule against Dezinho or the Sindicato, but his credibility was destroyed, although he appealed for years to defend himself. Eventually, Amorim was found guilty and was forced to retire.[50]

The arrest reminded Dezinho that without the command of the union, he would be more exposed. He began to think about finding a larger forum from which to advance in his career while the land struggle continued. He found it in the Workers' Party, getting himself nominated as their candidate for councilman in the forthcoming mayoral elections.

12

Death and Salvation

O N the night of November 21, 2000, right after the election of a
new mayor in Rondon, a nineteen-year-old man named Wellington de Jesus Silva stepped off the rear of a silver-colored Honda
motorcycle and began to walk hesitantly toward the house of Maria Joel and Dezinho, located in the Recanto Azul neighborhood. It
was suppertime, and the smell of rice and beans emanated from the
wooden cabins flanking the dimly lit, unpaved street. Piles of dirt and
saffron-colored sand were everywhere, because the municipal government was installing a sewer system along Paraguay Street, the road
where Dezinho and his family lived.

That night, Wellington wasn't paying attention to any of this. He was
wondering why he had agreed to do the "service" he was expected to
perform in the next hour.[1]

Unemployed in his hometown of Pau Brasil, a small and poor municipality in Bahia where he had found sporadic work in construction,
Wellington had arrived in Rondon less than a week before. His older
brother Rogério, who lived in town with some distant relatives, had
called him to propose work in a charcoal kiln, and Wellington had
accepted, immediately jumping on a long-haul bus.[2] In the period between his arrival and the night of November 21, Wellington had been
on Paraguay Street twice. He'd been guided there by his cousin Ygoismar Mariano da Silva, who also lived in Rondon and was the man
who had revealed to Wellington the real reason he had been called to

that town. Ygoismar was waiting in the Honda that night, and he had warned the reluctant Wellington that something bad might happen to him if he didn't complete his appointed mission.

With that threat still resonating in his mind, Wellington reached the picket fence surrounding Dezinho's cabin and, in line with the custom of the city, clapped loudly. "*Seu* Dezinho! Mr. Dezinho!" he said, raising his voice.[3]

If the young man outside was on edge, inside the house, all was calm. Maria Joel and eleven-year-old Joelina, her youngest daughter, were watching television.[4] It had been a day of relative and rare peace for the family. Dezinho had spent some time at home with the children, which was a special thing, because the previous two months of the elections hadn't been easy.[5] Maria Joel would refer to them as "a time of great tribulation."[6]

Dezinho had fully devoted himself to the campaign. At home, he had rehearsed his speeches atop a chair, in front of a mirror, a scene that made Maria Joel and the children laugh. He gave speeches in the town's poor neighborhoods and in the rural interior, his popularity leading some ranchers to conclude that Dezinho would receive staunch support.[7] However, as election day drew near, rumors spread that Dezinho would be murdered if he was elected councilman, and the prospects of victory became increasingly bleak. "People, including those in the occupied fazendas, were intimidated by the rumors indicating that if Dezinho was elected, he would be murdered the next day," Maria Joel recalled. Nobody really knew anything about a concrete plan or who specifically the instigator might be, but the strategy worked. Dezinho got only 238 votes, two ballots short of being elected councilman.[8] The results were a bitter blow.

But more opportunities seemed to lay ahead. By the time Wellington got to town, the talk in Rondon was that the mayor-elect, a man named Moisés Soares de Oliveira, who had won with the support of a broad coalition of progressive parties, might pick Dezinho as his secretary of agriculture, an ideal position for his shift from grassroots activism to politics that would also serve to influence the Sindicato.

This appointment had not been confirmed yet, but it seemed like a real possibility, because Dezinho had supported Moisés in the campaign, and the Workers' Party was a crucial ally in the coalition.

With that job possibility still on the horizon, Maria Joel heard Wellington's unfamiliar voice and asked Joelina to check who it was. (For security reasons, the children had been instructed to always verify the identity of visitors through the window.) After a fleeting inspection, Joelina described Wellington as "a young guy asking about Daddy."[9] Maria Joel then stood from the couch and went to talk with the visitor. It was fully dark outside, and in the barely illuminated area behind the fence stood a thin man she had never set eyes on before. Slight of build, he had short, curly hair and light-colored eyes. He was wearing a loose shirt and a pair of worn trousers.

"Good evening," Wellington said. "I want to meet with Mr. Dezinho. I need his help to get a pension for my dead grandmother."[10]

Maria Joel was used to her husband attending to all sorts of matters at any time of the day. "People routinely came to our home to ask Dezinho for help," she would say.[11] They rarely allowed strangers to enter their home, a place Maria Joel considered the safest for her family in that town, but if they felt that the person was trustworthy and in real need of help, he or she might be allowed in. Wellington was a total unknown, and he hadn't presented any credentials to Maria Joel or explained who had given him their address. But perhaps on account of his delicate, almost prepubescent countenance, which made him look painfully shy and insecure, or maybe because of his pronounced Bahian accent, she took pity on Wellington and decided to open the door. He just seemed to be an artless, needy boy. It was about 7:30 p.m. when he entered the cabin.

Wellington would have preferred for Dezinho to have been the one to open the door so he wouldn't have to go inside the house, but that didn't happen. Wellington had considered multiple scenarios after making the first contact, and he knew that Dezinho was the father of several children, but once he entered the house, he became unnerved by the presence of a preteen just a few feet from the burgundy couch where Maria Joel had invited him to sit and wait.

Though he didn't know it, there was someone else in the house—another man, but not the one he was seeking. Relaxing in the backyard, Joelson was getting ready for a shower. That afternoon, the couple's son had been playing soccer with his father, who had passed on to him a real passion for Brazil's national sport.[12] It had been an occasion to mend fences, because Joelson, now sixteen, was feeling angry about their constrained life.[13]

Still fresh in Joelson's memory were the details of a recent frightening occurrence. It had happened right before the political campaign and while Maria Joel and Joelminha were in another city for a medical issue. A group of six peasants had spontaneously shown up at Dezinho's home with an urgent message for him. "We've heard there's a plan to kill you tonight in your own home," they announced. "We're here to protect you." Dezinho gave full credibility to the account, having himself noticed some suspicious people near the cabin. Although Dezinho didn't let the children in on all the dangers he faced and often displayed a cavalier attitude toward the threats people came to alert him about, that day, probably because Maria Joel wasn't in town, he was candid with Joelson and Joélima, another of his daughters who had stayed home.

"Sweetie, some people want to kill Daddy," he said calmly. "Tonight I'll remain at home with some friends and with Joelson, because he's Daddy's little man, but I want you to sleep at the neighbors.'"

As Dezinho prepared with his comrades—all unarmed, according to Joelson—to respond to the threat, the union leader wrote a goodbye note that he gave to Joélima. "Deliver this letter to Mom if something happens to us. Tell her how much I love her."[14] In tears, Joélima followed her father's orders and went to a friend's house.

Maria Joel called home that day to check on things. Dezinho didn't tell her about the issue, but she sensed the great strain in her husband's voice. Joelson later said that Dezinho wandered through the house smoking one cigarette after another, thus fueling his own dread. In a normal place, an ordinary person would have simply called the cops, but they were in Rondon do Pará and Dezinho didn't trust law enforcement, so he instead crafted his own safety plan. One man was told to remain in the backyard all night and another was stationed next to the picket fence encircling the house. The rest would stay inside, volunteering "to die if they killed Dezinho," according to Joelson, who never truly understood why his father had decided for him to remain by his side. After midnight, exhausted by the stress, Joelson fell asleep.

Joélima didn't rest at all. At 5:00 a.m., at the first light of day, she ventured onto Paraguay Street to check on her father and brother. She found the lights of the cabin on and the door wide open. Inside were Dezinho, Joelson, and the peasants, all awake. There were also some neighbors who, informed about the threat, had joined the vigilante group. The anticipated pistoleiros had never shown up. A few days later, Dezinho called de Assis at FETAGRI and said, "If the death threats don't stop, this may be the last time we talk."[15]

"DEZINHO ISN'T HERE," Maria Joel told Wellington in her characteristically soft voice. "He's having dinner with a neighbor."

Momentarily flummoxed, the visitor asked if Dezinho would take long. When Wellington posed the question, he may have hoped for a yes so he would have a pretext to abort the mission and run. But Maria Joel, who thought that the matter might be pressing after Wellington had declined her offer to discuss it the next day at the Sindicato's office, shook her head and said to Joelina, *"Vai procurar o papai,"* instructing her daughter to fetch Dezinho.

The girl left, and Wellington started to show signs of tension. The glass of water Maria Joel had brought him shook slightly in his hand as the two sat waiting only a short space apart. Wellington was wearing a loose, short-sleeved beige shirt, which he was almost certain concealed the Taurus .38-caliber revolver he had tucked in his waistband.[16] But the barrel was pressing into his hip, and Wellington worried that Maria Joel might spot the wooden grip of the weapon protruding from the top of his trousers and bulging over his flat stomach.[17]

Unsuspicious for now, Maria Joel asked Wellington for details on the issue involving his grandmother. Wellington's accomplices had concocted a cover story in which he would mention a grandmother who had just passed away and the possibility of receiving a state pension. But they hadn't provided him with any further information, and Wellington knew almost nothing about the work his target did, so he didn't elaborate and responded evasively, attempting to mask his discomfort with some small talk.

"Will Dezinho take much longer?" he again asked Maria Joel.[18]

Only a few minutes passed before Joelina returned, announcing proudly that she had found her father and that he was on his way back home. At this point Wellington had to weigh his choices, the most urgent being whether he should wait for Dezinho inside the house or find a credible excuse to exit and carry out the execution in the open air. He decided to go with the second option, presuming it would be easier to escape and jump on the Honda with Ygoismar once he'd emptied the barrel.

"I'm going to buy a pack of cigarettes," Wellington blurted before standing from the couch.

This action made Maria Joel suspicious. "If you want to speak with Dezinho, why do you go to buy cigarettes now that he's coming?" she inquired.[19]

A fleeting stony silence ensued before he hurried to the door and replied he would be back in a while. On his way out, Wellington bumped headlong into Dezinho, the man he had been promised a bounty of 2,000 reais—about $600—to murder.

Maria Joel hadn't seen the gun, and she didn't think that Wellington could be a hired assassin, but she couldn't shake her distrust of the young man and went to the door to eavesdrop on the conversation. She witnessed a few seconds of the meeting, which took place a few steps from the doorway, in the area just outside the fence. She heard Wellington repeat the same spiel about his grandmother's pension, and momentarily satisfied, she turned and headed back toward the living room.

If Maria Joel didn't see in Wellington an assassin on a mission, something in the man's demeanor did raise Dezinho's suspicions, and the activist promptly threw himself on Wellington the instant he grabbed the gun.[20] Dezinho could have run, but this would have exposed his family, so he felt compelled to fight. Both men were about the same height, although Dezinho was at least thirty pounds heavier. They grappled and rolled before falling together into a deep ditch—one of the holes dug to install the pipes for the sewer system. Wellington had already emptied the contents of the revolver's barrel.

The thump of the shots resonated throughout the street, and Maria Joel rushed to the entrance, knowing something terrible had happened. She arrived in time to see the two men tumbling into the ditch.

"They've shot Dezinho!"

Her yells alerted Joelson and Joelina as well as dozens of neighbors, who rapidly assembled around the hole. At the bottom, Wellington lay pinned beneath Dezinho's body, the pair now still. The yellow tank top Dezinho was wearing, emblazoned with an ad for a congressional candidate of the Workers' Party, was soaked with blood.

"It wasn't me! I haven't killed him!" Wellington screamed.[21] "Someone shot him from a motorbike, he grabbed me, and we fell!"[22]

More and more people arrived at the scene and saw the man—whose narrative nobody believed—with his arms up, his shirt stained with blood, and his hair covered with sand. Scrambling out of the ditch, Wellington tried to make a break for it, but the crowd grabbed him. His accomplices—Ygoismar, who had been waiting on the Honda, and according to some sources, another man in a pickup—fled as soon as the commotion broke out.[23] The weapon was found in the hole, and then it was all over for the killer.

The air thickened with rage, and the situation began to deteriorate. Furious neighbors beat Wellington, pelting him with stones and sand. Someone slipped a makeshift noose around his neck, aiming to see *lex talionis* served right then and there.

Encircled by angry men and women pointing at him as if he were the feral beast in a fable, Wellington thought he was about to die.[24] Why had he shot that man? Why hadn't he aborted the mission when, walking toward Dezinho's house, he'd turned around and said to Ygoismar, who had now abandoned him to his fate, that he didn't have the guts to slay a man he knew nearly nothing about?[25] Those questions, now moot, resonated in his head.

To reassure his family in the most difficult times, Dezinho used to say that "not even a cannonball would finish him," but that night, seeing him motionless and bleeding, Maria Joel was filled with terror at the idea that a man she had allowed to enter their home had killed her husband.[26] With the faint hope that he might still be alive, even if severely injured, she ventured into the hole.

"Dezinho, don't die! Talk to me!" she screamed, her body racked with sobs as tears rolled down her face and dropped on her husband's chest.

Suddenly, the uproar coming from the crowd snapped Maria Joel out of her trance, and she noticed Wellington, his face swollen from blows, in the powerful grip of a neighbor named Magno Fernandes do Nascimento. She screamed, "Don't kill him!"

Some people thought her call was a pointless act of mercy. They were operating according to the commandment of an eye for an eye and a tooth for a tooth. But no matter how bewildered she was by the tragedy, Maria Joel realized that Wellington wasn't simply the man who had just killed her husband and ruined her life—he was also a potential informant. "He is the only one who can lead us to the mastermind," she uttered, convinced that the murderer was only an agent in a larger scheme perpetrated by the fazendeiros.

Soon Sergeant Claudio Marino Ferreira Dias appeared at the crime scene, where about a hundred people had gathered. The officer of the military police of the state of Pará had heard on the radio that "gunfire had happened in town," and he had rushed to the area.[27] Once on Paraguay Street, Marino met Wellington, who was "very dirty and injured," and promptly identified the victim as Dezinho.[28]

Murders were common in town, but this one certainly stood out. Not only had the city's most prominent activist been slaughtered, but the gunman had also been apprehended by bystanders. Marino would say that in ten years of service, he had "never witnessed the arrest of a pistoleiro in flagrante by locals."[29] Acknowledging the mounting anger of the crowd and the great number of Dezinho's supporters, notably in the rural areas, where the news would soon spread, Marino understood that, in the following hours, his "greatest concern as a law enforcement official would be to guarantee Wellington's safety, because the crime had deeply shocked the town."[30]

Although he was not the most senior officer (he had been based in Rondon for only fifteen months), Marino had carried out some important missions in the region.[31] He was the agent who had inspected Nunes's Fazenda Lacy to investigate the existence of slave labor and deforestation, for example, which had been reported by the Sindicato. In his nineteen-page report, Marino had concluded that contrary to the allegations of the Sindicato, the labor conditions offered by the King of Wood in his spread were nothing but exemplary.[32] On that occasion, however, Dezinho hadn't participated in the raid.

Marino had also interacted with Josélio several times, a man he would describe as a fazendeiro who was "a very good friend of [the police], someone who helped a lot in the city."[33] In a later interview, he would elaborate by arguing that Josélio helped the police to get state funding.[34] Though Josélio had been given several certificates stating that he had been "a friend" and had "provided services" to the police in Pará, Marino had received another type of help from him.[35] He would say that Josélio had twice informed him personally about violent crimes that were about to be committed in town—a murder and an attempted kidnapping.[36] Asked about how Josélio had acquired information about these crimes, Marino would say, "When you have power, things come to you. Usually, he who has power has access to information."[37]

The body of Dezinho was taken from the hole and placed in the bed of the police's pickup. Joélima arrived at the scene about that time. She couldn't believe that her father was dead. Just an hour earlier, Joélima had been at a friend's house when Dezinho had asked her if she wanted to accompany him home. She had said no because she wanted to play a bit more. Soon after, a girl riding on a bicycle had approached her to ask "if she was the daughter of Dezinho."

"Yes, I am," she had replied.

"They've just murdered your father."[38]

Joelminha, the eldest, would be the last to reach home. The eighteen-year-old, who was engaged to be married in two months, was at the high school when she was informed that something had happened to her father. Initially, her fiancé had said that "he had suffered an accident," leading Joelminha to think that her father had been "involved in a car crash while on his way out to the rural areas."[39]

She understood what had happened only when she found Maria Joel, sobbing, at the center of the crowd. "Where is my father? Mom, where is my father?" Joelminha kept asking. Then she saw two bloody handprints on her mother's blouse, immediately recalling a scene she had witnessed three days earlier, when she had heard her father saying in a speech that "he would shed his own blood for the good of the landless peasants if that was necessary."[40]

That night, Joelminha refused to accept what had happened. She went to the hospital where Dezinho's corpse had been taken for an autopsy. She didn't find his body in the morgue but "thrown on a weighing machine, like an animal."[41] Enraged, she leaped on Dezinho, "lifting him up while screaming to the hospital staff to treat him with respect."[42] For a second, she thought that despite the visible bullet holes, he wasn't dead.[43]

"I believed an elite of fazendeiros in control of everything had ordered doctors not to revive my father," she would recall.[44]

According to the autopsy report, Dezinho's cause of death was "acute blood loss as a consequence of internal hemorrhage."[45] Wellington had emptied his gun clip during the attack, but only three of the six bullets had struck Dezinho. Two had pierced him, and the third merely grazed his chest. One of the shots had entered the left side of the abdomen but missed the lungs and hadn't resulted in major injuries. The second was most likely the kill shot, penetrating his chest and causing critical damage to his heart.

Wellington was taken to a hospital to suture some face wounds and later was taken into custody. Marino called on all patrols in Rondon and nearby cities to be available should they need to "contain an attack" on the police station by supporters of the Sindicato.[46] As the officer prepared for a massive riot that would never materialize, he looked, intrigued, at Wellington, wondering who the guy was. Wellington's face seemed so familiar. Marino interviewed him in the cell, but Wellington remained silent when he was asked who had hired him. The officer realized then that he had seen the young man before, just hours earlier, during a series of traffic stops he had conducted on Rondon's main avenue. He recalled having stopped the Honda on which Wellington and Ygoismar were traveling because neither was wearing a helmet. Ygoismar had told Marino that he'd rented the motorcycle to go "to a fazenda," while Wellington had explained that "he had just arrived from Bahia to meet his relatives."

Those answers now opened a line of investigation. Was it possible that Wellington, who seemed so diffident, was actually a professional pistoleiro someone had hired in Bahia, about a thousand miles away, to come and take Dezinho's life?

And, more crucially, who had hired him?

Part III

After He's Gone

13

An Unusual Case

THE murder of Dezinho shocked Rondon do Pará. Bars, restaurants, and public squares were soon filled with people speculating on the how, the who, and the why—although the answer to this last question seemed rather obvious.

In the occupied fazendas, the news spread via small radio stations, and union members based in Rondon traveled to inform peasants in the interior. Before long, waves of shabby laborers and farmhands flooded the city, some in tears and others ready for a fight. Still, despite the fear of retaliatory riots or even an invasion of the police station where Wellington was being held, no acts of civil disobedience ever materialized.

The thrashing the killer had endured during his capture had left his face severely swollen, even impeding his speech. "He is being treated by doctors," an officer explained to reporters covering the assassination on the night of the murder.[1] "He will be interrogated as soon as he recovers."

Wellington was badly banged up, but what disturbed him most was an overwhelming feeling of uncertainty.[2] What would become of him now? Caught red-handed, he faced a maximum prison sentence of thirty years, and he feared what might happen to him during that time. Prison conditions, the *New York Times* wrote at the time, were "generally dismal throughout Latin America, with extreme overcrowding, torture and diseases like AIDS the most serious problems. But Brazil's treatment of its more than 200,000 prisoners stands out as extreme."³ A UN official would describe the conditions as "grotesque" and "subhuman," and they were indeed appalling, with recurrent and wild uprisings organized by incarcerated bosses of the powerful gangs that terrorized the country.⁴

Wellington was not yet part of this bleak system. According to official documents, prior to murdering Dezinho, he had no criminal record.⁵ In his hometown of Pau Brasil, he had occasionally worked in construction with his brother Rogério, who had called Wellington to propose the job at the charcoal kiln. Wellington had planned to save some money from the job and buy a small plot.⁶ "Mom," Wellington had said after talking with his brother, "I'm leaving for Pará to become a farmer."⁷

It was only after there was no chance to turn back that he'd realized the real job waiting for him was a murder, as Wellington would tell the police on the morning of November 22, when he provided his first statement, accompanied by a defense attorney named Adriana.⁸ Still bruised, Wellington told officers that he hadn't consumed drugs or alcohol before committing the crime. He was reticent, but he did give a detailed account of the few days he had spent in Rondon before pulling the trigger. Wellington explained that he had arrived from Bahia only five days earlier, "invited" by Rogério and by a cousin named Gilvaldo. Wellington also said he had stayed in the house of "an uncle" named Gilson. According to him, none of these kin had any involvement whatsoever in the murder. The one who had approached him "to carry out the service," he said, was Ygoismar, who had given Wellington the loaded revolver.

Slightly older than him, Ygoismar had taken Wellington to Dezinho's house twice to show him the target. The night of the murder, when he was already carrying the revolver, Wellington had approached the activist's home three times but had returned to the motorcycle to tell Ygoismar that he was going to turn down the assignment. "I can't do it," he had said. Ygoismar had replied that he must either complete "the service" or face the consequences. Fearing for his own life, Wellington had proceeded.

Before validating his statement with a messy signature—all Wellington was able to write—the murderer gave the first indication that he was unable—or perhaps unwilling—to reveal the identity of the crime's mastermind. "Other than Ygoismar, I don't know who contracted out the murder," he stated.

After testifying, the triggerman was briefly presented to the media covering the case. "I killed," Wellington muttered to reporters. He denied being a professional killer and expressed remorse for the homicide. His demeanor suggested an individual from an extremely humble background, who while devoid of a moral compass was not a professional assassin. Officers who had questioned him said that Wellington, transferred that day to a jailhouse in Belém, seemed overwhelmed by the situation.[9]

The way the execution had been carried out also suggested he was not an experienced murderer. A veteran assassin would never have chosen the victim's home as a place to commit the crime, because it made the venture unpredictable and risky. Another issue was the bounty. Even if he was an outsider coming from Bahia, Wellington should have figured out who Dezinho was before negotiating the price of the hit, which for a target of Dezinho's stature should have amounted to thousands of dollars. The sum he'd accepted signaled that either he was in urgent need of money or, more likely, he knew nothing about how to negotiate a killing contract of that kind.

Friends of Dezinho agreed that Wellington didn't seem to be a professional. Francisco de Assis, Dezinho's partner at FETAGRI and himself an activist with a price on his head, believed that Wellington had been caught in the net of a criminal organization hiring needy, rudderless people to commit crimes. "In Pará, the fazendeiro masterminded the assassination; an experienced pistoleiro was the intermediary, and often he hired a poor man like Wellington to commit the execution," de Assis explained. "The result was that poor men were paid miserable sums of money to assassinate other poor men, and then, if they were caught for the crime, they served the sentence that should have been imposed on the fazendeiros."[10]

MARIA JOEL'S CHILDREN would recall the day after the assassination as a time of overwhelming grief. Joelina, who had been the one to fetch Dezinho, was overcome with guilt. "I felt that I had somehow contributed to the death of my father," she said.[11] The public response to the tragic news also made it difficult for the family to get the space they needed to deal with their sorrow. Their house became a destination for hundreds of mourners—relatives, friends, colleagues from the Sindicato, left-leaning politicians, and state and federal lawmakers. Joelminha, the eldest, felt anger and frustration at witnessing her mother being compelled to speak of the crime again and again to those visitors.

"What for?" Joelminha declared. "I wanted it over. Why continue talking about it if he was already dead—if everything had already ended?"[12]

With the media interest in the case high, the murder promptly gained notoriety at the national level. Reporters noted that like many other Pará-based activists, Dezinho had been threatened over and over before he was murdered. In the previous five years, ten union leaders had been under threat of death and then assassinated in Pará, making the state "a national and international leader in violence in the rural areas," a leader of FETAGRI complained bitterly to a journalist.[13] The chief of staff of the government of Pará, reacting to such criticism, promised that he would "provide all the police help necessary to identify the instigators of the assassination [of Dezinho] and punish them."[14] Raul Jungmann, who had been appointed minister for agrarian reform back in 1996, just after the Eldorado Massacre, said that the federal authorities would "find all the legal means possible to put the culprits in prison."[15]

"This government," Minister Jungmann stated in a press release, "has a side: it is in favor of the landless and against latifundia."[16]

Brito also reached out to the media to comment on the murder—and he was candid about who he considered the number one suspect.[17] "The fazendeiro Josélio de Barros Carneiro must have been one of the organizers of the death of Dezinho," the president of the Sindicato declared.[18] He linked the assassination to the political situation resulting from the election of Moisés, seen as an ally of the union, and to the broader context of the Sindicato's struggle against grilagem. "It is said in town that the fazendeiros of Josélio's group had promised that, before Moisés takes office in January, all the people who lead the land occupations in the region will have been eliminated."

By that time, the relationship between Brito and Maria Joel had deteriorated. In the months before the murder, Brito had been increasingly uneasy about the role Dezinho continued to play in the background, ignoring Brito's position as chief of the Sindicato and, ultimately, "creating a movement of his own," as Brito put it. Weeks before the killing, Maria Joel had personally asked Brito to use money from the Sindicato to get Dezinho out of Rondon, because the number of threats was increasing, but he'd refused, arguing that the union had nothing to do with what Dezinho was doing on his own. Maria Joel replied that if Dezinho was murdered, she didn't want to see Brito at the funeral. They would never reconcile after that argument.

On the morning of November 23, some two thousand people, including federal and state officials as well as members of roughly thirty farmworkers' unions and civil rights groups, participated in Dezinho's exequies.[19] The mourners reached the town in caravans of cars and buses from throughout the state. By 9:00 a.m., a procession led by a pickup truck with Dezinho's coffin in its open bed left the town's central square and headed toward the cemetery. The BR-222—the former PA-70, and before that, the Nut Road—was blocked to traffic by peasants in shabby clothes, tension engulfing Rondon do Pará as protests unfolded.[20] A large black placard carried by supporters accompanying the body of Dezinho declared, "Ranchers of Rondon Killed Dezinho. The Fight Goes On. Agrarian Reform Now!" All the letters had been painted in white except for "Killed Dezinho," which was in red. As the corpse passed them by, neighbors, students, and shopkeepers joined the procession, and many businesses decided to close for the morning as a gesture of respect. The march passed "in front of the police station as a sign of protest," wrote a reporter.[21]

Maria Joel, wearing a casual sleeveless black blouse and loose olive-colored pants, had her hair pulled back haphazardly. She wore an absent, pensive expression, as if her mind was somewhere else even though she was standing on the vehicle's bed, right beside the coffin.

The tragedy had taken a visible toll on her. Eva recalled that her sister was "feeling as if, by taking the life of Dezinho, someone had also ripped off a part of her own body." The children, next to Maria Joel, were in tears.

The widow said her final goodbye to Dezinho moments before the interment. Surrounded by umbrellas held by supporters to protect her from the sun, Maria Joel opened her right hand wide and placed her palm over the glass window exposing her husband's face.[22] The poignant gesture seemed to express that she would never abandon her husband to oblivion—that she would fight on in his memory and would forever love him. Still, what would become of that devastated woman, a widow at only thirty-six, was far from predictable.

The police immediately launched their investigation. Because the case was sensitive, and probably because the authorities felt the homicide might remain unsolved unless the case was handled by an officer with some degree of independence and integrity, the task was given to the superintendent of the civil police in southern Pará, an officer based in Marabá named Walter Resende de Almeida. Well-built and in his early forties, Resende had worked for years uncovering crimes in the region. He was, in his own words, incorruptible.[23]

Significant human resources were allocated to the investigation, with a total of eighteen agents deployed in Rondon to sniff around and look for tips.[24] Resende believed that the murder had been "hired out" to Wellington, probably through a middleman, and he had little doubt that the motive was to be found in the larger context of the fight for land.

As law enforcement looked for suspects, one of the early theories considered by Resende was that the family of Hilário might have staged a vendetta, as many in town continued to think that the Sindicato had played a role in the rancher's death. Another early hypothesis was that Dezinho had been killed in retaliation for the occupations led by the union. Three estates had been occupied by that time: Fazenda Jerusalem (occupied in 1997); Fazenda Primavera II (1999); and Fazenda Tulipa Negra (Black Tulip), a 7,400-acre ranch whose takeover by some 150 families had been personally organized by Dezinho some months prior.[25]

Resende suspected the owners of these areas, but he also had in mind other fazendeiros who had previously clashed with the Sindicato. In his initial report, written a few days after the murder, Resende would mention Josélio, who, he wrote, "was known in the region."[26] In a later interview, the detective would explain that Josélio was widely known in police circles for both his clout and his controversial story. Given the flood of accusations made against him by Dezinho and Brito, it was reasonable to expect that Josélio would be one of the first persons interrogated, and Resende did follow that lead, but the detective would have to wait to hear from the fazendeiro. Coincidentally, Josélio wasn't in Rondon—he'd left town on the very eve of the murder.

Early in the probe, Shirley Cristina would inform Resende that the rancher had departed for Baixo Guandu on November 20.[27] "Mr. Josélio is out of town on a holiday trip to the city of Baixo Guandu (a fact publicly known in Rondon do Pará) where his father lives," Shirley Cristina wrote in a fax she sent to the police station after talking to Resende over the phone.[28] She never mentioned it, but Josélio's younger brother Chico had just been reelected mayor in Baixo Guandu; the trip, however, was apparently motivated by a birthday party for Josélio's father.[29]

On November 25, two days after the burial and with no suspects as yet other than the hapless Wellington, Maria Joel was called to give her statement. Now slightly more composed, she explained to Resende's team that days before the killing, a man named Francisco Martins da Silva Filho, one of the occupants of the Fazenda Tulipa Negra, had approached her husband to alert him about an imminent plot. "He said to Dezinho that he would be eliminated before January," declared Maria Joel, explaining that the meeting had taken place in her home. Although the account had been unnerving, Dezinho was used to receiving tips and warnings of that kind, so he hadn't taken any additional precautions following the meeting.

The widow recalled that Francisco had told Dezinho he had learned about the murderous plan against him from his own brother, a man named Pedro. In Maria Joel's words, Pedro was "a former policeman" who was said to have bonds of interest with fazendeiros. Resende would soon discover that, strangely enough, Pedro had been killed only two weeks prior to Dezinho's execution.[30] According to Maria Joel, Francisco had reported to Dezinho that Pedro had died as a result of *queima de arquivo,* or "burning the archives"—that is, because he knew too much about powerful people in town.

Resende dug into the issue. He soon found out that Pedro—whose full name was Pedro Alves da Silva—wasn't a former policeman but was actually a crook. Five-seven, with curly hair and thick eyebrows, Pedro was said to have worked as a debt collector, although he also did dirty jobs. According to Sergeant Marino, Pedro had participated in bank robberies, and rumors also indicated that, while working for fazendeiros, he had taken the lives of employees who threatened to sue their bosses over unpaid wages.[31] One of his main clients was allegedly King of Wood Nunes, and Josélio would later admit to having met Pedro once to try to find out who had robbed a truck owned by a friend.[32]

The momentous discovery for Resende occurred when he went to check the file on Pedro's execution to determine if there was a potential link with Dezinho's murder. To the detective's surprise, the police of Rondon hadn't opened an investigation into Pedro's killing, nor had they interviewed a single suspect or informant. The sole available piece of written information about the case was a one-page report in which Francisco stated that Pedro had been ambushed in a bar on the night of November 3.[33] Nobody could give Resende a reasonable explanation for why the crime hadn't been investigated yet, so the lead detective directed his attention to finding Francisco. He thought the man might have important information.

14

The Evidence Man

BY the time of Dezinho's murder, the bespectacled thirty-two-year-old Francisco Martins da Silva Filho was getting ready to flee Rondon.[1] As soon as his brother Pedro had been hit, Francisco had started to worry about the possibility that he would be the next man to fall.[2]

Who was Francisco, and why was he at risk?

Slightly shorter than Pedro but with his brother's wide eyebrows and curly black hair, Francisco had been born in Maranhão, although he'd been raised in Vila Rondon. In the 1970s his father, a peasant and staunch Christian, had moved with his wife and their nine children to the boomtown, where he earned his income as a logger. Francisco had a special relationship with Pedro, who was only a year older. He would recall their games in the jungle, when both kids used empty cans with strings as handmade pull toys. On Sundays, they all went to church. Despite their relative poverty—"I ate my first yogurt, a tiny luxury at the time, only at twelve," Francisco would say—they were happy.

The family was abruptly split when, Francisco said, his father realized his wife was having an affair. Henceforth, things went downhill for Francisco's mother, who in his words fell into a world of parties, alcohol, and fleeting affairs that would profoundly affect the upbringing of her two sons, who had remained with her. When she had no way to put food on the table, Francisco recalled, she forced Pedro and him to tell a judge that their father wasn't providing financial support after the divorce. "She made us travel twenty-six hours by bus to reach the town where the court was located and testify lies," Francisco would say bitterly.

Francisco said that eventually she managed a small brothel in Rondon. Francisco and Pedro spent many nights in the dingy whorehouse.[3] "That gave me some early insight into the underworld," Francisco would explain, claiming that many fazendeiros and their gunmen caroused in the brothel. Ultimately, Pedro broke with that life to follow his own path, moving as a young man to another city in Pará, where he apparently committed his first offense—check fraud—to pay for a course on how to become a private investigator. Francisco also relocated to Maranhão.

The brothers would cross paths again in the 1990s, when they returned to Rondon with their own families. Francisco was lucky enough to find his father still employed in the timber sector, working in a nursery owned by Nunes, who offered Francisco a job. "He was a rigid man and protested if he saw any of his employees doing nothing,'" Francisco would say of Nunes. "But I liked working with him. I started working at four or five a.m. and left the sawmill at ten p.m. He paid double for the extra hours." According to Francisco, he worked for Nunes from 1990 to 1995.

When Pedro returned to town, he also found a job in a sawmill. But it was in those years that Francisco first witnessed his brother carrying a gun. After their troubled childhood, they had no secrets from each other, so Pedro eventually confided to Francisco that he was joining Rondon's underworld to supplement his income. Occasionally, he would disclose some of the criminal assignments in which he'd participated. According to Francisco's recollections, Pedro was involved in crimes for at least three years before he—as well as Dezinho—was finally murdered.

Francisco quickly understood that Pedro had been killed to cover up murders involving police officers and the elite of Rondon. Because of this, immediately following Pedro's murder, Francisco began to fear for his own life; he himself knew a lot about those circumstances, and that placed him in a dangerous position. In fact, even before Pedro had been interred, Francisco was followed by mobsters, and only hours later, his wife, Maria, was almost run over. "Everyone in town knew Pedro and I were very close," Francisco would later say. "Those who had killed him knew that I also had confidential information, because Pedro explained many things to me, and so they came after me."

Reacting to those threats, Francisco convinced his entire family to flee. He settled with his wife and daughters, his mother-in-law, another of his brothers, and two of his sisters-in-law—including Pedro's widow, Regina, and their children—in a nearby town called Dom Eliseu.[4] He also looked for a gun on the black market. Francisco's ultimate plan was to buy time while he sold his home in Rondon. With the money, he wanted to start over in the state of Roraima, near the border with Venezuela.

The plan fell apart with Resende's investigation. Following the lead provided by Maria Joel, the detective went to Francisco's home, and even though Francisco wasn't there, Resende was lucky enough to find Maria and Regina, who were cleaning the house for sale. They were evasive when Resende asked them where Francisco was, and realizing that the two women were his only hope to contact a potentially crucial witness before he fled Pará, perhaps forever, the savvy Resende placed both Maria and Regina under arrest. The move, despite its dubious legality, worked. Once the women were at the police station, though Francisco's wife still attempted to mislead Resende, claiming that her husband was visiting relatives in another town, Regina relented and revealed that he was at his father's home.

A local officer was dispatched to the new address along with Regina, a strategy intended to compel Francisco to cooperate. Francisco had just finished lunch when he was caught by surprise.

"Your wife is at the police station," said the officer, a man Francisco believed accepted bribes in exchange for passing information to fazendeiros. The officer explained that Resende, the detective who had arrested Maria, was leading the investigation of Dezinho's murder. "You need to come with us," he told him. Reluctant but feeling he had no alternative, Francisco agreed. It was November 27, 2000, and that would be the last time he would ever set foot in his father's home.

The two men and Regina got in a car, and shortly afterward they arrived at the station.

"Are you Francisco?" Resende asked when he met Francisco, his voice deep and his tone uninviting.[5]

"I am, sir," Francisco replied.

"You must now tell me everything you know about the deaths of Dezinho and of your brother Pedro," he urged. "Your life is also at risk."

"Sir," he said, "we are in the wrong place to talk. This police station is not a place of policemen; it's a place of criminals. If you want to talk, let's meet outside."

Francisco deeply mistrusted the police of Pará, and Resende wasn't immune from that distrust. Francisco speculated the detective's ultimate intention was to learn what he knew about Dezinho before deciding whether Francisco should also be killed. "I was 98 percent sure that that would be my outcome," Francisco would later say. Still, he reasoned that, in order to get Maria out, he had few options but to cooperate with the superintendent.

Resende accepted the condition and allowed Maria to leave the police station with Francisco. Minutes later, the group met nearby, and they (Resende, Maria, Francisco, and another officer) traveled together to Dom Eliseu. Once there, they went to the house where Francisco's kin were staying in order to get "some evidence." Then the elusive Francisco finally provided a statement, one that Resende would later describe as "dynamite."[6]

Rondon do Pará, the witness said, was commanded by "a group of criminals comprised of approximately twenty people, whose leader [was] referred [to] with the code name the Judge."[7] The gang's members were the fazendeiros and the sawmill operators, those who controlled the lion's share of the region's land, businesses, and politics—men who, in Francisco's words, decided "who lived and who was murdered in town."

"That's what my brother Pedro told me before dying," he declared, referring to the group as a kind of criminal consortium of big landowners who bribed police officers so they could act with impunity.

The first name Francisco mentioned was Olávio Silva Rocha. Born in Bahia, Silva Rocha was a fazendeiro, the founder of the local radio station, and the owner of several gas stations. In 1988, he had been elected mayor and, in 1994, a member of the Brazilian congress.[8] In the recent elections, Silva Rocha had been a candidate for mayor.

The Brazilian Amazon in Pará state.
Heriberto Araujo

The Te-Chaga-U ranch, photographe
by the police at the time of the raid t
find human remains. *Police of Pará*

The party that raided the Te-Chaga-U, including Dezinho (*in striped T-shirt*
Zé Geraldo (*in red shirt*), and Gil (*in ski mask and white hood*). *Police of Par*

A story in *Jornal do Brasil* about the trial for the murder of Cavalcanti; Josélio's photo is left among the headshots.
Jornal do Brasil

Human remains found by the police at the Te-Chaga-U.
Police of Pará

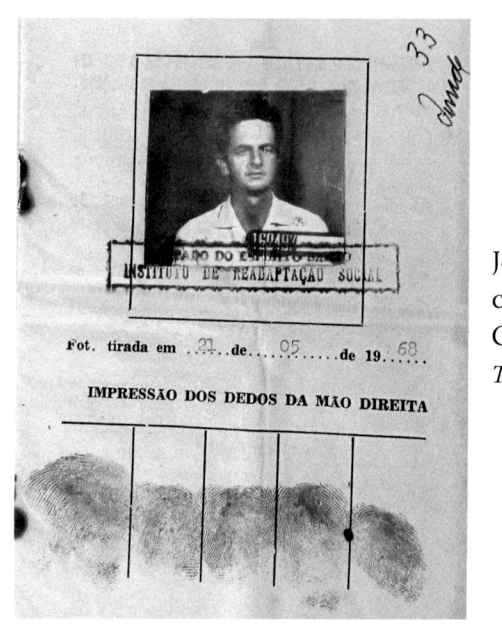

Josélio's prison identification card after his conviction for Cavalcanti's murder.
Tribunal de Justiça do Espírito Santo

A Provincia do Pará report of the chainsaw murder on September 10, 1975. It shows the letter delivered to the Reises, a picture of the victims, and Geraldo showing the clothes of his dead brother.
A Província do Pará

Dezinho addressing landless peasants during a demonstration in Rondon do Pará.   *Dezinho and Maria Joel's family archive*

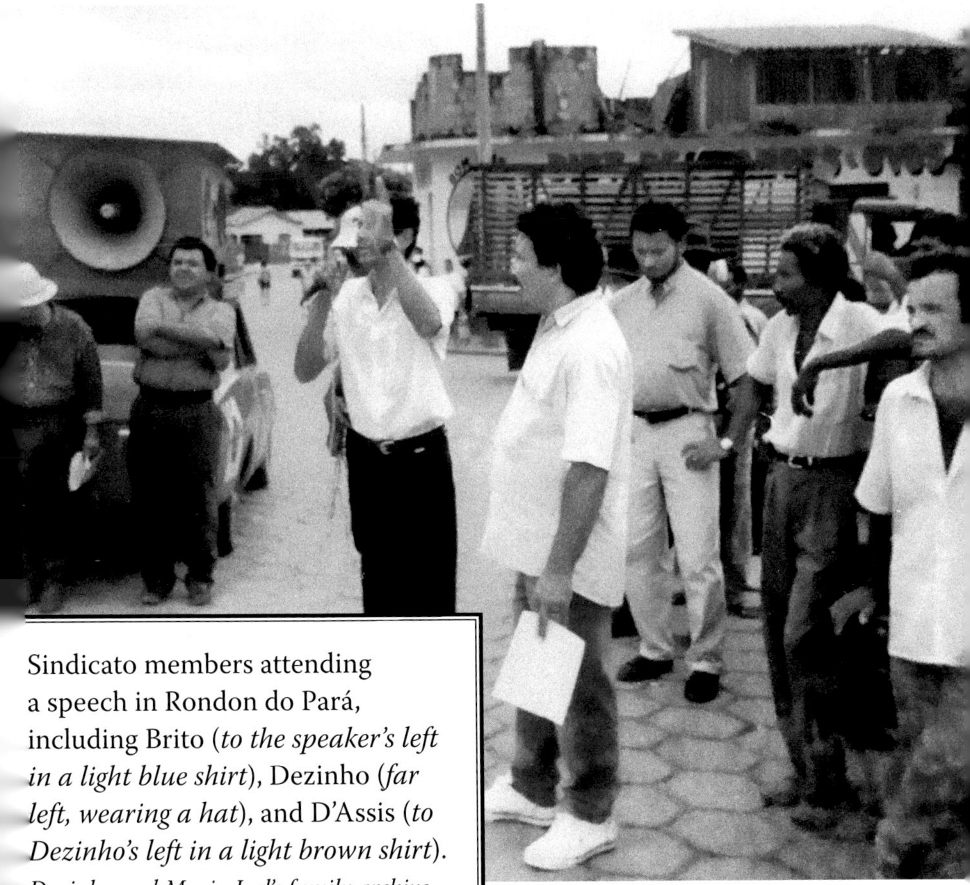

Sindicato members attending a speech in Rondon do Pará, including Brito (*to the speaker's left in a light blue shirt*), Dezinho (*far left, wearing a hat*), and D'Assis (*to Dezinho's left in a light brown shirt*). *Dezinho and Maria Joel's family archive*

Dezinho and his four children (*left to right*): Joelminha, Joelson, Joelina, and Joélima. *Dezinho and Maria Joel's family archive*

Maria Joel and Dezinho dancing a party in Rondon do Pará. *Dezinho and Maria Joel's family archiv*

Dezinho, Maria Joel, and their children at one of the last birthdays they celebrated together. *Dezinho and Maria Joel's family archive*

Josélio (*far right*) and other people photographed by the police of Minas Gerais, led by Detective Faria in the Te-Chaga-U fazenda.
Detective Antônio Carlos Correa de Faria

...monstration by MST members to protest the arrest of Dezinho ordered by ...ge Amorim in 1999. Dezinho is wearing a white T-shirt (*left of center*).
...inho and Maria Joel's family archive

Gunman Pedro Alves de Silva.
Francisco Martins da Silva Filho

The casket of Pedro Alves da Silva
before being buried in Rondon do Pará.
Francisco Martins da Silva Filho

PROCURADO

GOISNAR MARIANO DA SILVA
Qualquer informação ligar (0XX91) 326-1320
DEPOL RONDON DO PARA/PA

PROCURADO

ROGÉRIO DE OLIVEIRA DIAS
ou
ROGÉRIO DE JESUS SILVA
Qualquer informação ligar (0XX91) 326-1320
DEPOL RONDON DO PARA/PA

Wanted posters for Ygoismar Mariano da Silva and Rogério de Jesus Silva released by the police of Pará after Dezinho's murder.   *Police of Pará*

Rancher Décio José Barroso Nunes being interrogated by Brazilian lawmakers at the Brazilian Senate, 2006.
TV Senado

Maria Joel (*center, in a red shirt*) attending a meeting with President Lula. By her side, actresses Letícia Sabatella (*wearing a pink blazer*) and Camila Pitanga (*in white*).   *Salete Hallack*

Rondon do Pará - Agência ⬛

Sem-terra vão à rua para denunciar que mandante da morte de sindicalista continua em liberdade   *Fazendeiros reagem com tratores e caminhões, acusando sem-terra de ter assassinado pecuarist*

Fazendeiro é 'dono' de município

Delsão manda e desmanda em Rondon do Pará. Contrata pistoleiros, mata e sequer apaga pistas dos crimes que comet

AMAURY RIBEIRO JR.
Enviado especial

EXTERMÍNIO NO CAMPO RONDON DO PARÁ – Controlando a Justiça e a polícia, o fazendeiro Décio José Barroso Nunes, conhecido como Delsão, é praticamente o dono do município de Rondon do Pará. Lá ele manda e desmanda, sendo temido por todos. Seu poder é tal que ele não se preocupa de apagar pistas dos crimes de que é acusado.

O juiz titular do município, Paulo Amorim, foi suspenso pelo Tribunal de Justiça depois de ter sido acusado de envolvimento em fraudes com Títulos de Dívida Agrária (TDA) para beneficiar fazendeiros. A promotora, Josélia Barroso, está impedida de atuar na comarca desde 1996, quando a Polícia Federal encontrou um cemitério clandestino na fazenda do pai dela, Josélio Barroso. Os oficiais de Justiça, funcionários cedidos por cartórios ao fórum, são acusados pelo promotor substituto, Raimundo Ayres, de retardar o andamento dos processos em que Delsão está envolvido.

Hospital – O veredicto se confirmou quando Pedro, que bebia num bar da cidade, foi morto pelos pistoleiros de aluguel Francisco Veloso de Freitas (que é vereador em Bom Jesus do Tocantins pelo PSB), Luiz Carlos Aviz e Paulo Ferraz Rodriguez, em novembro do ano passado.

"No hospital, percebi que, antes de morrer, meu irmão ia me alertar para o perigo que eu corria, mas desistiu porque dois médicos ligados à quadrilha não tiravam os olhos dele. Pedro foi assassinado porque sabia demais. Dias depois, quando o Veloso começou a rondar a minha casa, vi que meu irmão estava certo e que também corria perigo", disse.

A advertência de Francisco a Dezinho não impediu a morte do sindicalista. Num ato de ousadia, mesmo após o recado de que estava marcado para morrer, Dezinho continuou a investigar as mortes de Juaraci Gomes da Silva, o Bodão, e de Piauí (cujo nome até hoje não consta dos autos do processo), que trabalhavam na Madeireira Barroso, de propriedade de Delsão.

Contratado na Bahia pelo intermediário Ygoismar Maria-

A viúva Maria da Costa mostra fotografia de Dezinho

Caminhões contra passeat:

RONDON DO PARÁ – Rondon do Pará, 21 de novembro de 2001. Comandados pelo presidente do Sindicato dos Trabalhadores Rurais do município, José Soares Brito, cerca de dois mil trabalhadores rurais saem em passeata. Exigem a prisão do mandante do assassinato do sindicalista José Dutra da Costa, o Dezinho, ocorrido há exatamente um ano.

"Pedimos a condenação dos fazendeiros assassinos", gritam em coro os manifestantes. A passeata dos sem-terra segue em direção à Avenida Central. Lá, é surpreendida por uma caravana de 50 tratores e caminhões. Delsão, acusado de ter sido o mandante da morte de Dezinho, comanda o passeio motorizado.

Acompanhado de seguranças, funcionários de suas madeireiras e amigos, Delsão tenta com o comboio neutralizar o impacto da passeata. Do carro de som alugado pelo fazendeiro saem palavras de ordem contra o movimento dos sem-terra. "O MST não quer terra. Só quer bagunça", grita o locutor.

Contra-ataque – Os f zendeiros partem para o co tra-ataque: os alto-falantes e faixas acusam trabalhador de ter matado o pecuarista J sé Hilário, em 1996. Os aut do processo, no entanto, con tam uma história diferente.

No depoimento prestado Justiça, o agricultor Antôr Piauí confessou ter matad fazendeiro. O motivo do crin José Hilário tinha agredi sua mulher. Piauí disse q acusou o sem-terra porque foi torturado na delegacia da cid de por pistoleiros e policiais.

Diante da iminência confronto entre as manifest tes da passeata e da carreat Polícia Militar entra em aç Sob o comando do tene Cláudio Moreira, 37 soldad armados com metralhador separam os dois grupos.

O lado direito da rua é servado aos trabalhadores. lado oposto ficam Delsão e amigos. No meio, os policia

Às 15h, os sem-terra volt para o sindicato. Imediatame te, Delsão encerra também reata dos fazendeiros. (A.R

A truck transporting crops through the BR-163, known as the "soybean highway," Pará state.
Heriberto Araujo

Soy silos on the Tapajos River, near the BR-163.
Heriberto Araujo

José da Silva, a settler, union associate, and former landless peasant on his plot in Rondon do Pará. *Heriberto Araujo*

Settlers and subsistence farmers transporting their crops to Rondon do Pará
Heriberto Araujo

Mayor Shirley Cristina at her inauguration ceremony, 2013; *third from right,* Councilwoman Joelminha.
Ricardo Tavares D'Almeida

Josélio de Barros and his wife, Shirley, at the inauguration ceremony of their daughter, Shirley Cristina.
Ricardo Tavares D'Almeida

Shirley Cristina and Judge Gabriel Costa Ribeiro, before he removed her from office.   *Archive of Judge Gabriel Costa Ribeiro*

Maria Joel receiving the Human Rights Award from Lula, 2007. *Dezinho and Maria Joel's family archive*

Maria Joel Dias da Costa in fro of the union building in Rondo do Pará. *Heriberto Araujo*

Lawyer José Batista Gonçalves Afonso at the archive of the CPT; behind him, files for court cases of murders and a poster of Chico Mendes. *Heriberto Araujo*

Décio José Barroso Nunes
Réu

Rancher Nunes at the jury trial for the murder of Dezinho, August 2019. *Tribunal de Justiça do Estado do Pará*

Maria Joel testifying at the jury trial of Nunes, August 13, 2019. *Tribunal de Justiça do Estado do Pará*

Key witness Francisco testifying at the jury trial of Nunes, August 13, 2019.
Tribunal de Justiça do Estado do Pará

Francisco Martins da Silva Filho, present day, at his home in Central Brazil. *Francisco Martins da Silva Filho*

Second, Francisco mentioned the wealthy Lopes family. He specifically accused three family members of being part of the criminal consortium: Manoel, whom Dezinho had once referred to as an associate of Josélio and instigator of homicides, and two of his several sons, Antônio, a doctor, and João, the former mayor of a little town in Bahia state, where he had known and befriended Nunes, the King of Wood.[9] Originally from Espírito Santo, the Lopes family had moved to the frontier back in the 1970s, investing in land, cattle ranching, and timber.[10] They had been founding members of the local logging association and the Associação.[11] They were heavily involved in politics.

The Lopes and Barros families had strong personal connections. Manoel Lopes was born in the 1920s in Baixo Guandu, Josélio's hometown.[12] One of his grandsons, Marcos, a son of former mayor João, had married Josélia Leontina, the district attorney.[13] Francisco never mentioned it, but another odd coincidence was that the first defense attorney of Wellington was a granddaughter of Manoel Lopes and also a daughter of João. Her name was Adriana, and she had assisted Wellington during his statement just hours after he took Dezinho's life.[14]

Francisco devoted some time to elaborating on the alleged role of the Lopeses in the gang. He said that Antônio Lopes de Angelo was a doctor and worked at Rondon's St. Joseph Hospital, where Pedro had died after being shot. According to Francisco, the gang met in the house of the doctor, who was a candidate for vice-mayor on Silva Rocha's ticket in the recent elections won by Moisés.[15]

The third name raised by Francisco was Lourival de Souza Costa and his alleged pistoleiro, a cowhand named Domicio de Souza Neto, who went by the name Raul. Less wealthy than Silva Rocha and the Lopes family, who had founded the St. Joseph Hospital in 1974, Lourival was struggling to retain control of his 6,600-acre Fazenda Santa Monica after Dezinho became president of the Sindicato. According to a confidential police investigation, Lourival was a suspected grileiro.[16]

"Who's the boss?" Resende barked after Francisco had reported all this.

Before replying, Francisco mentioned three other men who, by his account, were low-ranking thugs, including a council member named

Francisco Veloso de Freitas, who Francisco said had also participated in the murder of Pedro.[17] Then Francisco revealed who he believed to be the boss: it was Nunes, the King of Wood, whom gang members referred to as the Judge.

To prove that his information was reliable, Francisco provided Resende with names of victims and the locations where crimes had apparently been committed. He specifically described the particulars of four homicides involving both Nunes and his brother Pedro, who, by Francisco's account, had become a sort of private pistoleiro working to protect the man's interests. The witness, despite the tension of the moment, displayed an impressive recall of details.

As Francisco recounted, a first murder had taken place when a sawmill employee who had complained to Nunes about unpaid wages was allegedly removed from his home and shot to death, his body subsequently incinerated on a pile of tires beside the BR-222. (Resende would later discover that in one recent police report, Pedro and Nunes did indeed appear as suspects in the burning of a man near the highway, but ultimately no charges were filed.)[18] A second case, which had occurred just months prior to Pedro's murder, involved two brothers executed for stealing an expensive blade from one of Nunes's sawmills. The crime, Francisco claimed, had been committed by Pedro and another killer in a spot called Presa de Porco, an interior area of Maranhão where the thieves had attempted to hide. The fourth murder Francisco chronicled was that of a man named Juracy Gomes da Silva, who went by Bodão and was also a former employee of Nunes.[19]

When Resende questioned how Francisco knew of this, he explained that he had three sources: Pedro; Nunes himself, with whom Francisco had occasionally interacted even after he had stopped working in his sawmill; and some conversations that Pedro had recorded. The tapes, collected by Francisco after Pedro's death, were handed over to Resende along with dozens of photographs whose contents led the detective to conclude that Francisco's statement might be accurate. In one shot, Pedro posed bare-chested with a cap and holding a revolver, as if aping a mobster in some cheaply produced film. Other pictures were of rural areas; red marks highlighted the exact spots where Pedro had apparently buried bodies. The photographs signaled that Pedro had been gathering evidence to create an archive that could incriminate fazendeiros. The reason he did so—and whether he did it alone, or with Francisco—remained unclear, but one could argue that the secret archive was a powerful resource that could be held against the city's landed elite.

Finally, Francisco got to the core of the matter that interested Resende: how all those revelations connected with Dezinho's murder. Francisco explained that he didn't know anything about Wellington, the killer, but he did know that the criminal consortium wanted Dezinho dead. He knew, he said, because it was Pedro who had originally been assigned the job. Francisco also claimed to know the motive for the crime.

Nunes, explained Francisco, "along with other ranchers of the region, ordered my brother Pedro to kill the union leader Dezinho because he [Dezinho] had reported to several state agencies, as well as to unions, several crimes committed by the powerful group."[20] In other words, Dezinho knew about the murders Francisco had just described, at least some of them, and had reported them to colleagues at FETAGRI and apparently also to the police. Maria Joel, in her statement, had already mentioned that Dezinho had known about the incineration of the corpse near the BR-222.

According to Francisco, Dezinho's attempt to build a case against the King of Wood was a central motive for the murder. But there was yet another reason, Francisco declared: Dezinho's political ambitions and his alliance with mayor-elect Moisés.

As Brito had said to reporters, the election of Moisés had caused anxiety among fazendeiros, even more so following rumors that Dezinho might be selected for a top post in his administration. Francisco recalled that one day during the campaign, while he and his brother were with Nunes in his office, Pedro had said that "any candidate could win the election, but not Moisés. If he ever won, he would not take office, because he [Pedro] would show him who" was really boss. This was apparently an allusion to Nunes's bravado. Subsequently, Nunes himself had reinforced the threat. "[Moisés] may win, but he won't take office," Nunes allegedly stated, according to Francisco's deposition.[21]

Dr. Angelo, whose ticket with Silva Rocha had been defeated, would tell the police that Nunes had contributed to their campaign.[22]

If, by Francisco's account, the motives seemed clear and matched Resende's early assumptions that Dezinho's murder was intended to put an end to the activism against large landholders, an unsolved question remained: why had Wellington, and not Pedro, been the one to kill Dezinho?

As Francisco would later explain, Pedro had attempted to kill Dezinho twice, once in Rondon and another time at a gathering of activists in Marabá, but he had never found an opportunity to shoot him and get away. But why hadn't he persisted? Francisco claimed that he had prevented his brother from murdering Dezinho to try to keep Pedro from being killed himself. Informed of Pedro's deadly intentions, Francisco told the detective, he had convinced Pedro to desist, because, in Francisco's words, if Pedro had carried out the assassination of the activist, his brother "would have been the first suspect to be arrested." In Francisco's reasoning, Pedro's eventual detention in Rondon would have led to his murder by members of the criminal consortium and corrupt police officers to prevent his becoming a snitch. In the end, Pedro had been killed in what Francisco believed was an attempt by Nunes to "burn the archives" before taking Dezinho's life.

Confirming Maria Joel's statement, Francisco told Resende that, in the interlude between Pedro's murder on November 3 and Dezinho's assassination on November 21, he had gone to the house of the activist to warn him about the imminent threat against his life. The two men had met for less than half an hour, and Francisco had revealed to Dezinho the existence of the secret archive.

"Take care," Francisco had said. "My brother Pedro was killed, and you'll be next."[23]

"I'm aware," Dezinho had replied.[24] "Your brother tried to kill me several times. But I trust in God." Then he'd asked Francisco to examine Pedro's audiotapes to find out whether they contained "something to further incriminate Nunes and other fazendeiros."[25] Francisco had listened to some excerpts, confirming that the tapes contained sensitive information, but later, when he noticed that he was being watched, he'd decided not to further involve himself in the issue and fled to Dom Eliseu.[26] He was preparing to sell his house and relocate near Venezuela when Resende had found and arrested Maria and Regina, forever thwarting Francisco's plans to get out of town.

IT WAS LATE into the night of November 27 when Resende finished his questioning. Although Francisco's statement, if proven reliable, would be crucial for the investigation, there was no concrete evidence for either the murders the witness had just described or the existence of the alleged criminal gang. At this time, the sole support for those claims was Francisco's statement.

The detective also wondered why the witness was providing such information. "I didn't go to talk to the police," Francisco would say. "I was found by the officers and left with no alternative but to reveal what I knew." Even if those claims were true, it seemed evident to Resende that the shrewd Francisco was also pursuing some sort of self-serving agenda, the most evident being to avenge the murder of his brother, which Francisco attributed to Nunes. It was also possible, Resende speculated, that Francisco was attempting to use the archive to blackmail the ranchers at a crucial time. But there was at least one other reason Francisco had chosen to talk—his links with Dezinho.

Francisco wasn't a member of the Sindicato or any other landless organization, but he had joined the occupants of the Fazenda Tulipa Negra to fight for a plot. That experience had given him a positive view of Dezinho, a man Francisco had initially thought was "a person who encouraged invasions" before changing his opinion, describing the activist as a "simple, poor, and tenacious" man who pursued a cause to benefit the community. Francisco would admit he particularly appreciated Dezinho because when he'd approached him to join the Tulipa Negra just months prior to Pedro's murder, the union leader hadn't discriminated against him even though he was aware that he was the brother of a gunman. (Dezinho might have had an ulterior motive for that, perhaps to gain the pistoleiro's sympathy and access to privileged information on fazendeiros, but researchers have noted that landless organizations like the MST were open to all kinds of people, recruiting even "'hopeless cases'—drug addicts, emotionally disturbed street kids and violent criminals—and turn[ing] them into productive, fulfilled citizens" by teaching them how to work the land.[27])

But in this tangled story of treason and murky alliances there was, above all, one crucial question: who was Francisco really? The essence of his account, though potentially plausible, raised many underlying questions, such as how had he *really* gotten to know so much about Rondon's underworld? Francisco told the detective that Pedro had passed him information, but the detailed testimony and the multiple connections he was able to establish among the alleged members of the criminal consortium suggested that the knowledge came from something other than mere hearsay. Had Francisco ever been part of the gang and for some reason decided to become a snitch? Had he ever aided Pedro in committing felonies? Francisco would say "he had never committed a crime," although he admitted to having occasionally worked with Pedro, but only when his brother was collecting debts in arrears.

Despite his reservations, Resende found the statement credible, and Francisco became a key witness in the case. That meant he was now in even greater danger, so the detective immediately arranged with the state authorities to have Francisco and his family enter a witness protection program. Three days later, Francisco, his wife, and their children were transferred by law enforcement to an undisclosed location in central Brazil.[28] They were told to cut off all ties with friends and relatives in Rondon.

Immediately after Francisco issued his statement, Resende worked to reinforce the case against Nunes so he could eventually ask a judge to issue arrest warrants. The detective had to tie up at least three loose ends. First, he had to dig into the specific motive that had prompted the King of Wood to subcontract the execution of Dezinho. According to Francisco, Dezinho may have been building a case against Nunes, and the election of Moisés had caused great worries among the establishment, but even so, why had the rancher—who up until then was mostly known for his business drive rather than for his alleged criminal behavior—ordered the hit?

Second, Resende had to find a link between Nunes and Wellington—between the suspected instigator and the triggerman—and then uncover whether the assassination had been plotted just by Nunes or, as Francisco claimed, by the Lopeses and Lourival as well.

Seeking information on these issues, Resende called several people to provide statements. The same day Francisco was interrogated, another team of police had questioned Olivandro, the son of the dead rancher Hilário and the man who, according to Brito, had been driving the car the day he was almost abducted. Olivandro showed no love for Dezinho, but he denied that the killing was a personal vendetta.[29] He also denied having participated in Brito's kidnapping attempt.

Dezinho's entourage also testified. Attorney Carlos Guedes, who worked at CPT and had been providing legal assistance to the Sindicato, recalled that Dezinho, shortly before his death, had named Nunes as one of the men using gunmen to intimidate him.[30] Guedes said that the activist had been informed by the authorities that a massive swath of land containing both the Fazenda Tulipa Negra, which belonged to a rancher named Kyume Mendes Lopes, and Nunes's Fazenda Lacy had been the object of land-grabbing schemes. Brito, also interrogated, confirmed that the Sindicato was planning actions on the Lacy ranch.[31] "We wanted to occupy the fazenda," Brito declared, "as it is only used to harvest timber."[32] Once again, he alluded to the possibility that Josélio, who had not been mentioned until then by Francisco, was somehow involved in the murder.

Previous court files showed that Nunes was seriously worried about the potential loss of the Lacy, his most valuable asset. When rumors in 1999 had signaled that the spread might become a target of the Sindicato, he had filed a complaint against the union, his attorney successfully requesting that the judge issue a restraining order. Judge Amorim—the same judge who had ordered Dezinho's arrest and was later removed—had ruled that an occupation of the Lacy, no matter the suspicion that the spread was the result of grilagem, would be illegal and threatened Dezinho and Brito with heavy fines if they attempted it.[33] Nunes asked the judge to impose a financial penalty of nothing less than $2.8 million if the threats of "invasion" by the Sindicato persisted.[34] The judge opted for a much more modest deterring sum, ruling that, for every single "threat of invasion," the Sindicato or its leader would have to pay an $11,000 fine.[35]

While Resende seemed to be making quick progress discovering the motives for the crime—all of them apparently pointing at Nunes as the instigator—his investigation was failing to uncover any links between the King of Wood and Wellington. The press argued that Ygoismar, the man on the motorcycle, and Rogério, Wellington's brother, had been hired to kill Dezinho after Pedro had refused to do it. For some reason never officially established, Ygoismar had subcontracted the crime to Wellington, netting himself some 15,000 reais for the murder—about $4,500, a sum much closer to what the execution of an activist with Dezinho's profile was likely to cost in the region.[36] This fact was never confirmed, because Ygoismar, a middleman in the scheme, according to Resende, was never found.

Nine days after Dezinho's death and forty-eight hours since hearing from Francisco, Resende had yet to connect all the dots. He still had to further investigate the potential involvement of the other fazendeiros. But the detective had enough evidence to write his first report and request a warrant to arrest Nunes. He wrote that Nunes was "a powerful man with a stable financial situation, an owner of sawmills and fazendas, which surely will allow him to escape to an unknown place as soon as he knows that his group is being dismantled."[37] Nunes, Resende wrote, "could also intimidate witnesses, as has been happening with Francisco," so he urged the judge to take "urgent measures." In his opinion, "the group commanded by Delsão [Nunes] . . . must be plotting other deaths."[38]

The district judge of Rondon, a woman named Iacy Salgado Vieira dos Santos, issued a warrant without delay. It was early in the morning on November 30 when Nunes was arrested and taken to Marabá for questioning.[39] The arrest came at a particularly rocky time for the businessman because, less than two weeks earlier, Nunes had signed a divorce agreement with his ex-wife, dividing his wealth and splitting up his family.[40] The rancher had managed to keep control of most of the sawmills and fazendas, valued at several million dollars, but in exchange, he would pay his ex-wife some $600,000. He was also forced to hand over the family's beautiful villa with a 15,000-square-foot garden and a swimming pool. Eventually, the divorce would take a toll on Nunes's relationships with several of his four children.

In Marabá, Nunes gave a long deposition in which he admitted that he had known Pedro for four years, claiming to have hired Francisco's brother to "make some collections of arrears."[41] Nunes specifically mentioned an instance in which Pedro had investigated the sudden disappearance of a man named Vicente Paulo Favoreto, a client suspected of having faked his own death to avoid settling a large debt, and another in which he paid Pedro's expenses during a two-month work trip in Maranhão. Nunes also acknowledged having a long-lasting friendship and business links with the Lopeses. Although he admitted that two of the murders attributed to him by Francisco were indeed of current or former employees, Nunes nevertheless denied having ordered or committed any crime, insisting to Resende that he knew nothing about them or about the murder of Dezinho.

Detective Resende didn't believe him and transferred Nunes to a prison in Belém while he continued his investigation. The businessman was placed in a "special cell" that his attorneys had managed to get for him through a motion.[42] (Brazil's penal code states that college graduates are entitled to be held in cells separate from common inmates so they don't get mixed up with dangerous gang members, but Nunes had no university degree.)[43] Before entering the jailhouse, Nunes, wearing a short-sleeved white shirt open to his hairy chest and a pair of chinos, talked to reporters covering the unfolding case.

"I'm innocent, and I'll prove it," he said, accompanied by three attorneys.[44] He was visibly tense and upset. A journalist wrote that, as he fired off questions, Nunes "started to lose control," replying "even before the questions were concluded." He was adamant that he hadn't had "the slightest contact" with Dezinho, although he didn't refrain from expressing his disdain for him. "I think that he [Dezinho] shot a lot and shot everywhere, right?" Nunes said, suggesting that the union leader had made many enemies and his fate was somehow inevitable.

Many thought that the case was heading toward a rapid resolution when footage of Nunes entering the prison was broadcast by Pará's news stations. Still, not everyone in Rondon believed Nunes was guilty. Sergeant Marino, the officer who'd arrested Wellington and said Josélio was a friend, believed that Resende was either being misled, falling prey to confirmation bias due to Francisco's account, or, even worse, intentionally accusing the businessman for his own purposes.

"It was a *cabala*," Sergeant Marino would say of Resende's investigation.[45] (In Portuguese, *cabala* refers to the kabbalah but also means "a conspiracy to harm someone.") "Not even the best police in the world solves a homicide in five days."[46]

In Marino's opinion, Nunes was a fair man whom "people in town spoke well of."[47] Marino instead suspected Lourival, saying he was acting in defense of the Fazenda Santa Monica, also suspected of grilagem and rumored to be a top priority for occupation by the Sindicato. He claimed that the fazendeiro, visibly frustrated, had once admitted to him that unless the police did something about the issue, he would hire a gunman. Supporting this version of the events was testimony attributing the ownership of the revolver used by Wellington to "Raul," one of Lourival's employees.[48] According to eyewitnesses, the fazendeiro had also been spotted twice near Dezinho's house on the eve of the crime.[49]

Nunes didn't spend much time in custody. On December 13, less than two weeks after his arrest, an appellate judge in Marabá accepted the defense attorney's application for habeas corpus on the basis that there was not "sufficient" evidence to support his preventive detention.[50] In his ruling, the appellate judge never cited Francisco's testimony.

The application had followed a rather odd process at the appeals court. The initial motion had been filed on December 11. Theoretically, applications for habeas corpus were randomly assigned to appellate judges through lotteries, and the first motion had landed on the desk of a hard-line judge described as "incorruptible."[51] Within hours, though, Nunes's attorneys had decided to withdraw the application, only to resubmit it forty-eight hours later, on December 13. The case then ended up in the hands of another judge, who almost immediately accepted the request.

"It was difficult to accept that the justice system was failing us only thirteen days after a murder as predictable as that of Dezinho," Maria Joel would say. She believed that Nunes was the instigator and assumed that some bribery was involved in his release.

The day Nunes was freed, masses of his workers celebrated the ruling by shooting off fireworks and honking car horns. "Our City Is Empty Without You," read a large banner held by Nunes's supporters.[52] Agribusiness and loggers' associations joined the propaganda campaign by releasing statements depicting Nunes as "innocent" and describing his companies as pillars of the town's economy.[53] The father of Bodão, the worker whom Francisco claimed Pedro had killed, told reporters that the King of Wood had nothing to do with the crime.[54]

Maria Joel and her children would face further legal setbacks. Months later, the same appellate judge who had accepted the habeas corpus application would rule unilaterally that the prosecution of Nunes be suspended until the five audiotapes provided by Francisco to Resende were transcribed by law enforcement and incorporated in the case.[55] It would take years—literally—to carry out this process, which was completed only after judges warned uncooperative officers that they might be charged with felonies if they continued to ignore their orders. The delay—caused by a lack of both manpower and technology, according to the police of Pará—stalled the case.[56]

As Nunes was being released from custody, Maria Joel was visited by Dezinho's brother, Valdemir. He had once kneeled before his now-deceased brother, asking him to leave the Sindicato and return to Maranhão, and now he was back in Rondon with the same mission. Valdemir, worried about his sister-in-law's fate and that of his nieces and nephews, intended to convince Maria Joel to leave as soon as possible. He had become a councilman in Maranhão and said that he would help them to start over.

But Maria Joel refused to move. In a media interview published four days after Nunes was freed, the widow made it clear that she wasn't willing to throw in the towel so soon.[57] She said firmly, "I will continue Dezinho's fight."

15

A Cause Larger Than Death

SCHOLARS studying the traumatic consequences of violent deaths in bereaved families have noted that although murder has forever "fascinated mankind," until the late 1970s, "the scientific record [was] virtually silent on how homicide affects the relatives of the victim."[1] When psychologists and researchers began to study the impacts of such deaths, they found that relatives of murder victims are frequently engulfed in a sense of victimhood "by the multiple losses endured: loss of a family member, loss of illusions of safety and invulnerability, loss of a sense of trust in the surrounding community, and loss of a belief system."

Maria Joel endured that woeful inner turmoil in the aftermath of the crime, often seeking solace in her children. At night, she joined them in their room, all lying in a group embrace, each of them silently wondering what their life would now be like. In her innermost thoughts, Joelminha, the eldest, hoped for an existence "freed from constraints and prohibitions."[2] "A normal life, because now he's dead," she thought, not without remorse.[3] As a consequence of trauma, the young woman had stopped eating, her weight dropping to seventy-seven pounds.

Joelminha wasn't the only family member wishing to put it all behind her and look to the future. Maria Joel's mother, who had never approved of Dezinho's involvement in the Sindicato, now pressed them all to deal with the pain in silence, turn the page, and start over. "Look what happened with Dezinho. Let it go, Maria Joel, let it go!"

she insisted, supported by a group of aunts and uncles also living in Rondon.

Hearing those words, Maria Joel remembered her early days in Rondon, when her mother had described the land as one of abundance. She thought about the bus trip with Joelminha and Joelson from Maranhão and how much she had hated the ugly boomtown with its sandy streets and uninviting atmosphere. How strongly she had sensed—almost *known*, she now realized—that their life there, despite her mother's enthusiasm, would lead to their downfall. If she had written that letter to Dezinho suggesting another place to try their luck, maybe things would all have been different—or maybe not. Anyway, it wasn't her mother's fault that things had turned out this way. They were a humble family of sharecroppers, and like thousands of others, they had simply been looking for a better future on the frontier at a time when the government said the Amazon was "a land for people, for people without land."

But in making the crucial decision about the family's path forward, Maria Joel decided that she would not bow to fear and would not follow her mother's advice. This time she would follow her own intuition— and her aim was to find a way to make the killers pay for the murder of her husband. That was exactly why, almost instinctively, she had protected Wellington from death on the night of Dezinho's murder.

Maria Joel's resolve to have the assassins prosecuted, sentenced, and jailed wasn't motivated just by the pain of the loss. It was also a question of memory and dignity. For nearly a decade, her family had lived a life of isolation and restriction because Dezinho had said that one must believe in something in life and fight until the bitter end to achieve it. Even if Maria Joel and her children hadn't necessarily agreed with him, accepting defeat now and just leaving town would make it, as Joélima, the middle daughter, would put it, "as if Dad was murdered twice."

Thus, Maria Joel decided that the family would remain there and resume their lives in Rondon. Undoubtedly she would pay a personal price for that choice, and not just the one Maria Joel's mother and many others in her entourage feared—the risk of further executions of family members. Remaining also meant that it would be harder to reach a point of healing, where the emotional suffering caused by their loss might dwindle from ever-present to intermittent. In that town, they would always be seen as the widow and the fatherless children of the slaughtered activist.

But Maria Joel hoped that, by staying, she would win a symbolic moral battle with the people who had gambled that killing Dezinho would end the movement he had built. She would show them that she would not surrender to fear.

"We weren't going to run away; we weren't the ones who had done anything wrong," Maria Joel would declare.

Mayor Moisés decided to help the family out by giving Maria Joel a job in one of the town's public kindergartens. The employment, she would later admit, would become almost therapeutic, in addition to providing crucial funds to support her family. Surrounded by joyful children, she earned an income and received respite from her thoughts for several hours each day. Maria Joel determined that her children also needed space to deal with the trauma, so she sent them to Bragança, where Dezinho had often hidden from gunmen.

Some months passed, and peasants and rural workers continued to show up at their home, often bringing Maria Joel rice, corn, manioc, fruits, and meat. She began to hear from them that they wanted her to fill the vacuum left by Dezinho. "People told me, 'You must continue; it must be you,'" Maria Joel said.

The lobbying had a real impact on the widow. "She had trouble sleeping, because she was consumed with doubts, wondering whether to follow in Dezinho's footsteps," her sister Eva would say.

Her mother, anticipating that Maria Joel might get into the same trouble as her dead son-in-law, suddenly announced that she was moving back to Maranhão. "She came and told me she was suffering. She argued that it was too painful for her, so she was leaving," Maria Joel recalled bitterly. Other brothers and sisters who disagreed with Maria Joel's decision to build a new life without breaking with the past also distanced themselves, skipping protest marches organized by her, the CPT, and FETAGRI. Only Eva and her children would truly remain by Maria Joel's side. Joelminha finally accepted that she would have to once again put on hold her desire for "a normal life" for the sake of the family's unity and her father's memory.

On November 21, 2001, the first anniversary of Dezinho's assassination, Maria Joel and the CPT organized a demonstration that brought together hundreds of people protesting a lack of consequences for the evildoers and what one British author called "an unwieldy and highly bureaucratic judicial system that is heavily biased in favor of the big landowners."[4] "Down with Latifundia. Agrarian Reform Now!" read the banners held high by union members as they walked throughout Rondon.[5] When the demonstrators reached the highway bisecting the town from east to west, they ran into another small protest group— one backed by dozens of bulldozers, rigs, and tractors—supporting the fazendeiros. "Landless Don't Want Land! They Want Trouble," read a white banner hanging from a large truck. Sources suspected that the vehicles belonged to Nunes and that the demonstrators were his employees. The fazendeiro staunchly maintained he was not involved in the crime and was now engaged in a counter campaign to clear his reputation.[6]

Tensions rose, and about forty military police in riot gear were dispatched to set up a cordon separating the groups.[7] Maria Joel had planned to take the stage, but she was warned that it was too dangerous. Instead, Batista, representing the CPT, addressed the crowd along with de Assis, Dezinho's close friend at FETAGRI. Unfazed, they made a blunt announcement: the following day, the Sindicato would take the names of any family willing to participate in future occupations. "The next day," recalled de Assis, "the line of people waiting to register their names at the Sindicato was so long that it reached the other side of the block from the union building."[8] Eventually, some 1,500 families would enlist in those years to take part in occupations.[9]

"They mistakenly thought," Maria Joel said of the consortium of fazendeiros she blamed for the murder, "that by killing Dezinho, they would wipe out the movement. But that didn't happen."

Less than a month later, *Jornal do Brasil* ran a seven-part series under the title "Extermination in the Countryside." A reporter had previously spent over two weeks investigating "the serious consequences of the absence of the state" in southern Pará. The impunity was indeed appalling. Data from the CPT showed that of over 1,200 land-related murders committed in Brazil between 1985 and 2000, only about 10 percent had ever reached trial, while a mere 0.5 percent had led to the instigator being sentenced.[10]

One of the interviewees in the *Jornal do Brasil* series was Josélio de Barros. The fazendeiro denied that laborers had ever been subjected to debt bondage on his fazendas, although after the Te-Chaga-U bust seasonal employees did claim they were not paid.[11] "What we need in the region is a qualified workforce. Those tramps [*vagabundos*] don't want to work," he declared, his dislike of the workers quite obvious.[12]

Despite Resende's suspicions, Josélio hadn't been targeted in the investigation of Dezinho's murder. Josélio had returned to town five weeks after leaving for a "holiday" on the eve of the killing, and once he was back in Rondon, the detective had interrogated him. "I'm not a member of any criminal group. I'm a friend of other fazendeiros, but just to promote the development of the region, not to commit crimes," he'd stated, denying being part of any gang.[13] Josélio had admitted "two contacts with Dezinho" in his life: one in court, probably for the Te-Chaga-U case, and another during the mayoral elections. He claimed that, at this point, he owned 42,000 acres in four fazendas. "I have valid deeds for all of them," he said. At sixty-five, he had become quite wealthy. The Fazenda Serra Morena, still under his control despite the issue of the fake deed, had a current market value of 10.6 million reais, some $3 million.[14] After Josélio gave his statement, Resende decided not to investigate him further.

Resende had also questioned the other cattlemen mentioned by Francisco—the Lopes family, Olávio Rocha—but found no evidence sufficient to press charges against any of them. The district attorney of Rondon agreed, so Nunes remained the official suspect at that time. Maria Joel and her family, however, believed that a consortium led by Nunes was behind Dezinho's death and that Josélio might have had an important role in the crime, a belief shared by many at the Sindicato. No evidence gathered in the investigation supported this hypothesis. Still, an odd episode was revealed by Sergeant Marino that hinted at the possibility that Josélio knew, right before Dezinho's murder, that Rondon was in for a crime spree and it was perhaps a good idea to get out of town for a while.

According to Marino, one of the two crimes that Josélio had warned him about while he was based in Rondon was the murder of "a detective named Pedro."[15] Marino explained how the information had been passed to him: some days before the murder of Francisco's brother, Josélio had been driving around in his car when he had spotted the sergeant on the street. Josélio had pulled over and told him that "there was a plan to kill a detective named Pedro." Based on that statement, Marino thought Josélio was indicating a policeman named Pedro, and since there was no officer in Rondon named Pedro, Marino didn't pursue the lead.

Francisco would counter that law enforcement was aware beforehand that his brother was slated for death. His assumption was that the police knew that Pedro was about to be killed but didn't prevent the murder from happening because some officers and fazendeiros were involved. Francisco never mentioned Josélio to Resende, but later he said that Josélio was "more dangerous than Nunes because he had no problem at all with committing murders."[16]

For the series, *Jornal do Brasil* was also granted an interview with Nunes, who was candid about his views on the issue of land activism and his innocence. "I employ eight hundred people on my ranches and in my sawmills. The criminals are these landless people," he said in his office.[17] A subsequent encounter between the journalist and Nunes went less smoothly. While the reporter was at the courthouse accessing some files, the King of Wood suddenly appeared. "You're very inconvenient," Nunes shouted before pushing the reporter. "Get out of my way. Only young women are after me. You'll get into trouble!" he warned. Olávio Rocha, also there, stepped in to prevent a fight.[18]

The journalist didn't quote Maria Joel in the pieces he filed, but a photograph of her that ran along with one of the stories encapsulated the widow's transformation. Wearing a sleeveless striped blouse and a pale pencil skirt combined with a pair of rubber flip-flops, Maria Joel stood holding a large banner with a picture of Dezinho and a slogan written in all caps: "THE FIGHT FOR THE AGRARIAN REFORM IS LARGER THAN DEATH."[19] She looked calm, determined, poised.

By then Maria Joel was planning to take command of the union, still under Brito's leadership. She believed that "the fight initiated by Dezinho had to continue"—that the movement had to produce some tangible achievements. Until then, the Sindicato hadn't won a single inch of land, and Maria Joel agreed with those in the union and the CPT who argued that her leadership might have a real impact. Unsurprisingly her daughters and son strenuously opposed the idea. They were willing to join the quest for justice but wanted no more activism.

"When I told them that I might be the next president [of the Sindicato]," Maria Joel recalled, "they said, 'Mother, we've already lost Father; we don't want to lose you, too. We don't want to become orphans.'"

Still, the children were now young adults, and after the death of Dezinho, they were no longer kept under lock and key, which critically eased domestic tensions. Joelminha had overcome her eating disorder and gotten married, while Joelson was starting to see the controversial choices made by his father in a different light. "When we were young, my sisters and I didn't understand why my father was doing a job that caused so much pain to the family," he would explain. "But later we realized how important my father's work was. Laborers who had escaped from the fazendas continued to come to the Sindicato asking for help. They needed us."

Maria Joel's plan to eventually succeed Dezinho soon prompted the gossipy Rondon to question her ability to lead the movement at such a critical time. When the idea began to circulate at the Sindicato, some of the townspeople treated her like a lunatic, revealing sexist attitudes and gender bias. "People said, 'What the hell does that woman want to do in the Sindicato? The husband already died—he who was a man who fought for land and never managed to produce a settlement. Go take care of your children! If a man couldn't do it, how will a woman do it?'" Maria Joel recollected.[20] She was called "the crazy woman," "the woman who doesn't love life," and "the woman who doesn't take care of her children."[21]

These preconceptions and slurs were hardly a surprise in a country long known for its traditional views on gender roles and its trivialization of all forms of violence and abuses against women, including murder and rape.[22] In the Brazil of those days, women were routinely discriminated against in all aspects of life, including in social movements that, like the Sindicato, claimed to be among the most progressive.[23]

However, Maria Joel fought back against the prejudices and ultimately convinced most of the members of the union that she was the right person to lead. On August 9, 2002, she was elected president of the Sindicato, receiving the support of 170 union delegates, against Brito's 131 votes.[24] The new leader named her sister Eva director for women's policies and her close friend Zudemir dos Santos de Jesus—a short and skinny woman with a wide smile—director of social issues. For the first time, the Sindicato's board held equal numbers of women and men, and the former "Sindicato of Dezinho" became popularly known as "the Sindicato of the women."[25] The name printed at the entrance of the office was changed to add *female rural workers*, becoming Sindicato dos Trabalhadores e Trabalhadoras Rurais de Rondon do Pará.

Soon the little town would discover that the changes Maria Joel was about to initiate with the help of CPT lawyer Batista and FETAGRI would go well beyond the symbolic.

16

The Law of the Gun

LESS than two months after the election, Maria Joel and her al-lies chalked up a major achievement. High-ranking officials at INCRA, the federal colonization agency, met in Marabá and ordered the seizure of four occupied fazendas in Rondon do Pará, concluding that the areas had been the object of grilagem, or at least that the supporting documents provided by the owners were not legal.[1] In total, the land at stake was 35,000 acres. INCRA announced that about 285 occupant families would be settled in plots.

The proceedings leading to the expropriation had been initiated months before Maria Joel had become president, but the process had been stalled and, in the case of some ranches, hadn't even begun, so the decision was largely credited to the role of the Sindicato. Eva described her sister's actions: "Maria Joel entered the premises of INCRA in Marabá and said to the officers, 'If you had gone ahead and seized the land-grabbed fazendas before, my husband would have not been assassinated.' Then a long silence ensued among the officers. . . . She had become a very strong person."

The creation of four settlements proved that Dezinho's strategy of occupying areas in order to achieve some land reform, though extreme and in the end tragic for him and his family, had ultimately been effective. INCRA recognized that the activist had been a central agent of change by naming the first of the four settlements after him: the former Fazenda Tulipa Negra, previously claimed by Kyume Mendes

Lopes, was transformed into the Projeto de Assentamento José Dutra da Costa (Dezinho's full name).[2]

"People cried around me," declared Maria Joel, recalling the day she announced to the occupants that their struggle had finally borne fruit. "They cried in gratitude. It was a sad moment for all of us, because we had gained the plots only after Dezinho had been murdered. But I felt that he was there with me, that he was present, and that we had succeeded in achieving something that he had wanted so much."

Most peasants still lacked proper homes, electricity, or schools for their sons and daughters, but it was a promising start. Not only had Maria Joel managed to get them land and valid deeds but she had also persuaded INCRA to commit substantial federal funds to improve rural roads, build cabins, and provide access to low-interest credit for peasants to buy seeds or tools. Over time, some $2 million in loans would be provided by INCRA, a milestone for families who had previously struggled even to open a bank account or to buy a motorcycle in installments.[3]

"Maria Joel is a star put before us to be able to follow her," said Eliane dos Santos, a mother of six who had occupied one of the fazendas while Dezinho was alive and who was ultimately given a 120-acre plot by the government.[4] "We would be nothing without Maria Joel. Before having my own plot, where I produce oranges, bananas, avocados, manioc, and so many other products I sell in town, I worked as a maid, and I often had to resort to charity to be able to feed my children. Today I can say to someone, 'Hey, come have dinner at my house.' I have both a home and a plot. For me, it's a major achievement." In the new settlements, many expressed the same pride at what they considered a crucial feat that had not only improved their lot in life but also reinforced their confidence in themselves and in the movement.

The "Yes We Can" moment gave Maria Joel immediate respectability and prestige. The news spread like wildfire in the interior, the name of Dezinho's widow being on the lips of everyone everywhere in that swath of jungle. Advised by FETAGRI and the CPT, Maria Joel initiated propaganda campaigns in an effort to maintain the momentum.

Ads financed by the Sindicato were aired on local stations, saying, "Maria Joel calls all landless workers in Rondon to register with the Sindicato to have a piece of land." The brief announcements, broadcast over local radio stations reaching a vast interior inhabited by laborers, farmworkers, cowboys, and peasants, prompted waves of rural dwellers to join the Sindicato, which would soon reach the milestone of 5,000 members.[5] The new targets were the Fazenda Santa Monica—claimed by Lourival, who had been investigated for Dezinho's murder but who at this point hadn't been charged with any crime—and the Fazenda Fé em Deus (Faith in God), which was claimed by another big landowner suspected of land-grabbing and later accused of treating laborers like slaves.[6]

While Maria Joel defended the same ideals as Dezinho, people noticed that her command of the union wasn't exactly like her husband's. Although less charismatic, she was more democratic, paying attention to her advisors and making crucial decisions in a less impulsive and confrontational way. For instance, in order to minimize the risks of violence in the new occupations, Maria Joel instructed the occupant families, about two hundred, not to settle directly on the fazendas. Instead, they were advised to build their shacks in a nearby area held by a smallholder who sympathized with their cause and had agreed to have the landless relocate there temporarily.[7] Once the occupant families were settled there, the Sindicato informed the media and INCRA about the situation, drawing the attention of journalists and top officials to the Fazenda Santa Monica. CPT lawyer Batista was a crucial ally in those carefully crafted pressure campaigns, which took place as Luiz Inácio Lula da Silva and his Workers' Party led the polls to win the presidency of Brazil in the forthcoming elections.[8] (Lula did indeed win the 2002 presidential elections.)

The Sindicato's successes and, at the national level, Lula's real chance to finally win the presidency immediately put the fazendeiros on the defensive. Some landowners with sketchy deeds attempted to infiltrate the movement through sham petitioners posing as landless peasants. Others, hearing rumors that their estates might be occupied, showed up at the union and warned of retaliation. Calm and unintimidated, Maria Joel responded that "land stolen from the state had to be seized." Her children witnessed some of these tense meetings in which Maria Joel built herself into an image of a strong, determined person.

Josélio didn't meet with Maria Joel, but 2002 was a year of renewed concerns for the fazendeiro. Weeks before Maria Joel was elected president of the Sindicato, the rancher again approached ITERPA, this time through his daughter Shirley Cristina, to request that Pará's colonization agency issue a certificate stating that Josélio was the legitimate owner of the Fazenda Serra Morena.[9] Josélio's earlier request, back in the late 1970s, had resulted in ITERPA's president publishing a note in the official gazette informing the public that the deed was a fake.[10]

Now, Pará's colonization agency again refused to issue the certificate.[11] Weeks later, on April 15, 2002, the Barroses addressed ITERPA once more, admitting for the first time that they weren't the legal owners of all the land they had been controlling for decades.[12] They asked ITERPA for the chance to fix the problem by buying some 12,000 acres that were in fact owned by the government. They claimed to have been unaware of ITERPA's earlier declaration that their deed was a fake, and Shirley Cristina requested "special conditions" to buy the land.

On September 10, 2002, with the issue still unresolved for the Barroses and less than a month after Maria Joel had been elected president, another murder caused a deep shock in town. Magno Fernandes do Nascimento, the man who had seized Wellington when he had attempted to flee after shooting Dezinho, was executed by two men who showed up at his residence around 3:00 a.m.[13] On the night of Dezinho's murder, Magno had provided a statement to the police, and he was considered an important eyewitness regarding Wellington's participation as the triggerman.

Law enforcement managed to uncover the identities of at least three men involved in the crime, including two young triggermen who, hours before killing Magno with two shots in the head, had been drinking and got involved in a fight in a dingy bar. But detectives found no leads or reliable sources who could explain why it happened or suggest a specific person as the instigator. The thirty-nine-year-old former serviceman appeared to have had no known enemies or debts, and he hadn't been a member of the Sindicato.

Still, relatives suggested that the execution might have been linked to Dezinho's case. According to a family member, Magno's wife had recently left him because she could no longer bear the threats against her and her husband.[14] Apparently, those threats were related to both the role Magno had played in preventing Wellington from getting away and the bragging he had been doing about this act to friends in bars. The case would ultimately remain unsolved, but years later, many continued to believe that Magno's hired assassination was one of the earliest actions intended to intimidate Maria Joel and knock out a witness in Dezinho's case. Two trails of evidence sustained this hypothesis—leads never really pursued by the police of Rondon. One was that, according to the mother of one of the triggermen, her son worked in a fazenda run by Nunes.[15] Second, a sister of Magno would tell the police that she'd been informed by acquaintances that the gunmen, never arrested by the police, had escaped capture by hiding on Nunes's land.[16] However, according to court files, the fazendeiro was never interrogated or even considered a suspect.

But the event that would definitively set off alarms at the Sindicato and in Maria Joel's family took place less than a year and a half later: on February 7, 2004, Ribamar Francisco dos Santos, the union's treasurer and the coordinator of the landless families camped near the Fazenda Santa Monica of Lourival, was executed.

A short, chubby man with black hair and chestnut-colored eyes, Ribamar was murdered by killers who shot him from a motorbike as he was leaving his home at dinnertime. As had happened with Dezinho, the crime was witnessed by one of his daughters, who was inside the house clearing the table after dinner.[17]

The murder of Ribamar was understood as an action intended to get Maria Joel to take a step backwards. Just weeks prior to the killing, Maria Joel had met in Marabá with the United Nations special rapporteur on extrajudicial summary and arbitrary executions. In that meeting, she had explained her story as well as her current work to promote agrarian reform to the UN official. She had also reported rumors that "someone close to her" would be killed as retaliation.[18] "I was very anxious about these rumors, because I thought that the target would be one of my children," Maria Joel would say. She had also received threatening phone calls in which she was advised to begin organizing her own funeral, because soon she, too, would be executed to put an end to her work at the Sindicato.

As with Magno's case, the police uncovered the identities of the pistoleiros who had killed Ribamar, but they never found them or established the motive for the crime, at least officially. Court files show that law enforcement had two theories about the murder.[19] One was that Ribamar might have been killed by the owner of a construction company that was supposed to be building houses in the INCRA settlements but wasn't carrying through, prompting Ribamar to ask the authorities to halt further payments. The crime was perhaps a vendetta by the businessman involved, the police reasoned. The second theory connected the crime to Lourival. According to Maria Joel and other members of the Sindicato, Lourival had been heard in town saying that if the Fazenda Santa Monica was finally "invaded" by the occupant families, he would retaliate with blood and then would commit suicide.[20] On the eve of Ribamar's execution, Lourival had apparently been spotted near Maria Joel's home.[21]

Lourival was interrogated before securing a preventive habeas corpus to evade detention, his attorneys arguing that no reliable evidence linked him to the crime. Finally, no charges were pressed against the rancher. Josélio was also questioned, because Maria Joel had told the police that the fazendeiro was pressuring those who sympathized with the Sindicato not to provide assistance for further occupations.[22] In a short statement, Josélio categorically denied any participation in the death and argued that he was the victim of a defamation campaign. "My name always comes up in this kind of case," he admitted. "I believe the people heading the Sindicato are persecuting me."[23] He was not investigated further.

By this point, his concerns about the outcome of some of his fazendas were obvious.[24] Josélio worried about the potential implications of Maria Joel's leadership, but the wider political context in Brazil also unnerved him. Two years earlier Lula had been elected president with the support of movements like the MST and FETAGRI, and now the government, for the first time in the hands of a former union leader, was expected to return the favor with a thorough agrarian reform program in the country. The seizure of large areas of land for which Josélio lacked proper deeds seemed like a real possibility.[25]

Josélio's anxiety over the issue was further amped up on November 30, 2004, about nine months after Ribamar's murder. Three officials from INCRA showed up at his fazendas and requested access to take GPS coordinates, prompting the fazendeiro to suspect that the federal agency, in collaboration with the Sindicato, might be preparing a legal action to seize some fazendas.[26] He and his daughter Shirley Cristina postponed the inspection twice. When the time finally came, a discontented Josélio accompanied the three members of the INCRA team, among them a woman, and threatened them with potential attacks if his estates were afterward "invaded."

"I won't do anything against the girl, because she's a woman," Josélio warned, according to the later statement given by the civil servants, "but you," he said, referring to the leader of the team, "I'll look for you wherever you'll be!"[27]

Now sixty-nine years old, Josélio was still seen as the legendary strongman of Rondon, but his physical decline was evident and his influence had dwindled. In fact, the Barroses, in addition to using lawyers to request that ITERPA allow the family to remain in control of the land for which they lacked proper documents, were now seeking to secure their power through politics.[28] Shirley Cristina had just made her first attempt to be elected mayor, and her father had been the main donor to her campaign.[29] She failed, ending up with the least votes in the 2004 mayoral elections, but many anticipated that she would not give up her political ambitions.

THE POLICE NEVER found out who had hired the pistoleiros to kill Ribamar, so the murder would remain unsolved and unpunished. When some form of institutional reparation came, it was from INCRA, which, just before the end of 2004, created three new settlements in

Rondon amounting to 44,000 acres. In total, 325 families would be given plots.[30] It was another milestone for the Sindicato, albeit preceded by blood and suffering, and Maria Joel warned the authorities that the pending land reform had to be carried out unconditionally, not just as a response to the execution of campaigners.[31]

The loss of Ribamar happened at a critical moment for Maria Joel. She had received threats before, but she had assumed that, as Dezinho's widow, she was somehow too obvious a target to actually be attacked.[32] But the killing of a close colleague convinced her that she wasn't untouchable.

She responded to the crime by increasing the pressure on the authorities. A first move was to petition the government of Pará to be included in a program to protect threatened activists. Her application was accepted without delay, which meant that Maria Joel now would be escorted day and night, both inside and outside her home, by one or even two plainclothes officers.

At the same time, the widow explored new strategies to pursue justice and bring the case of Dezinho before audiences abroad. On November 30, 2004, Maria Joel, the CPT, and two Brazilian nongovernmental organizations lodged a complaint before the Washington, D.C.–based Inter-American Commission on Human Rights "against the Federative Republic of Brazil for the alleged violation of the rights [of Dezinho] to life, physical integrity, personal liberty, a fair trial and judicial protection."[33] The legal move would not have immediate consequences, because the commission might spend years examining the case before ruling on whether the petition met the criteria to be formally investigated. But to internationalize the case and prosecute the Brazilian state meant renewed public attention with which to pressure state attorneys and judges in Pará.

Coincidentally, Darci Frigo, one of the Brazilian attorneys and land activists assisting Maria Joel with the complaint at the commission, had recently received an award from the Robert F. Kennedy Memorial Center for Human Rights.[34] Through his mediation, the American NGO and some two dozen international partners—including the California-based Food First and the Oakland Institute as well as the Seattle University

School of Law—would submit an amicus curiae to the commission some months later. In the 37-page document, the advocacy groups summed up some of the burdens faced by Dezinho while leading the Sindicato, including his "arbitrary detention" by Judge Amorim, the multiple death threats made by ranchers, and the overreaching grilagem. Josélio and Nunes were named several times as the masterminds of killings.[35]

"José Dutra da Costa's [Dezinho's] life and death tragically illustrate the absence of the rule of law in Brazil," the amicus curiae read. "*Grilagem* in Pará, and the fundamental negligence of the government that permits it to flourish, are underlying causes of Dezinho's death."

On the domestic front, the Sindicato approached Lula's government to take urgent action against the reigning lawlessness. In a letter co-signed by the CPT, FETAGRI, and several NGOs, Maria Joel wrote to the secretary of human rights, Nilmário Miranda, on January 5, 2005.[36] She referred to the unsolved cases of Magno and Ribamar and attributed to Nunes a series of crimes, from grilagem in the Fazenda Lacy to illegal deforestation and debt bondage. Maria Joel also requested that the federal police investigate the King of Wood and that the prosecutors give priority to the case of Dezhino's murder, which was still stalled in the courts in Pará and now was the basis of international litigation against the Brazilian state.

O Globo caught wind of the letter and dispatched a reporter to Rondon, which again became the subject of headlines in the national papers. The journalist filed a long piece titled *"Um Brasil sem lei,"* "A Brazil Without Law," in which the city was described as the fief of "grileiros of large areas of land that set up groups; issue orders to kill, threaten, enslave, intimidate; and remain unpunished, in the style of old feudal lords."[37] Nunes was depicted as a mobster facing "more than 500 labor lawsuits, most of them for cases of slave labor in the pristine forest" of his fazendas. The paper also reported that investigations against Nunes for his "extensive criminal curriculum" had physically vanished from the archives of the local courthouse, a practice that *O Globo* described as not rare in Pará, where the files of at least eleven court cases involving hired assassinations had recently "disappeared"— and, with the documents, the chances of prosecuting the suspects.

Maria Joel again posed for a picture to run with the story. Three years had passed since she had appeared alone in the photo in the *Jornal do Brasil*. This time, she was surrounded by dozens of peasants raising their sickles and hoes. Referring to the criminal consortium of fazendeiros whom she accused of spreading terror through murders, Maria Joel said to the reporter: "I'm small to go up against an organization of their size. . . . But like my husband, I'll continue this fight."[38]

The letter and the story by *O Globo* captured the attention of Secretary Miranda.[39] As he was promoting a new national program to protect the defenders of human rights at the time, Miranda decided to go to Pará as a kind of response to the media criticism. Early in February 2005, he traveled to Rondon with another of Lula's ministers and talked with Maria Joel, heading afterward to Belém to meet state authorities and other land and environmental campaigners.[40] In the state capital, Miranda was unexpectedly approached by a petite blue-eyed seventy-three-year-old American Catholic nun named Dorothy Mae Stang.

"She came to ask for protection for an activist in the region of Anapu," Miranda recalled, referring to a logging boomtown located some 300 miles west of Rondon.

Dorothy had been born in Dayton, Ohio, and had moved to Brazil in the 1960s as a missionary of the Sisters of Notre Dame de Namur. In her early years, she'd lived near Vila Rondon and met Father Fontanella, the Italian priest deported in 1976.[41] By the time she met Secretary Miranda in Belém, the nun had made her name as an environmental activist in Anapu, where she was promoting a sustainable settlement project in which homesteaders were given a plot under the condition that they cleared only a fraction of it. One of Dorothy's mottoes was *A morte da floresta é o fim da nossa vida*, "The death of the forest is the end of our lives."

Only a few days after meeting Miranda, Dorothy was brutally executed, and her story would be blasted around the world. The homicide would have profound consequences for the fate of the rainforest, at a turning point due to the successful introduction of soybeans, a crop rapidly conquering global markets from China to Europe.[42] Unexpectedly, Dorothy's tragic death would also have an impact on Maria Joel's life.

17

Land or We Burn the Jungle

I N early 2005, just weeks before Dorothy's murder, deforestation in a remote region of Pará was causing increasing concern among federal officers from Brazil's environmental protection agency, IBAMA. The area was located some 500 miles west of Rondon and lined the BR-163, a 1,100-mile highway cutting through the state of Mato Grosso and the neighboring Pará from north to south.

After satellite data exposed levels of illegal clearing and burning beyond anything seen in years, agents of IBAMA and the federal police were deployed to contain the destruction, which was particularly intense on the border of Mato Grosso and Pará. In that area, the savannah transitioned into thick Amazonian rainforest to the north. At the time, it was one of the most promising agricultural frontiers of South America for soybeans, a crop for which Brazil was now the world's largest exporter.[1]

The story of the BR-163 and the settlement process spurred by its construction during the dictatorship harkened back to the history of Rondon and its foundational PA-70.[2] Initially, fortune hunters from all walks of life had founded hamlets along the road. They made a living by harvesting wood and engaging in wildcat gold mining in the surrounding forests, inhabited by the Kayapó Indigenous group. Almost everyone was armed, according to accounts from the time, and strongmen ruled the roost through extreme violence.[3]

In the late 1990s, the region experienced a drastic transformation with the arrival of the soybean.[4] Agribusiness became the engine of an unprecedented economic boom that enticed multinationals, investors, and new waves of migrants to towns on the frontier. In only a decade, the area under cultivation doubled in Mato Grosso, the flat landscape of which was more suitable for combine harvesters than the undulating geography of Pará.[5] The real estate market hit new highs, with plots near the BR-163 seeing a five- to tenfold increase in price.[6] Predictably, the scramble for land led to deforestation and grilagem in tracts of the jungle controlled by both the Kayapó and the government.

The wider global context at that time only exacerbated the conflict over the frontier. The world market for soybeans was rapidly growing, especially due to increasing demand in China, while the United States, one of the world's largest soybean producers, had suffered droughts in the Midwest that limited the output and pushed prices up.[7] A third factor influencing both the price and the global supply chain of soybeans was a frightening dystopian disease causing the deaths of livestock and people in Europe.[8]

The scientific name for this was bovine spongiform encephalopathy (BSE), but it became known as mad cow disease. A cow who contracts it—after a long incubation process lasting from two to five years—shows signs of unusually nervous and violent behavior and has trouble getting up or walking.[9] The human variant of the disorder is called Creutzfeldt-Jakob disease, and cross-contamination is thought to take place when humans eat certain parts of a cow infected with BSE, like tissue from the animal's brain and spinal cord, which is often used for hot dogs and taco fillings.[10] BSE is undetectable through the testing of live cattle, and after early outbreaks occurred on British farms in the mid-1980s, the disease spread quickly and quietly through herds of cows, mostly through infected meat and bone meal used to supplement cattle feed with protein content.[11]

The European Union, acknowledging the risks of "one of the most significant animal diseases in the 20th century," took a radical step in 2001 to tame the transmission: they banned the use of animal-protein-based ration for livestock.[12] "The resulting shortage of protein for animal ration," wrote a group of scientists, was "filled primarily by soy meal, whose amino acid composition and nutritional properties are superior to those of other vegetable meals and oils."[13]

In cities along the BR-163, like Sinop and Sorriso, former frontier outposts now turned capitals of Big Ag with a passion for imported pickup trucks and fancy restaurants, the world's hunger for soybeans was understood as a historic opportunity for development. Soybeans became known as the "wonder crop" or the "showcase crop" by media outlets like *Forbes*, and the *Financial Times* would later write that soybeans were "the crop of the century."[14] The BR-163 was referred to as the "soybean highway" because it connected croplands in Mato Grosso to the shores of the Amazon and the Tapajos Rivers, waterways located farther north in Pará and accessible in less than a day under normal circumstances. The basin was thought to have the natural conditions to become a fluvial export corridor—like the Mississippi river system in the United States—if investments were made in infrastructure. Priority number one was the paving of a 650-mile tract of the BR-163, which remained a treacherous sandy road filled with potholes.[15]

Improvement of the road was an old demand of the agribusiness caucus to solve the recurrent stranding of hundreds of big rig trucks while they were transporting produce to ports on the Tapajos River, and Lula's administration had signaled it would carry out the project.[16] Politically, the plan made sense for the government because, despite the opposition of environmental groups, it would bolster the support of the agribusiness caucus, "the largest bloc in Congress since Brazil's redemocratization."[17]

Soon after the project was officially announced, serious problems surfaced that made Lula's administration change its mind. The areas next to the BR-163 became hot spots of destruction, with some areas registering a 500 percent increase in deforestation.[18] Ultimately, officials were prompted to halt "the paving until they could formulate a forest-management strategy for the region," wrote Scott Wallace in *National Geographic*.[19] In the meantime, logging licenses were suspended, and officials reviewed dubious land claims. In swaths of jungle of great biodiversity, further destruction was prevented by creating two natural parks in which most activities were banned.[20]

But environmental offenders decided to fight back, and a general climate of lawlessness spread in that region, which was difficult to patrol due to its vastness.[21] Groups of landowners and loggers blockaded tracts of the BR-163 for weeks, leaving hamlets devoid of basic supplies like diesel or drinking water.[22] Buses were burned and corporate planes were prevented from landing in small airports. Local political elites even threatened the federal authorities with irreversible disruptions.[23] Fazendeiros were said to be stockpiling arms and ammunition and hinted at the possibility of violence and pollution of waterways.[24] The leader of a loggers' group warned that "blood would flood" the area unless they could continue operating, no matter that, as a top official later pointed out, "the main problem with sawmills along the BR-163 is just one—it is all illegal."[25]

Precedents made the threats seem credible. Months earlier, an office of IBAMA had been torched after a raid was launched against illegal sawmills.[26] In another incident, a top civil servant working for an environmental fund had been kidnapped and held hostage until plans for creating new protective areas were abandoned by the mayor of a town crossed by the "soy highway."[27] Hence, on February 3, 2005, the government of Lula, pressured by rebellious protesters led by a senator named Fernando de Souza Flexa Ribeiro, who was later found to have great appreciation for Josélio, canceled some of the bans and restored logging licenses.[28] Environmental groups warned that surrender to extortion was a disastrous move.[29] Much sooner than anyone had expected, Lula's government would have to backtrack due to the reverberation of a murder that made global headlines—that of Sister Dorothy.

On the rainy morning of February 12, the silver-haired missionary was walking through a jungle path near Anapu when she was approached by two men known to have links to grileiros.[30] In the heart of that towering jungle, they asked her whether she was armed, to which Dorothy replied by removing her Bible from her bag and declaring, "This is my only weapon."[31] Then they shot the American nun six times.[32]

Her assassination was the culmination of years of threats.[33] Once she went to have dinner at a restaurant, and a fazendeiro who recognized her "came over to Dorothy, shook his fist in her face, and said in a menacing voice, 'I know who you are. One day, we will get you.'"[34] For years, Dorothy had been depicted by her enemies in Pará as an agent of the United States and even as a member of the Bush family pursuing an obscure ulterior motive in the resource-rich Amazon.[35]

"I know that they want to kill me, but I will not go away," Dorothy once declared.[36] "My place is here alongside these people who are constantly humiliated by others who consider themselves powerful."

Secretary Miranda—one of the last high-ranking federal officials Dorothy met with in her life—said he had perceived in the American nun that "she had come to accept her own death as a kind of martyrdom" to benefit the environmental and social movement she had built in Anapu.[37]

News of the homicide circled the planet and put the Amazon and Lula's policies on the spot. The *New York Times*, in a blunt editorial, wrote that the "ghastly crime," which soon was determined to have been contracted by two fazendeiros, was "a message" for the Brazilian president: "Land-grabbers are telling the government that they run Pará."[38]

Feeling compelled to respond, Lula acted with unprecedented assertiveness. He dispatched 2,000 army troops to Pará to take control of the situation, and in order to defuse international criticism, his government restored and expanded the banning of illegal activities along the BR-163, the paving of which was suspended indefinitely.[39] (It would be concluded only decades later by order of President Jair Bolsonaro.)[40] Additionally, researchers wrote, "a presidential decree subsequently established a moratorium on the granting of land titles and logging permits in a 14.6-million hectare region along the highway" where illegal clearing had caused great damage to the environment.[41] The set of measures "shut down the land-speculation market overnight on the contested frontier."[42]

Less than a year later, the federal authorities would consolidate the environmental protection by creating a string of new preserves along the "soy highway" adjacent to the preexisting Indigenous lands of the Kayapó. Spanning an area of almost 16 million acres, these preserves would become, in the words of a scholar, a "protected forest corridor unique in scope and ecological diversity in the Amazon and in the world."[43] The government would also provide funds to IBAMA to fight offenders. From Brasília, top officials monitored deforestation using cutting-edge satellite technology and remote sensing. When spots of deforestation were discovered, almost in real time, IBAMA officers trained in jungle combat were deployed to the rainforest to arrest land-grabbers, chainsaw operators, and corrupt officials. This made IBAMA a nightmare for environmental offenders.[44]

Soybean multinationals joined the conservation efforts by announcing that they would track via satellite the source of the soybeans and would not purchase any grown on jungle areas deforested after July 2006.[45] The deal was called the Amazon soy moratorium and would play a crucial role in taming deforestation.[46] In 2005, the Brazilian Amazon had recorded deforestation of over 7,300 square miles, the third largest area of rainforest ever leveled in a single year.[47] Five years later, the rate had been slashed by almost two-thirds.[48] With this achievement, Brazil earned a reputation as "a global leader in its efforts to reduce deforestation and emissions" and proved to the world that the rise of its Big Ag wasn't incompatible with the preservation of the rainforest.[49]

THE MURDER OF Dorothy and its immediate political implications represented an opportunity for the environmental and social movements in Pará. Maria Joel, who had fleetingly exchanged words with Dorothy months prior to her death, found herself unexpectedly in the media spotlight as Brazilian and foreign media outlets reached her to comment on the case of "another predictable murder like that of Dezinho," she said.

"If they killed a famous American activist, guess what they'll do to us poor, unknown Brazilians, in the heart of a devastated forest," Maria Joel said to O Globo.[50] In an interview with the conservative O Estado de S. Paulo, she warned Lula's government: "If things don't change, and now, you [the media] will soon be back here to cover [the news of] my death."[51] *Folha de S. Paulo* wrote that although hundreds in Pará had been threatened for their activism, Maria Joel was the only campaigner receiving police protection in the state; the newspaper story included a photo of the activist walking, escorted, through the streets of Marabá.[52] She also talked with Henry Chu, the Brazil correspondent of the *Los Angeles Times*, who in the aftermath of Dorothy's murder visited Rondon and wrote a long piece titled "Light Shines on State Called Brazil's Heart of Darkness."[53] Maria Joel told the American journalist that despite the threats, she would not give up.

CPT lawyer Batista, sensing that the political context was favorable, launched a public campaign to revitalize the case, which was still stalled in courts. Through the mediation of a priest based in Rio de Janeiro named Ricardo Rezende, a man with a long history of defending farmworkers, Batista reached a group of Brazilian actors who had recently founded a small organization called Movimento Humanos Direitos. It campaigned to give visibility to victims of human rights abuses, often organizing meetings between celebrities and top political authorities that were publicized and received the attention of the media and society. Two actresses named Camila Pitanga and Letícia Sabatella, then at the height of their careers, learned of Maria Joel's plight and agreed to participate in a rally in Rondon.

On April 10, less than two months after Dorothy's murder, both stars marched beside Maria Joel and her children, who were wearing white T-shirts stamped with photographs of Dezinho and Ribamar. Throngs of spectators participated in the public event, never having imagined that two of the best-known actresses from soap operas watched by virtually the whole country would ever show up in that forgotten corner of Brazil, much less for the widow of Dezinho. "Let's protect Maria Joel!" they yelled to a multitude of some 2,000 people.[54] Maria Joel said that day, "Justice could be faster and lead to trials for those responsible for the death of my husband. How long will we have land widows?"[55]

About a month later, on May 19, Maria Joel got the chance to tell her story to the most powerful man in Brazil. Pitanga and Sabatella managed to set up a meeting in Brasília with Lula, and the leader of the

Sindicato was invited to join them.[56] Wearing a spotless red blouse, she sat right next to the president of the nation and asked him to order the federal police to carry out an independent investigation into the murder of Dezinho. She wanted the authorities to fully uncover the group of landowners who, in Maria Joel's opinion, had plotted the scheme along with Nunes. "I don't want to die, Mr. President," she said. One of the celebrities read a letter written by Maria Joel's children imploring Lula to take measures to prevent their mother from becoming the Amazon's next martyr.

Father Rezende, also present with some other campaigners, addressed Lula to reinforce the urgency of Maria Joel's case. "Mr. President, in the past, other people who received death threats came to this office asking for help and were later murdered," said the priest, referring specifically to the case of a member of the CPT.[57] "What can be done for Maria Joel?"

Lula gave a deep sigh and answered, according to eyewitnesses, "I'm the president, but I don't have the power to prevent a death. If in the United States they killed Kennedy, how could I prevent the murder of an ordinary citizen here in Brazil?"[58] Still, aides said that the federal police, commanded by the minister of justice, would look into the case.

In the meantime, Dorothy's killers had been arrested, and the plot behind her death had been uncovered. The murder had been hired by two fazendeiros because they wanted to land-grab large areas in the sustainable settlement project in which Dorothy had been working with homesteaders. The five men involved in the crime—two triggermen, a middleman, and the two ranchers—would ultimately be convicted and receive long prison sentences. Those who actually shot her were tried and jailed less than a year after the crime, almost a record for the standards of Pará. Putting the two fazendeiros behind bars would take much longer.

Soon after telling her story to Lula, Maria Joel would have her own opportunity to see justice served—or so she thought.

18

Amazonian Justice

O N October 10, 2006, Wellington de Jesus Silva was arraigned on charges of murder before a Belém-based court.

The trial of the man who had taken Dezinho's life was being held in the state capital after Batista had successfully filed a motion requesting the transfer from Rondon. The argument the CPT lawyer had presented to the court was that the plaintiff could not get a fair jury trial in that town ridden with violence.

The judgment was crucial for Maria Joel, and she wasn't alone on that day in which she would cross paths with the man who had misled her, entered her home, and killed her husband. Hundreds of supporters traveled to the state capital and camped at the entrance of the courthouse, a landscaped square of mango trees located in downtown Belém, not far from the shores of the muddy waters of the Pará River.[1] Many of these supporters were members of the Sindicato and FETAGRI, and they carried banners displaying Dezinho's picture to pressure both the jury and Raimundo Moisés Alves Flexa—a bespectacled judge who had presided over some of the trials in the case of Dorothy's murder—to hand down a tough sentence. Even the actress Sabatella flew to the state capital to attend the trial.

Everyone was eager to hear Wellington's testimony and follow the unfolding of the case. Many wondered what Wellington's position would be, whether he would reveal the whole scheme behind the murder. The defendant indeed had good reason to confess. He had been caught in the act by the townspeople and there was ample evidence proving that he had been the shooter, including the gunshot residue found on the clothes he was wearing the night of the murder.² It seemed extremely unlikely that Wellington would be acquitted by the jury, so people thought his attorneys—public defenders, because he had no way to pay for a private lawyer—would advise him to plead guilty, cooperate, and disclose the name of the instigators.

The defendant was without handcuffs when he was brought to the courtroom. The now twenty-five-year-old man wore an oversize yellow shirt and looked tiny and fragile when he entered and sat in the large wooden chair placed in front of Judge Flexa. Although six years had passed since the murder, Wellington's physical appearance was almost the same as the day he'd been arrested—slim and clean-shaven, with short hair, and utterly introverted. But his mien was different. Wellington now looked gloomy and evasive, and his face had lost the guileless air that had once made him seem a reluctant outlaw.

The big news of the day was revealed soon after Wellington was presented to the judge. Commotion erupted in the courtroom as Américo Lins da Silva Leal, the trial lawyer who had defended one of the fazendeiros in Dorothy's case, presented himself as Wellington's new defense attorney, replacing the lawyers who had been handling his case.

The son and grandson of judges, Leal specialized in winning jury trials for seemingly undefendable landowners and loggers charged with felonies.³ Five-foot-seven and slender, with a disheveled gray beard and hair that made him look like a slimmer version of Karl Marx, Leal was widely known for his use of courtroom theatrics. One of his defense strategies was to depict victims of land-related murders as evil and unlawful. (He had said of Dorothy that she had "the DNA of violence, the DNA to kill" just because she was American.)⁴ Leal's hard-hitting speeches had resulted in the acquittal of the three high-ranking police officers leading the Eldorado Massacre back in 1996.⁵ (Many years later, a higher court reversed the controversial verdict rendered by the jury and jailed two of the officers.)⁶

If until that point Wellington's conviction had appeared to be inevitable and pleading guilty was the defendant's sole chance to persuade Judge Flexa to be lenient in his sentencing, the involvement of Leal made things less predictable. State attorneys were in fact dumbstruck by the coup de théâtre and its potential implications. "[Leal is] paid by whom? The defendant is a person who lacks the resources [to hire him]. Someone is financing it, and that is the same person who paid Wellington to assassinate Dezinho," a prosecutor said.[7]

Leal's immediate request to Judge Flexa was to postpone the trial for at least a week. He argued that because he had just taken over the case, he hadn't had enough time to construct his defense. Judge Flexa agreed, and the trial was postponed for a month. Maria Joel and Batista suspected that the strategy was intended to defuse the public pressure placed on the jurors by the presence of both the protesters and the actress Sabatella.

If that had been Leal's calculation, the tactic was a failure. On the new date of the trial (November 13), the mobilization grew in number and muscle. About a thousand peasants and activists arrived in Belém to stage a vigil. Over a hundred wooden crosses, many bearing a photograph of Dezinho, were stuck in the grass of the square in front of the courthouse, a powerful image captured and broadcast by the media.[8] A large banner read: "It Is Not Enough to Condemn the Pistoleiro. We Want the Fazendeiro Decio José Barroso [Nunes] in Prison!" This banner had been hung from the streetlights by the "Dorothy committee," a watchdog group that monitored the trials of the nun's killers and supported Maria Joel's quest for justice.[9] This time, Sabatella could not attend the trial, but Pará's governor-elect, the Workers' Party member Ana Júlia Carepa, showed up to support the widow.[10]

Judge Flexa announced the opening of the trial, and a court officer read the charges against Wellington. A few minutes later, Leal revealed what his—astonishing—defense strategy would be: a new narrative of how the events had unfolded that night of November 21, 2000. First and foremost, Leal claimed that Wellington had never been involved in any homicide; he was instead just a passerby whom Dezinho had grabbed after being shot by someone else. This explained, according to Leal, why both men had fallen into the hole. Wellington supported this thesis by reversing his previous self-incriminating statements and depicting himself not only as the victim of a miscarriage of justice but also as a powerless and voiceless citizen whose most fundamental rights had been violated.

"The police beat me because they wanted me to confess to a crime that I did not commit," the defendant told the jury.[11]

The central piece of evidence presented to the jurors to support the new account—and therefore Wellington's plea of not guilty—was Leal's own interpretation of the trajectory of the gunshot wounds in Dezinho's body. Leal said that there was absolutely no doubt that the shooter that night was left-handed, which, according to him, made it impossible that the murderer could have been Wellington, who was right-handed. Leal and Wellington attempted to convince the jury of this narrative by reenacting the crime in the courtroom, with Wellington playing Dezinho and Leal the unknown killer.[12]

Before concluding his testimony, Wellington said that he didn't know who might have shot the activist or why the revolver was found in the hole. He also claimed that he didn't know Nunes.[13] Wellington's mother, called to testify as an "informant" by Leal, insisted that her son had no criminal record and was a good boy who never lied to her. "He doesn't know anything about the crime. . . . I don't know why he's in jail," she said.[14]

Maria Joel seethed with indignation as she observed the unfolding of a strategy intended to cast doubts over Wellington's guilt and mislead jurors about what had really happened that night. Perhaps, the widow would say later, it would have been better to let the vigilantes lynch Wellington the night her husband was murdered so the culprit "would now be in a hole underground like Dezinho."

The defense strategy, if it was ever truly intended to get Wellington acquitted, was a total fiasco. The testimony of Maria Joel, who chronicled in detail how the killing had taken place, as well as evidence such as the gunshot residue on his clothes, led the seven jurors to determine that Wellington was guilty beyond a reasonable doubt. He was convicted of first-degree murder, and Judge Flexa, noting that the gunman had acted "like a coward" and had shown no repentance, sentenced him to twenty-nine years in prison.[15] The prosecutor in the case said that Leal's defense strategy had been "suicidal" for the defendant.[16]

The media reported that Nunes's defense attorneys had attended the trial, sitting "in the front row of the courtroom, but gave no interviews."[17] Some commentators believed that the King of Wood might have paid for Leal's legal services with the aim of keeping the defendant under control and making sure Wellington didn't implicate him. In their opinion, this explained Leal's odd defense strategy; if Wellington hadn't been the shooter, he couldn't know who had hired him for the crime.[18] "It was evident that Wellington hurt his case because he wanted to benefit the fazendeiro," declared an attorney and human rights activist.[19] Later, Leal would deny this hypothesis, insisting that he had done his best to represent Wellington.[20] He would also refuse to disclose how much he had been paid or who had hired him for the jury trial of Wellington, but he admitted that "for a job like that, he would charge no less than 100,000 reais," some $50,000 at the time.[21]

Wellington kept silent while he listened to the tough verdict and was then escorted out and transported back to the jailhouse in a retrofitted paddy wagon.[22] For the more than ten hours of the trial, he hadn't exchanged a single word or glance with Maria Joel.[23] Wellington's mother and sister, also present in the courtroom, sobbed and refused to comment on the verdict.

Dressed completely in black, Maria Joel reacted to the verdict by slowly rising from the bench and embracing her daughters and son. Joelson's arms circled his mother and sisters powerfully, and the family sobbed on one another's shoulders. There was sorrow in their weeping, but also relief. "Nobody can decide over the life and death of people and remain unpunished," declared Maria Joel, visibly satisfied.[24]

Outside the courtroom, hundreds celebrated the conviction with joyful tears and congratulatory hugs. Surrounding Maria Joel, who felt reinvigorated and was preparing for reelection as leader of the Sindicato, friends and farmworkers sang and waved the flags of peasant unions. It seemed like a historic date in the fight against impunity in Pará.

Much sooner than anyone could have expected—after less than twenty-four hours, in fact—Maria Joel's sense of accomplishment was overshadowed by a decision that showed true justice was far from achieved. On November 14, the same day Wellington's picture and the news of his sentence splashed across the front pages of newspapers in Pará, the district attorney of Rondon, a woman named Lucinery Helena Resende Ferreira, addressed the judge with her written closing arguments in the pretrial proceedings in the case of Nunes.[25] Her controversial request was that the charges against the fazendeiro be dismissed.

The prosecution of the King of Wood had been resumed after the audiotape transcripts were completed in mid-2004, almost four years after being turned over by Francisco to Detective Resende. No conclusive evidence incriminating Nunes in Dezinho's murder had been found in the recordings, although he was indeed referred to as "the Judge" by some speakers taped by Pedro. In light of the situation, the judge presiding over the case had prompted both the district attorney and the plaintiff to file their closing arguments before eventually deciding whether the case deserved to reach trial or whether the charges had to be dismissed (*impronunciado*).

District attorney Ferreira had picked the second option. In the closing arguments she filed a day after Wellington's conviction, she wrote that there was "no certainty or even enough evidence" that Nunes was involved in the crime.[26] Ferreira gave no credibility to the "isolated statement of Francisco," supporting her motion to dismiss with the findings of a report produced by the federal police after Maria Joel had met President Lula.

An agent instructed to look into the case had indeed concluded that there wasn't sufficient information to single out Nunes as the sole mastermind.[27] He wrote that the King of Wood owned "massive lands" with deeds that presented "doubts about their legitimacy." He also said of Nunes that "his name had a long history of conflicts in Rondon do Pará." But the officer noted that nothing had really been proven against the businessman. The agent therefore concluded that the entourage of Hilário might have planned the assassination of Dezinho in association with rancher Lourival, the owner of the Fazenda Santa Monica. To reach this conclusion, the officer minimized the revelations of Francisco incriminating the King of Wood, relying instead on the statements of several members of the family of Nunes's second wife. They claimed to have seen Olivandro, Hilário's son, with Wellington on a bus coming from Bahia and also to have been approached by Lourival after the crime in order to help the suspected middlemen— Ygoismar, Rogério, and other outlaws—flee the region.[28]

Batista presented his counterarguments to prevent Nunes being dismissed on the grounds of such disputable hearsay evidence provided by the defendant's kin. In a forty-page motion, he quoted Francisco's statement extensively and cited the unsolved murders of Nunes's former workers, which had never been duly investigated by Resende. He also mentioned an excerpt of a transcribed phone call taped by Pedro in which an unidentified man said to another, "You know that Delsão [Nunes] likes to kill, right?"[29]

Now it would be up to Rondon's judge to examine the legal arguments of both sides and decide whether Nunes's case should go to trial. Many in Maria Joel's entourage suspected that a strategy had been carefully crafted to have the case shelved right after Wellington had been sentenced.

THE GOLDEN AGE of Nunes's entrepreneurial success took place in the years between Dezinho's murder and Wellington's conviction. Over the course of those six years, Nunes had consolidated his position as the main fazendeiro of Rondon and one of the largest landholders in southeast Pará, with a series of investments that proved his ambition and also showed how little the criminal prosecution had affected his appetite for business. By 2006, on the eve of turning fifty, Nunes was an owner or partner in no less than seven sawmills and four fazendas.[30] He had become the largest source of employment in Rondon, with some 800 people working in his companies and up to 3,000 benefiting in the form of indirect or subcontracted jobs.[31] When federal officers later searched official databases to find out how many cars and trucks he owned, the "number of vehicles under his name [was] so large" that, according to a confidential report, the computerized system could not process the results.[32]

His business empire had expanded well beyond the wood sector. He owned butcher shops, supermarkets, a coffee-processing plant, and several apartments in Pará and Minas Gerais.[33] His latest pet project was a large and modern meat plant, where Nunes, now pursuing vertical integration in his cattle business and aiming to export meat cuts worldwide, slaughtered the livestock from his fazendas and processed the meat and leather. Opened in 2006, the meat plant, later recognized as meeting international standards and therefore being allowed to export to Asia, had cost several million reais, according to sources from the sector, with a person in his entourage putting the figure at some 20 million ($4 to $5 million). Still, the crown jewel of his portfolio continued to be the Fazenda Lacy, with its declared value of 107 million reais, or about $50 million.[34]

Nunes was proud of his achievements and described himself as a tireless entrepreneur, but many suspected that, like Josélio, Nunes's success story involved a mix of hard work and a dark side that had become obvious only after Dezinho had shone the spotlight on him.[35] Public records indicated that Nunes was an alleged repeat offender who had been either indicted or fined for many offenses, including illegal deforestation and trade of undocumented timber.[36] He and his companies were also suspected of using several front men to impede the authorities from seizing assets for owed taxes and fines.[37]

Some considered him to be the fazendeiro who had become the ultimate successor of Josélio in Rondon, employing his immense wealth and fearsome reputation to maintain power and influence. "In an earlier time, Josélio de Barros was perceived as a kind of patriarch with ascendency over the fazendeiros," declared a Rondon-based district attorney who'd had access to privileged information for years.[38] "Nothing happened without his consent. But Josélio was now old, and he didn't carry out any more activities"—meaning threats—"against rural workers. He was the past, not the present. . . . By that time, Nunes had become the most feared person in Rondon."

The source, who dealt with some of Nunes's cases, stated that the King of Wood resorted to intermediaries and fixers to buy people, claiming that he had been approached by one of his defense attorneys to shelve Dezinho's case. "Nunes believes that everyone has a price," the prosecutor quoted Nunes's attorney as saying to him. The source also reported that district judges and Rondon-based officials feared the consequences of enforcing the rule of law against Nunes, especially after the revelations made by Francisco. "When I started to work on Dezinho's case, the local judge told me, 'Careful, this is a very dangerous case. We're not safe here,'" recalled the prosecutor. Those fears were echoed by officers at IBAMA and other federal agencies, who spoke of the personal risks incurred by raiding fazendas and sawmills in Rondon.[39]

A myriad of labor cases against Nunes and his companies had also contributed to depicting him as an unscrupulous and hard-nosed entrepreneur. While he was spending millions building his meat plant and developing his farming and logging projects at the massive Fazenda Lacy, officials inspecting his businesses found multiple irregularities.[40] He also faced a flood of litigations from unhappy employees, a legal quagmire that had begun while Dezinho led the Sindicato and only grew over the years.[41] Nunes was even summoned to testify in Brasília before a parliamentary commission of inquiry investigating rural conflicts and labor abuses. On that occasion, the entrepreneur admitted—with his shirt open to his torso in the Brazilian congress— that he was being sued by workers; he argued that this was somehow normal.[42]

In Rondon, the Sindicato remained the pivotal organization helping some of those workers. "The practice of debt bondage was still common in many fazendas, with trucks of migrant workers brought to the region and delivered directly to ranches, where they didn't receive payment," explained Zudemir, one of the union directors. Nunes, she said, was routinely accused of failing to pay employees. Eventually, task forces created by the federal government to fight modern slavery would free Nunes's employees from working up to seventeen hours a day. According to official records, they were not allowed to leave their workplaces without prior written authorization from managers.[43] One of Nunes's companies would be placed on a "dirty list" of employers deemed guilty of engaging in modern slavery.[44]

Launched under Lula in 2004 and published by the federal government, the "dirty list" punished corporations by both damaging their reputations and blacklisting them from access to subsidized loans.[45] It was an assertive strategy to try to tame contemporary forms of slavery in the country, where the modernization of agriculture and the country's rapid economic development had not eradicated debt bondage.[46] "Those who enslave in Brazil are in agribusiness, many of them producing [farm products] with the use of high technology," said a member of an NGO investigating the appalling issue.[47] "The cattle, for instance, get first-class services: balanced diets, vaccination with computer-controlled systems, and advances in breeding by the use of artificial insemination, while seasonal laborers live without the right to decent water, food, and shelter; they are beaten and humiliated; and they can't return to their homes."

In Rondon, according to the CPT, hundreds living on fazendas continued to be victims of those schemes.[48]

IN DECEMBER 2006, Maria Joel's work defending social justice was recognized by the Brazilian state. She was bestowed with the Human Rights Prize of the presidency of the republic, the greatest honor awarded by the federal government to defenders of human rights. Maria Joel had received the award in the Dorothy Stang category, which credited citizens for their work defending civil rights. The prize came while she was still waiting for the ruling that would decide the outcome of Nunes's criminal conviction and weeks before she was re-elected president of the union.

In Brasília, Maria Joel met President Lula for a second time at the Planalto Palace, where she gave a speech honoring Dezinho's memory. "Some succumbed and fell by the wayside, but they made the history of our Brazil," she declared.[49] Then she urged Lula—who had been re-elected two months earlier with the largest number of votes ever received by a candidate for president in Brazil—to undertake true land reform. "Every man and woman has the right to life," she said. "By enacting agrarian reform [in the country], I know that the deaths in the rural areas will end, and no one will be above the law."

In response, Lula gave Maria Joel an effusive hug accompanied by a charming smile. The scene, captured by press photographers, seemed to indicate that both were on the same wavelength—the two of them, the man and the woman of the people, shared personal dramas, social struggles, and a career that showed it was possible to break from the pack and spark change. But Maria Joel couldn't have felt more estranged from the man who had once represented a political model for her husband. "At the beginning, he was a great hope, later, a big disappointment. In my meetings with him, Lula promised that some things would be done [in favor of farmworkers], but later, the opposite was done" by his government, she would complain.[50] Many during those years would feel frustrated by the policies privileging agribusiness carried out by Lula and his administration.

Receiving the prize didn't help Maria Joel in court. On March 26, 2007, the district judge of Rondon ruled that "in light of the complete absence of evidence of the responsibility," the case against Nunes was dismissed.[51] Batista filed an appeal in Belém and published a press release accusing the authorities of fostering impunity with that decision.[52]

Less than two weeks later, on April 24, a man showed up in the Sindicato. He had been assigned, he would say, to murder Maria Joel.

Even inside the union building, Maria Joel was still escorted by the police who had been guarding her since the assassination of Ribamar, but the man had been brought by an acquaintance, so he was able to access her office. Presenting himself as a pistoleiro, Luiz Gonçalves da Silva asserted that he had been hired to take her life for a payment of 2,000 reais. He nonetheless claimed that he didn't want to kill her and had shown up to request 300 reais, some $140 at the time, so he could buy a bus ticket and leave the city.

Maria Joel, displaying nerves of steel, seemed to accept the offer but pretended she didn't have the cash at hand and asked the man to come back the following day. He agreed. Luiz and another man accompanying him returned to the Sindicato and were arrested by the police as soon as they got their hands on the banknotes. In the brief conversation the pair had with Maria Joel, they claimed that they had been hired by Nunes.[53] According to Maria Joel's family, both men were later found dead in a dump in an action seemingly intended to "burn the archives" of that particular episode.

The same day the outlaws were arrested, district attorney Ferreira, who now wanted Lourival tried, filed a motion with her counterarguments to Batista's appeal.[54] She played hardball, and for some reason lambasted both the CPT and Batista for their media campaign to have Nunes tried.[55] She was now adamant that "there was no way to prosecute" him and underlined that Francisco's statement was riddled with "inconsistencies," although she didn't present a single example in her motion. Furthermore, she said, Dezinho and Maria Joel didn't inspire much trust, either. She argued that the testimony of the widow had little credibility because she had become the president of the Sindicato. Regarding Dezinho, Ferreira wrote that he had become the nemesis of fazendeiros in Rondon not for his activism and fight against grilagem, but instead "because he was suspected of having been the instigator of the assassination of the cattleman José Hilário," whom she described as "a pioneer in the region."

Batista appealed the dismissal of the case in higher courts. But because both the prosecutor and the district judge of Rondon considered Nunes innocent, his appeal appeared to have little chance of success in the multi-judge panel court now tasked with ruling on the issue. Soon, even more controversial decisions from the judicial system would shock Maria Joel and the lawyer.

In December 2007, only eight months after having his conviction confirmed by an appeals court, Wellington was granted permission from a judge to spend Christmas with his family. He was supposed to be out of the jailhouse for about ten days, but Dezinho's assassin would never return to his cell.[56] Maria Joel believed that the same people who had paid the attorney Leal, and perhaps someone at the Rondon district attorney's office, had now bought a judge to give Wellington a legal way to leave the prison and vanish forever, effectively granting him impunity.

"Only offenders who demonstrated that they were reintegrating into society could leave the prison, and Wellington was not," the widow would complain.[57] The case had reverberations in Pará's legislature, where legislators suspected the plot was intended to kill the assassin, burn that archive, and bury, once and forever, the case of Dezinho.[58]

By the time he got authorization to leave the jailhouse, Wellington had already served a sixth of his prison term, which, according to the law, entitled him to enjoy an amelioration of the incarceration conditions if certain requirements—like good conduct and, of course, the certainty that he would not flee—were met. Early reports produced by the prison's director showed that Wellington had displayed good behavior.[59] However, more recent reports had noted that he had not worked in jail and in general had expressed no will to rebuild his life.[60]

WELLINGTON'S ESCAPE WAS a momentous episode in Maria Joel's struggle for justice. "We came to think that everything was over," middle daughter Joélima, previously a vocal supporter of campaigning for justice, would say. Maria Joel's four children were now unanimous in support of this idea and wanted their mother to give up, accept defeat, and move on with her life.

The four fatherless children were outraged at the dramatic twist in the case, which not only instilled in them a profound distrust of the justice system but also sustained the fear that the same man who had taken their father's life could now use his freedom to kill Maria Joel. Fueling their indignation and paranoia was the situation Maria Joel found herself in due to the safety protocols imposed by the program to protect threatened activists. Their mother was under escort at all times, even when she simply stayed inside her house, the constant presence of the police curtailing her right to privacy and freedom.[61] Zudemir, one of Maria Joel's best friends, would say, "Under this kind of police protection, with so many hours spent indoors due to the threats, you don't live; you vegetate."

The children were not exempt from this sense of oppression. For instance, on the anniversary of Dezinho's murder, Joélima used to wear a T-shirt with a photograph of her father and was routinely approached by random townspeople urging her to remove it. "They told me, 'Are you crazy wearing *this* here in Rondon?!'" Joélima recalled. "I answered: 'For God's sake, he's my father!'"

The situation Maria Joel and her children were in illustrated the deep divide in Rondon over their story. Many townspeople appreciated the family's achievements—the creation of the settlements, the raising of awareness of issues like grilagem or deforestation, and the Sindicato's involvement in fighting social injustices and the absence of the rule of law. But others criticized Maria Joel for pursuing a strategy to guarantee continual media exposure. They also defended the fazendeiros, acknowledging that the elite were the economic engine of the town.

The economic and political clout of large landholders had in fact grown over the years. As was true in many other former boomtowns, employment in the wood sector of Rondon had continued to shrink, further damaging the local economy. In 2008 alone, some 250 jobs were lost in the sector, a considerable number for a town of a few thousand residents.[62] The perception among breadwinners was that siding with Maria Joel and the Sindicato could lead them to be deprived of job opportunities. "People thought that it was preferable to do away with Maria Joel rather than [face] the possibility of Nunes being tried and jailed, because his companies would close, and there would be no work," Joelminha recalled. "We were treated like villains because people thought we were contributing to making Rondon infamous," she said.

The great influence of the establishment in politics, especially in the municipal government, left Maria Joel's family even further out on a limb. The family had close ties to the Workers' Party in Pará, and Maria Joel had maintained some of Dezinho's previous links with state legislators and members of congress, but in Rondon the control of city hall remained in the hands of the same agribusiness elite suspected of having an interest in taming the actions of the Sindicato, especially those intended to provoke further seizures of land suspected of grilagem.

This wasn't a feature unique to Rondon. In many former and current Amazonian boomtowns, local governments were in the hands of ranchers, gold prospectors, and loggers suspected of dubious behavior. They controlled city halls with the backing of voters who supported the idea that if it weren't for the booming development sustained by the continual depletion of resources, the economy would collapse. This explained the limited political success of social activists and ecologists in a region so widely associated with the environment as the Amazon; the world cared about the fate of the forest, but the immediate concern of local breadwinners was getting a job.

The 2008 mayoral elections left little doubt about who the main political players were in Rondon do Pará and how antagonistic their backers were to the Sindicato. Three candidates were competing for power. The first contender was former mayor and former congressman Olávio Rocha, who was participating in his third mayoral election. Second was Dr. Antônio Lopes de Angelo, now on a separate ticket. Both men had been accused by Francisco of participating in the alleged criminal consortium led by Nunes, although no evidence had sustained the accusations and the police had never pressed charges. Last but not least was Shirley Cristina, who was running under the flag of the Brazilian Social Democratic Party (PSDB) of former President Cardoso, a powerful party that, in Pará, would end up electing the same candidate for governor (Simão Jatene) three times.

The elite of Rondon do Pará had a great interest in city hall because it had become a strategic platform to influence the town's affairs. Most of this clout stemmed from the control of a skyrocketing budget that in large part came from federal and state transfers and had as a wider context Brazil's economic takeoff and Lula's expanding social programs. In only a decade, the budget of the municipality grew threefold, and the allotment for staff grew at an even faster pace.[63] Laws regulating the proper use of public money were strict, but in little towns like Rondon, the mayor was in a unique position to influence the use of incoming funds. This was particularly true because city hall controlled many jobs, either directly or through subcontractors. All too often, those jobs rewarded loyalty and support in elections over talent or qualifications, fostering a sense of widespread clientelism.

Against all odds, Maria Joel would find in this complex political environment the will to overcome all the legal setbacks and continue with her mission.

19

Sink or Swim

IN those early months of 2008, as the children of Maria Joel lobbied their mother to quit the Sindicato and move on, the Workers' Party of Lula approached the union leader with the proposal that she be their candidate for mayor and compete for power against Olávio, Dr. Antônio, and Shirley Cristina.

The family had not had much luck in politics, so this proposal was initially received with great caution. Dezinho in 2000 and Eva in 2004 had failed to be elected council members, and Maria Joel posited that attempting to take control of city hall was likely to expose her to even greater dangers, as had happened with her husband. Also, although the Workers' Party was in power, it had never elected a mayor in Rondon.

But as she brooded over the issue, Maria Joel realized that the election was also a potential opportunity to show the townspeople that her family, too, was ambitious and politically influential and that she also had a development project for the town. A run for mayor would also give Maria Joel a leg up on a political career in the ranks of the Workers' Party. This was something that Dezinho had considered for himself before being slain.

In hindsight, one of Maria Joel's daughters would say, "My father had been killed because the fazendeiros didn't accept that an immigrant from Maranhão, in their view destined for nothing"—that is to say, destined to remain forever an underpaid laborer—"had taken the reins of the union, was trying to make a revolution, and was even attempting to become a politician, which involved more power. . . . When my

mother was given this opportunity," she went on, "we came to understand that, despite its inherent risks and challenges, this was the only
possible choice for us, enabling us to continue with our work and use
any political power we could gain to show people in Rondon that we
were not going to accept defeat" after the legal setbacks.

That view ultimately prevailed, and Maria Joel accepted the challenge, which, in the context of Wellington's escape and the dismissal of
Nunes's case, seemed like doubling down.

By March, Maria Joel announced that she was stepping down from
the presidency of the Sindicato to focus on her campaign. To the surprise of many, Maria Joel's running mate was Janilton Silva Rocha, a
son of Olávio. He represented the Brazilian Labor Party (PTB), which
had changed much from the Vargas era but now had formed a local
coalition with the Workers' Party. The campaign manager would say
that Janilton's profile allowed Maria Joel "to expand the political base,"
because he had the support of voters in the urban area, which represented some 70 percent of the electorate, while she was stronger in
the interior.[1] The widow would admit that accepting Janilton as her
running mate—a decision made by the party, she argued—wasn't easy.

"The early talks between us were very tense," Maria Joel declared,
"but Janilton kept saying, 'I'm my father's son, but I'm not like him.'"
Still, the odd political alliance seemed to work, as early polls indicated
that Maria Joel was in second place, only a few votes behind the front-
runner, Olávio.[2] "We believed she had a real shot at being elected,"
said a director of the Workers' Party in Rondon.[3] A local reporter who
covered the campaign said that Maria Joel had the full support of the
party and its local electorate, but many in Rondon doubted she could
receive enough votes to beat the fazendeiros.[4]

According to data from Brazil's top electoral court, Maria Joel was
the sole candidate in Rondon who didn't declare either campaign con-
tributions or official expenses.[5] Without capital of her own to invest—
her declaration of assets consisted of a home valued at $10,000—she
depended on the money transferred from the party's committee, which
was barely enough to organize rallies or produce any merchandise.
Shirley Cristina received some $100,000, with the PSDB being the

largest contributor and Josélio the most important individual donor, according to official reports.⁶ Dr. Antônio declared about $90,000, with most of the money coming from his own pocket; Olávio, $50,000. In May, a sudden twist in Nunes's case emboldened Maria Joel. Against all odds, the appeal filed by Batista had prevailed, and the court in Belém had overruled the previous ruling dismissing the case against Nunes, reactivating the prosecution against the fazendeiro. The former ruling, wrote one of the appellate judges, had ignored "several evidences" indicating that Nunes "was the instigator of the crime, through payment or promise of payment" to the triggerman.⁷ That didn't signal that Nunes had been found guilty of anything but instead meant that, contrary to district attorney Ferreira's motion to dismiss, there was reasonable evidence pointing to Nunes as the instigator of Dezinho's murder, and thus he should be tried by a jury.

Her spirits renewed, Maria Joel launched her campaign with a display of political muscle. On August 20, about six weeks before election day, the governor of Pará traveled to the little town to participate in a rally for Maria Joel's campaign, whose slogan was "Transform Rondon." Hundreds of townspeople dressed in red, the color of the Workers' Party, joined the event, where Maria Joel, all smiles, promised that her number one priority if she were elected mayor would be "to universalize social rights."⁸

Escorted by plainclothes officers and by her children, who devoted themselves to the campaign and proved once again that the unity of the family ultimately triumphed over individual choices and disagreements, Maria Joel crisscrossed the sprawling and overpopulated periphery. In neighborhoods like Jaderlândia or Recanto Azul, she met with the families of unemployed woodworkers, seasonal laborers, and rudderless young men. Those were areas ridden with illiteracy, malnutrition, abuse of children and women, prostitution, and unwanted teenage pregnancies.⁹

Land reform and upward mobility through family farming was Maria Joel's main proven achievement, but many in those neighborhoods, especially the generation born and raised in the city, refused to become smallholders, so her plan was to undertake a set of public works in order to provide jobs and stimulate the sluggish local econ-omy.[10] For teenagers who had dropped out of school, she wanted to fund training programs so they could learn a trade, get a regular in-come, build a family, and not resort to drug trafficking, a new curse across the Amazon. In the coming years, Rondon would score one of the worst per capita intentional homicide rates in the country due to gang violence.[11] "If there were land for all, things would be different. The new generations get lost in this perverse system that provides no opportunities. If the land had been distributed in a fair way, perhaps some drug trafficking would still exist, but not like today," Maria Joel would say about the issue.

Joélima would recall that in those neighborhoods of sandy streets and rickety cabins, located not far from where the family had first set-tled, acquaintances kept a photograph of Dezinho in their home, the framed picture hung on the wall as if her father were a family mem-ber. Her older sister Joelminha nonetheless had a down-to-earth view of things and wondered what those people would really do for them now that Maria Joel needed their support in the form of votes. Politics in that town were complicated, and the obvious qualities of the can-didate were often undercut by money—"by the power of money," as Joelminha put it. Reporters who covered the election would recall that it was commonplace for some candidates to send food baskets (*cesta básica*) to families living in poor neighborhoods to buy their support.

Thirty days ahead of the vote, a new poll showed Maria Joel in sec-ond place, "neck and neck with Olávio," according to the campaign manager. At the closing event of the campaign, several thousand peo-ple joined her for a march. Sweating and filled with optimism, Maria Joel believed she had a real shot. But the outcome would be much different. On October 5, 2008, Olávio was elected with 7,499 votes, 34.6 percent of the valid ballots.[12] Shirley Cristina was second with 6,571. Maria Joel came in third, with 3,878 votes, which was over ten times what Dezinho had received as a candidate for councilman but still accounted for only 18 percent of the electorate. Eva, running for a seat as councilwoman, also failed to be elected. The results were "a shock" for the family.

Looking back, Joelminha would say, "That night on the eve of voting day, many neighborhoods slept with flags of the Workers' Party on their windows but woke up flying another flag. That night, wherever there was a red flag, a vote was bought, money was offered."

Courts would prove that she wasn't completely wrong.

THE 2008 MAYORAL elections would be the most contested of Rondon's history up to that point, and only after almost two years of legal battles between Shirley Cristina and Olávio would the town's mayor finally be decided.

The litigation process had started in the wake of the election, when Shirley Cristina and a branch of the office of the state attorney that investigates crimes committed during campaigns (like illegal fundraising schemes or the use of public money to benefit a specific candidate, often incumbents) sued Olávio for "abuse of economic power" and for vote-buying, requesting that he be removed from office.[13]

In September 2010, an electorate court found Olávio guilty of illegally donating to voters some 1,500 gallons of oil during the campaign, a violation of the electoral law. He was removed from office, and although he appealed, the ruling was later upheld by Brazil's top electoral court.[14] Shirley Cristina, the candidate with the second-most votes, subsequently took office, becoming Rondon's first female mayor.

The removal of Olávio was hardly an isolated case in the country. According to O Globo, in the first quarter of 2009, when the elected mayors were sworn in, a mayor was ousted from office every sixteen hours.[15] "Voter buying dramatically influences municipal elections in Brazil," stated two researchers from MIT and the University of California who studied the country's voting system.[16] The prevalence of this offense, noted the experts, was especially high in the most undeveloped regions, like the Amazon and the northeastern states, where clientelism was so entrenched and people remained so poor and dependent that they voted for whoever bought them essential things like food or a botijão, a pressurized bottle of liquefied petroleum gas used to cook.[17] It costs about $20.

Shirley Cristina's court battle represented an unprecedented open confrontation with another political faction of fazendeiros and with Olávio's Party of the Brazilian Democratic Movement, one of the most influential despite its internal divisions. But for the daughter of Josélio, that mattered little, at least for the time being. By becoming mayor, she had succeeded in finding a platform to advance her political career, stalled until then. Now she aimed high, envisioning an eventual move to state legislator and perhaps even a leap into national politics if she did a good job as mayor.

Like her father, Shirley Cristina wasn't a great fan of the Sindicato or of the landless movement, which anticipated tough times for the union when she took office. "We have our rights violated when our lands are invaded," Shirley Cristina once complained during a meeting with one of Lula's ministers.[18] She believed that the occupation of ranches was the wrong way to promote agrarian reform and only created "an industry of the invasion" that didn't translate into agricultural output.[19]

There was little doubt that, as Shirley Cristina argued, agribusiness was now a driver of the national economy. During Lula's two terms, Brazilian agribusiness exports grew threefold, from $24 billion to $76 billion, making Brazil "an agricultural superpower," according to Colin Powell, then U.S. secretary of state.[20] But a nationwide agricultural and livestock census offered an interesting picture of the sector that didn't accord with the assumptions of Shirley Cristina and many other big landowners, who downplayed the role of smallholdings in Brazil's agriculture. Official data showed that some 4.4 million Brazilian family farms of less than 250 acres produced 70 percent of the food that Brazilians ate (rice, beans, corn, milk, pork, and poultry).[21] Yet those small farmers held only 21 percent of the country's agricultural lands, while an elite 1 percent of landholders owned a staggering 44 percent of the country's arable lands, or some 360 million acres.[22]

Another study found that smallholders were able to achieve such productivity despite the fact that state support for large-scale industrial agriculture—mostly subsidized loans absorbed by soybean, cattle, sugar, cotton, and corn producers in multi-acre fazendas—was seven times larger than the amount received by small landowners.[23] Later censuses would confirm these trends shaping Brazil's contemporary agriculture: a sharp and growing land inequality and concentration combined with robust state support for export-led producers of monocultures.[24]

After her political defeat, Maria Joel returned to the presidency of the Sindicato, but she was no longer in charge of the union by the time Shirley Cristina became mayor. On March 28, 2009, Maria Joel was elected coordinator of FETAGRI in southeastern Pará and stepped down from the presidency of the union.[25] The new position was based in Marabá—a ninety-minute drive from Rondon—and involved coordinating some nineteen farmworkers' unions spread across the region.[26] For her, moving to Marabá meant putting some distance between herself and the city that had so marked her life. Some advised Maria Joel to take a break from her quest for justice and social rights. Away from her painful memories, they argued, she could once again embrace the joy of life.

Part IV

The Downfall

20

The Widow Must Fall

WITH Maria Joel's new job at FETAGRI, the Sindicato was confronted with the challenge of filling the vacuum she left behind. Initially, her vice president took over, but ultimately Maria's Joel's successor would be her sister Eva, who won the union's next election, held in February 2011. However, Eva soon had to resign due to poor health, and Zudemir, Eva's former vice president, found herself commanding the Sindicato during the years Shirley Cristina was the town's mayor.

"For us, it wasn't an easy time," she recalled. "Could anyone ever imagine us, from the union, sitting to talk with her, who doesn't like rural poor and was the daughter of one of the men who first chased Dezinho with pistoleiros? In the end, we had meetings with her," she said. "But it wasn't easy."

Like her predecessors, Zudemir received death threats, and one day a man attempted to break into her home with the apparent intention of murdering her. She reported the case and was placed under police protection for eight months.[1] Two of Zudemir's daughters decided to emigrate to the United States through smugglers who helped people cross the U.S. border.[2] From Florida, one of her daughters would later say that she had finally escaped from a town "where people are killed for almost no reason." Eventually, Zudemir, emotionally traumatized, would also quit the union and leave Pará.

Meanwhile, Maria Joel got her first chance at FETAGRI to prove that she could deliver at her new job. In mid-2011, FETAGRI, in alliance with the MST and other landless organizations, launched a protest action in Marabá against the government of President Dilma Rousseff, who had been elected to succeed Lula months earlier. In southern Pará, the strategy of occupying land to push the authorities to redistribute it among the rural poor had resulted in 70,000 families receiving plots in some 500 settlements.[3] "There are more settlements here than throughout [the rest of] Brazil," a leader of FETAGRI would say.[4] However, thousands were still occupying fazendas of dubious provenance, and the movement was critical of Rousseff's government, accusing it of ignoring their demands while grilagem continued unabated. Funds to grant subsidized loans to family farmers—a key issue to improve the productivity of those already settled—also appeared to be a low priority for Rousseff, an economist and technocrat less sensitive than Lula to the requests of left-leaning rural organizations.[5]

Early in May, hundreds of demonstrators launched the mobilization in Marabá by setting up a protest camp near the INCRA offices. Maria Joel was one of the leaders and was charged with negotiating their demands with the officials of the federal colonization agency. As the talks concerning the possibility of devoting more funds to agrarian reform unfolded, taking weeks because the negotiations involved the green light of top authorities in Brasília, the protest gained force with the continuous arrival of peasants from across the region.

Then news of a double homicide sent a chill through the protesters. A couple of prominent Amazon activists—José Cláudio Ribeiro da Silva and his wife, Maria do Espírito Santo, both in their fifties—had been murdered in northern Pará on May 24 by gunmen riding a motorbike. Hiding in the bushes, the pistoleiros had shot them in a forested area before cutting off José Cláudio's ear as proof of his assassination.

"The grisly death of the couple," wrote a *Vice* reporter, "harked back to other violent moments in the history of the Amazon," such as the murders of Chico Mendes or Dorothy Stang.[6] The case quickly gained public attention because, at least since 2001, both environmental activists had been receiving death threats from illegal loggers trespassing on the reserve where the couple of campaigners lived. José Cláudio had even "predicted his own death six months before it happened," wrote Tom Phillips in *The Guardian*, referring to a public talk in which José Cláudio had said, "I could get a bullet in my head at any moment."[7]

President Rousseff instructed her justice minister, José Eduardo Cardozo, to order the federal police to carry out a full investigation.[8] The assassination also reverberated in the Brazilian congress, where a comprehensive and controversial piece of legislation called the Forest Code (Código Florestal) was being discussed to regulate the use of land and unify environmental requirements for farmers operating across the country.[9] A congressman from the Green Party took the floor to honor José Cláudio and Maria, "those heroes who were assassinated."[10] Lawmakers from the agribusiness caucus, as well as cattlemen and soy producers following the debates from the gallery in the congress, responded with jeers.

The double homicide was splashed across the front pages of the national newspapers and was discussed extensively on TV shows, but the notoriety of the case didn't prevent violence from erupting just as the Forest Code was entering the crucial phase of discussions in the Brazilian congress. In fact, the killing of José Cláudio and Maria would be just the first of a series of attacks that some understood as a concerted terror campaign to influence the outcome of the new law.[11] (After long debate, the Forest Code would eventually favor environmental offenders by declaring an amnesty on illegal deforestation and land clearings committed before 2008.)[12]

Just three days after the murders of José Cláudio and Maria, on May 27, another threatened land activist, Adelino Ramos, was murdered while he and his family were selling their products in a food market in the Amazonian state of Rondônia, near the border with Bolivia.[13] That same day, another union leader, this time in Maranhão, barely escaped an assassination attempt by triggermen.[14] Twenty-four hours later, the dead body of another peasant was found only seven kilometers from where José Cláudio and Maria had been executed.[15] The man was said to be the sole eyewitness to the crime, having accidentally seen the faces of the murderers, and so it appeared to be a crime intended to "burn the archives."[16]

The terror campaign spread fast. The offices of the CPT were flooded with anonymous calls, threatening priests known for their advocacy for environmental protection and peasants' rights. "I'm calling to tell you," said a menacing voice over the phone to a CPT member in Acre, another Amazonian state, on June 3, "that you better warn your friends at the CPT that people died in Pará and Rondônia, and now it is the time for Acre and Amazonas [states]."[17] Just one day earlier, hooded thugs had stopped a car near Marabá to execute another peasant, subsequently cutting off his ear, and about a week later, yet another farmworker was murdered with two rifle shots aimed at his ear.[18] The violence seemed uncontrollable, and it frightened and outraged demonstrators camped near the INCRA offices. The leaders decided then to expand the actions of the protest movement to make sure their voices were heard.

By mid-June, with the leaders demanding that the president of INCRA come to Marabá to negotiate and that President Rousseff take determined steps to protect the lives of activists, Maria Joel and her colleagues engaged in acts of civil disobedience. They entered INCRA premises—a fenced complex with several one-story offices and a large grassy yard—and set up a new camp. They also cut off avenues and blocked highways, preventing vehicles from accessing a strategic bridge over the Tocantins River that connects the south and north of eastern Pará. Five to eight thousand activists and peasants caused "chaos" in the region, according to the media.[19] (With forty-six days of uninterrupted demonstrations, this would be the longest protest movement staged by farmworkers in Brazil in 2011.)[20]

"We won't move away from here until some of our demands have been met," Maria Joel bellowed at INCRA officials. As the mobilization escalated, the policeman protecting Maria Joel was growing anxious. If malefactors were planning to execute some iconic Amazon-based activists, she was a potential target, the officer reasoned. Indeed, an uncontrolled mob of protesters might be the perfect context for a pistoleiro to get close to her without being noticed. The officer didn't forget that Wellington remained at large.

One Friday, with protesters taking a break before engaging in a new week of demonstrations, Maria Joel called Joelson to inform him that she was heading home for the weekend.[21]

"Mom," Joelson answered, his soft voice sounding distressed, "don't come home."

"What's going on?" she inquired.

"We didn't want to tell you so you wouldn't worry," Joelson replied, "but we're being watched—for some days already. There are two guys in front of our home." He explained that he had monitored the presence of the strangers through the video system mounted on top of the fourteen-foot walls encircling the home, where an electrified fence had been installed to prevent criminals from breaking in.

Maria Joel hung up, a knot in her stomach, and informed the policeman protecting her. "Let's go to a hotel," he proposed, and they checked into a small inn in Marabá with the intention of being out of public sight for twenty-four to forty-eight hours, after which they would decide what to do next. Shortly afterward, Maria Joel was trying to calm herself in her room when she was startled by a knock on the door.

"Maria Joel," the policeman muttered through the door, "quick, let's go! They've discovered where we are staying!"

The officer said he had noticed the presence of suspicious men in the hotel lobby who said they were looking for Maria Joel. He determined that the place was no longer safe and took her out through a back door leading to the garage. "Let's go to Rondon," the officer said. Maria Joel agreed.

They jumped in the car and took a branch of the Trans-Amazonian Highway that connected Marabá with the BR-222, then headed east. Night engulfed the rolling highway, which was flanked by fazendas. The car sped along the deserted road, the atmosphere in the vehicle chilly, a stony silence broken only by the roar of the engine. Maria Joel called her son again. "I'm coming back home," she said. "We're being followed."

After about an hour, they passed the Fazenda Serra Morena of Josélio, the meat plant of Nunes, and several dirt roads leading to the interior—to Vila Gavião and the settlements. The vehicle finally reached the western entrance of the city, where Maria Joel called Joelson for the third time.

"Mom, they're still here," he reported. "Wait for a while before coming. I'll text you."

Maria Joel and the policeman spent about an hour moving here and there through the town, not stopping the vehicle for longer than a few minutes at a time. Rondon was calm and quiet except for some young people drinking on the terraces of bars. Joelson finally called back. "Come now, Mom. They've left."

The car approached the house, and Joelson opened the steel garage door using the electric system. The rollers screeched as the mechanism slowly lifted the heavy sliding door. Maria Joel wondered if she should get out and run toward the house or wait until the garage door was fully opened and the car could move into the garage. She decided to get out; Joelson came toward her.

"You're home. It's okay," he said, hugging his anxious mother.

After a few days, Maria Joel, proving yet again that she would not surrender to fear, returned to Marabá. The men skulking around the house seemed to have vanished, and she still had to take part in the negotiations with INCRA—which had yielded to the protests and had announced that it would increase the budget for family farmers fivefold.[22]

A few days later, as Maria Joel was preparing to give a speech at a public square, the policeman approached her, clearly unnerved. "We're not safe here," he announced. "We're being watched again. We need to leave immediately."

Batista, who was present that day with other colleagues of the CPT, had a talk with the officer. The man was on edge. He told Batista that he suspected the murder of Maria Joel could be imminent, as he believed that hired gunmen had infiltrated the protest camp. The officer retrieved a semiautomatic rifle and got Maria Joel settled in a pickup before driving again to a different hotel. Suspecting they were being followed, the officer drove recklessly. Finally, Maria Joel called a member of congress named Beto Faro—a former FETAGRI leader and acquaintance of Dezinho's—to ask for help. The Workers' Party lawmaker immediately offered his home in Brasília as a place for her to stay.[23] The widow flew to the capital city that night.

One morning a few days later, someone called Faro's home on the landline. The lawmaker's maid answered.

"Hello?"

"Maria Joel?" said a man.

"No, this is not her. Who's speaking?"

"Tell her that we know where she's staying."

With that, the line went dead.

By midnight, Maria Joel had relocated to another apartment. A few days later, she flew to Rio de Janeiro to go deeper into hiding. Fearing her children's phones could be tapped, she didn't even dare to call them.

THE WIDOW WOULD never find out who was behind the death threats or what the real intention of the call had been. Was it related to her involvement in the protests in Marabá and the Forest Code, or was it connected to the prosecution of Nunes and the possibility that the rancher would be put in pretrial detention? The arrest of the King of Wood was now a real possibility because he had lost a series of appeals in state courts to have his case dismissed. By the time of the protests in Marabá, Nunes's defense attorneys were appealing to the Supreme Court to keep the businessman out of jail.[24] Still, Maria Joel had no real evidence to support her supposition that the threats might have originated in Nunes's entourage.

Whatever the case, the dread left scars on her. Now in her fifties, Maria Joel felt physically and emotionally exhausted. For almost two decades, she had been afraid for her life or for the lives of her family members, and her escape from Marabá to Brasília and later to Rio de Janeiro harked back to the unpleasant memories of Dezinho's hurried departures—which in the end hadn't saved him. "It was as if we were reliving the same situation as my father—the threats, the impunity, the fear of death. She didn't have enough strength to come back to work; my mother was finding it really hard to deal with the terror campaign," Joélima would say.

In Rio, Maria Joel stayed with friends and colleagues of the movement, meeting lawyers and activists who had helped her with the complaint filed at the Inter-American Commission on Human Rights. Father Rezende realized that she needed help. He reached out again to actresses Pitanga and Sabatella to set up a public support campaign. Several actors gathered in a film studio and recorded a series of interviews for a 25-minute documentary titled *SMS for Brasília, SOS for Dona Joelma* [Maria Joel].[25] The video, in which a crestfallen Maria Joel appeared surrounded by some of the country's top celebrities, was published on social media. *Fantástico,* one of the most popular and highly watched TV news programs in Brazil, broadcasted an excerpt of it nationwide along with an interview with the widow.[26] She was in tears and appeared fragile.

One of the big names of cinema that would speak for Maria Joel was Wagner Maniçoba de Moura, who would later earn his ticket to global celebrity with his interpretation of drug lord Pablo Escobar in the Netflix series *Narcos.* The son of an army man, Moura was born in Bahia, in an interior town called Rodelas, where people—he would recall later with indignation—saw it as "normal" that farmhands worked in the fields in exchange for food.[27] Moura campaigned for Maria Joel in order "to protect her and raise awareness about the murders of rural workers, activists, and human rights defenders in Pará and other regions of deep Brazil, where representatives of the State and the Law are often the same criminals who enslave and kill the most vulnerable," he would say.[28] Eventually, Moura would become an ambassador for the International Labor Organization in the fight to eradicate debt bondage and child labor.[29]

The public campaign worked to attract the attention of the authorities, and on July 7, a meeting was held with Minister Cardozo in Brasília to request urgent measures from the federal government.[30] A close aid to Rousseff, whom Cardozo would defend during his later impeachment process, the minister of justice was nonetheless noncommittal about taking specific actions to ease Maria Joel's plight. In his meeting with the widow, also attended by Joélima and a dozen activists and federal lawmakers, Cardozo argued that there was "no point in seeking only the punishment" of those who poisoned her life.[31] Instead, he said that he would tackle the "cause" of the problem, which, in his view, was the "land issue." Cardozo admitted that it was a "historical problem . . . not easy to solve."

Maria Joel listened carefully, and then she took the floor. "Not long ago," she said, her soft voice gaining strength, her gestures denoting profound disappointment, "my son was told in town, 'Hey, your mother is going to die, she must die! Because she is hampering the [economic] prospects of Rondon do Pará.'" Cardozo and the others listened in silence. "Mr. Minister, for a son to hear that after his father has already been murdered, and now his mother is on the front line . . . What is this?!" Maria Joel met with other top aides to President Rousseff during that visit to the capital, but she was skeptical that she would find any real help. Still, having the support of friends and celebrities, which gave her access to federal authorities and the national media, reminded her that she wasn't alone and, equally crucial, that her story hadn't been forgotten.

Back in Rondon, the widow, reinvigorated, organized a series of protest acts with the CPT on the eleventh anniversary of Dezinho's death. Pitanga, Sabatella, and other actors and lawmakers participated in a march on November 26.[32] Dressed in orange T-shirts, the celebrities walked holding hands with Maria Joel and her children, who wore white T-shirts stamped with Dezinho's picture. Hundreds gathered to see the celebrities who, one by one, took the stage to praise Maria Joel. Mayor Shirley Cristina didn't meet the delegation.

Representatives of both the federal and the state governments also showed up to deliver, for the first time, a formal apology for the murder of Dezinho. The admission of guilt was part of a broader agreement signed by Maria Joel, her lawyers, and the Brazilian authorities within the framework of the complaint lodged before the Inter-American Commission on Human Rights. That organization had decided three years earlier that the case of Dezinho was "admissible with respect to potential violations of Articles 5 and 16" of the American Convention on Human Rights—that is, that the dead activist's "right to humane treatment" and "freedom of association" might have been violated.[33]

The Brazilian government had until then contended that there was no reason for an international body to intervene because the case was being properly handled by the Pará legal system.[34] But when the commission signaled that it might open a case because rights had potentially been violated, the government, anticipating it might lose the legal battle, proposed a friendly settlement. At the end of 2010, Maria Joel negotiated a comprehensive deal in which the family would receive a pecuniary reparation of 50,000 reais, some $25,000. The widow would also be entitled to a monthly pension of the same amount as the official minimum wage, which varies every year but at the time was about $350.[35]

"We could have asked for more money," she would say. "But we earmarked a series of social projects for the INCRA settlements."

In fact, most of the family's requests were aimed at improving the living and working conditions of the community. Maria Joel and her children wanted the government to build artesian wells, bring electricity to the houses of peasants, and improve the roads leading to the settlements.[36] The family also demanded an investigation into the dubious origin of four fazendas that remained occupied but had not yet been seized. Last, Maria Joel wanted the Brazilian state to present a formal apology for her husband's death in a public ceremony and to build a monument to Dezinho. As was true of other demands in the agreement, the commitment to build a monument would ultimately never be fulfilled.

21

"Load the Trucks"

THE day Brazil formally admitted responsibility for the murder of Dezinho, few people knew that Rondon do Pará had recently been the object of an undercover intelligence-gathering investigation into Nunes, who had obtained a habeas corpus from the Supreme Court to avoid pretrial detention.[1]

The mission, carried out by two officers of the federal police, aimed to create an inventory of the properties and assets of the wealthiest man in town. The person who had ordered the investigation was Jônatas dos Santos Andrade, a robust, five-foot eight federal judge with thick lips and a laconic style. He had nothing to do with Dezinho's case.

Judge Jônatas worked in a Marabá-based court that dealt with labor disputes. According to him, Nunes had more than 500 outstanding cases brought by former employees of his ranches and sawmills.[2] The King of Wood owed 3.26 million reais (some $1.9 million dollars) in unpaid wages, compensation, social security, and other benefits to former employees, according to court files.[3]

Judge Jônatas was sick of both Nunes's repeated violations of the labor law and the rancher's dilatory tactics to dodge his financial obligations. He would say he had given Nunes many chances to settle up. In addition to his formal requests, the judge recalled having met the fazendeiro "two or three times" in his office to talk about the issue, giving him six months to find a solution. Initially, the rancher offered assets that were not owned by him, said Jônatas, but later, he simply ignored the judge's rulings.

Labor rights in Brazil are governed by a complex series of laws and regulations, but fundamental rights like the maximum number of working hours, applicable annual minimum wages, and the right to join a union are contained in articles 7 and 8 of the constitution. If an employer breaks the law and fails to pay what is owed to an employee even after a judge has ruled he must, the labor courts can enforce the sentence by distraining capital or assets, often from bank accounts, so plaintiffs can be compensated.

That was precisely what Judge Jônatas had attempted after Nunes ignored his warnings. But then he found out that balances in Nunes's accounts were extraordinarily low while the real estate (ranches and sawmills) and chattels (vehicles, machinery, livestock) thought to be his were nominally owned by third parties, thus confirming the judge's suspicions that Nunes had resorted to front men. Some fazendas and trucks had been transferred to Nunes's children when they were as young as ten years old, according to court files.[4] Police databases also showed that assets were held under the name of his second wife, his sister-in-law, and a net of acquaintances who had worked in the past with the King of Wood.[5] Land, the core of Nunes's wealth, presented an even larger legal challenge to confiscate.

"We were not sure," said an assistant of Judge Jônatas, "whether Nunes was the legitimate owner of the land that he claimed to own."[6]

Prosecutors had in fact begun to question Nunes's ownership of the Fazenda Lacy, the spread Dezinho had believed to be the product of grilagem. In 2009, Pará's attorney general's office, positing that a great share of the 270,000 acres of the fazenda was actually owned by the state, had pressed charges against Nunes for land-grabbing and defor-estation.[7] Prosecutors accused him of participating in a series of scams involving forgery and even of setting fire to a cartório near Rondon to burn archives—this time, literally—so forged deeds could replace the incinerated ones and subsequently be registered in another notary office.[8] The case involved high financial risks for the rancher, with prosecutors requesting a fine of $14 million. If found guilty, Nunes would also lose the Lacy, which was his biggest moneymaker through the sale of timber to domestic and foreign markets including the United

States, Germany, Spain, and Denmark.[9] Refuting the claims, Nunes resorted to the same strategy he'd always followed: hire expensive attorneys to present floods of documents and a series of appeals, effectively stalling the case for years in higher courts. Meanwhile, he continued conducting business.

Wondering how he could target him, Judge Jônatas finally resolved to instruct the federal police to conduct an investigation of Nunes's possessions. The agents, acknowledging Nunes's clout over the economy of Rondon, employed a ruse to carry out the mission. They showed up at businesses they believed were controlled by Nunes with the cover that they were looking for fugitives from justice.[10] For days, they interviewed managers and overseers, gathering from the sources they spoke to (who had no idea of their real purpose) such background information as who owned that specific sawmill, ranch, and shop.

Containing forty-four pages of locations, photographs, GPS coordinates, names, and taxpayers' and companies' identification numbers as well as the plate numbers of several dozen vehicles (from $100,000 big rigs to $1,000 dilapidated sedans), the report of the federal police detailing the net of corporations and assets linked to Nunes would be crucial for Judge Jônatas to plan an impactful seizure operation intended to show the King of Wood that, in his court, breaking the law had consequences. The formula found by the judge to make the "rule of law prevail" in that corner of the Amazon forest was to go after vehicles and cattle.

ON THE MORNING of February 16, 2012, a leaden sky hung over Rondon do Pará. Clouds had gathered, and the vivid color of the jungle was muted.

A team of over sixty people had arrived in town the previous night and checked in at a hotel. They were federal police, staff of the federal labor court of Judge Jônatas, and a group of subcontracted truck drivers and cowhands. Even an auctioneer, a *leiloeiro*, had been hired for the occasion. Ten large livestock trailers used to transport cattle as well as five rollback tow trucks were also brought by order of the judge, who had also chosen to join the group himself.[11]

The party woke early, around 3:00 a.m., although the tension had prevented some from sleeping at all that night. Before daybreak, the group split into three squads, heading simultaneously toward either a sawmill or a fazenda mentioned in the police report. Two patrols were stationed at both the entrance to and the exit from Rondon to control the movement of vehicles. In the meantime, Judge Jônatas remained in town to coordinate the mission and make sure everything went smoothly. That day, he was dressed in a pair of light blue jeans and a black sport coat over an untucked white shirt, and his lank jet-black hair covered the nape of his neck and part of his ears.

An official from the labor court named Frank was leading one of the teams, which was comprised of truck drivers and a couple of federal officers. Frank had a warrant to seize every vehicle in the parking lot of one of Nunes's ranches with a plate that matched those in the police report.[12] The spot wasn't far from the meat plant of the fazendeiro.

A second team was commanded by a woman named Alma, who had almost two decades of work experience confiscating assets. Alma and her colleagues headed toward a sawmill located within the fenced boundaries of another fazenda of Nunes. She was also targeting vehicles, but hers was a more hazardous mission, because the location was remote. As Alma would later admit, it was an "extremely risky job, because there was no [phone] signal connection or safety" in the place, and so she had been instructed by Judge Jônatas "to leave the ranch before 3:00 p.m., before it got dark, so as to be exposed to the fewest risks possible."[13]

A man named Jarbas commanded a third group on a ranch where some of the best of Nunes's cows were pastured. The rancher's herd, estimated at several thousand animals, was composed of two different breeds. One was the Nelore, which, according to a Brazilian cattlemen's exporting association, "stands out for the high reproductive capacity of its dams."[14] The second was the crossbred Tabapuã, which "has a hornless character, which favors its handling and weight gain." Accompanied by the cowhands brought by Judge Jônatas, Jarbas was responsible for seizing the cattle.

Frank was the first to reach the target. He presented himself to the security staff at the gate and announced his intentions. He and his men accessed the enclosed area without trouble and headed toward the parking lot. Everything seemed to be proceeding just fine.

Alma used GPS coordinates to reach her spot, along with eight other people, including two federal officers holding semiautomatic rifles and wearing black bulletproof vests. A guard opened the gate of the ranch, and Alma immediately felt unnerved. "In that location, there were no fewer than 160 men and apparently not a single woman other than me," Alma would recall. Besides, the overseer was reluctant to obey her orders. The police finally instructed the man to put aside his macho attitudes and follow Alma's instructions or face being arrested. She then told the overseer that she would be taking all the big rig trucks, livestock trailers, pickups, sedans—everything—matching the data and description previously provided by the police. Soon the work began.

The real problem surfaced at the third location, where the cattle had to be seized. In a commercial beef herd, especially one made up of different breeds like Nunes's, the animals don't all have the same market value. The price of a cow is established by multiple factors like age, size, weight, body proportions, and the investment possibilities that it offers. For instance, a specimen of pedigreed breed and high-quality semen or ova is priced much higher than an average dairy producer.

The sale of embryos is a booming business in Brazil.[15] "Elite cattle" are often sold in exclusive auctions covered live by specialty agricultural cable channels and attended by wealthy fazendeiros looking to improve their herds through genetic manipulation.[16] Ranchers have compared some cows to pieces of art for both their outstanding beauty and the potential return on investment, and the market prices of some cattle occasionally reach astonishing heights.

In 2010, for instance, one Nelore specimen—the snow-white, two-year-old, 2,100-pound Parla FIV AJJ—was sold for 3.8 million reais, over $2 million at the time.[17] This majestic cow, a superstar producer, would almost double her market value a few years later and become the most expensive cow in Brazil, if not on the whole planet.[18] The breeder and seller, a well-known family of fazendeiros operating in São Paulo and

Mato Grosso, would later be suspected by top IBAMA officials of leading a scheme involving grilagem, massive deforestation, slave labor, and money laundering along the BR-163, the "soy highway."[19] The media would refer to this family as the biggest destroyer of the Amazon jungle of their time, although the family denies any wrongdoing.[20] Federal prosecutors pressed charges against several members of the family for environmental damage, but judges later dismissed some of the cases.[21] As of this writing, the lawsuit is ongoing.

Nunes didn't have million-dollar cows, but he had spent a lot of money to improve the genetics of his herd and had even built a laboratory to perform in vitro fertilization.[22] Therefore, as the officer commanding the seizure of the cattle realized, not all the specimens had the same value. The question was which cows should be confiscated to cover the debts Nunes had with his former employees.

"We gave a deadline to the ranch's staff to hand over documents on the specifics of the cattle so we could make a confiscation as fair as possible," explained Frank. "We waited for two, three hours, but they didn't handle it." The ranch's staff, officials concluded, were attempting to delay or even hinder the operation.

By mid-afternoon, Frank, Alma, and Jarbas reported what was going on to Judge Jônatas. Much of the mission had progressed successfully, with a caravan of confiscated trucks and vehicles emerging from the jungle and reaching Rondon do Pará, but the officials were still struggling to confiscate the cattle. "I understood that things were not going well. At that point, we hadn't loaded a single cow into the trucks," the judge would say.

Born in the city of Santarém, on the shores of the Amazon and the Tapajos Rivers, Jônatas was a man of progressive ideas who believed in social and environmental justice. Before moving to Marabá, he had worked in Parauapebas, the mining boomtown that had developed with the Carajás mine. There he had gained a reputation as a resolute judge when he ruled against Vale, one of Brazil's largest companies, for flouting the labor law in its operations to mine iron ore.[23]

"My determination to resolve cases causes me a lot of conflict. I've always been determined to counter injustice," Jônatas would say of himself. His tenacious and disciplined character had been molded, he would later admit, during the years he had spent in the army. "I spent the day in the lush jungle, marching, and then at night, I studied law. This experience trained me to endure psychologically straining situations."

That afternoon of February 16, 2012, when he heard from his staff that one part of the operation was at risk of failing, Judge Jônatas decided to hit the road. Once in the fazenda, he told his crew to proceed with the confiscation without knowing the specific market value of each cow.

"Load the trucks," he said.

Nunes's staff realized that the order would inflict great economic damage on the businessman, as court officials would now simply attribute an average weight and price to each animal, regardless of its characteristics. The officials calculated that to reach at auction an amount close to the sum Nunes owed, some 900 specimens should be seized. Later, when he was asked whether it had been an unfair decision, Judge Jônatas would reply, "You act carefully only when people cooperate with the law."

Shortly afterward, the trailers were loaded, and the cattle set off for Rondon, posing a new challenge for the officials: where to store the confiscated assets, which would end up clogging Rondon's city center and main avenues. Alma suggested that the vehicles she had previously seized be transported to Marabá, but what to do with the cattle? Caged in the trailers and without proper food and water, the animals risked dying in the tropical heat. As an official recalled, "Cattle eat, shit, and graze, and the judicial system doesn't have a framework to sustain this for a long time. So we had to sell Nunes's cattle right away, in Rondon." Judge Jônatas had anticipated part of the problem, which explained why he'd hired cowhands and an auctioneer with a portfolio of potential buyers. Still, the cattle had to be held somewhere while the auction, organized over the phone, was completed and the cows finally reached the fazendas of their new owners.

One facility available was the exhibition center where ExpoRondon, the annual rodeo and livestock fair the Barros family had promoted throughout the decades, was organized. The venue—a large enclosed open-air area with corrals and lawns—was owned by a union of fazendeiros presided over by João Malcher Dias Neto, Josélio's son-in-law and the husband of Mayor Shirley Cristina. Judge Jônatas contacted him about the potential use of the space, but Malcher refused to cooperate, arguing that the venue "was owned by the union, and the union worked to preserve the interests" of fazendeiros.[24] It wasn't the first time Malcher had sided with Nunes; previously, he'd signed a flattering statement about the King of Wood in which he claimed that Nunes supported low-income families.[25] The document was used by Nunes's lawyers in his defense in the case of Dezinho's murder.

In the meantime, according to the account of Judge Jônatas, Mayor Shirley Cristina showed up. "She was very angry," the judge recalled. "Why is this happening?" she said, according to the judge's account. "I'm also a rural producer, and I defend the interests of the fazendeiros in Rondon," the mayor argued, also taking the side of Nunes.

The judge reacted calmly but firmly. "Do you also owe 3.5 million reais in unpaid debts?" he responded. Promptly, he signaled the federal officers with him to come closer, sending a veiled warning to Josélio's daughter—either she would calm down or she would be arrested. Shirley Cristina, according to Judge Jônatas's later account, toned down her bluster, but she still tried to change the fate of the cattle by calling the state governor, who was from her same party. Judge Jônatas responded by calling the president of Pará's judicial system to inform him of the ensuing operation and its legal foundations. After a few minutes, visibly frustrated because she had apparently been defeated in the power struggle, Shirley Cristina left.

That wouldn't be the only time the mayor would upset members of the justice with her pushy ways. At that point she was involved in a mounting legal fray with Gabriel Costa Ribeiro, the newly appointed district judge of Rondon. Judge Gabriel, then in his mid-thirties and working for the state courts, complained that Shirley Cristina behaved as if she wanted to hinder his work by reducing the number of public servants available at the courthouse. The Brazilian bar association supported Judge Gabriel, releasing a harsh statement in which it accused Shirley Cristina of "hindering the functioning of the local courthouse" and behaving in ways "that return our society to the stone age, when justice was made with one's own hands."[26] The mayor and the young judge would clash further over the following months.

After the unpleasant encounter with Josélio's daughter and son-in-law, Judge Jônatas issued an order requisitioning the exhibition center.[27] Nunes, who hadn't shown up, attempted a last-minute delay by sending "several boxes with documents" apparently containing the specifics of the cows. But Judge Jônatas refused to halt the ongoing auctions. "This comes too late," he said to one of Nunes's men.

Some 900 beasts were sold on February 16 and the morning of February 17. The largest buyer of the cattle, based in Rio de Janeiro, purchased 733 bovines. The second largest bidder, who got 159, was a man living in Rondon, which triggered speculations that Nunes, in a desperate attempt to minimize his losses, might have bought back his own livestock through a middleman.[28] Some 900,000 reais, or about $450,000, were raised.[29] In the ensuing months, nineteen vehicles— including Volvo, Mercedes-Benz, Volkswagen, and Ford heavy trucks as well as Mitsubishi, Toyota, and Chevrolet pickups—were sold for 2.04 million reais, about a million dollars.[30]

The day after the seizure, Nunes's attorneys lodged an appeal to reverse the confiscation, arguing that Judge Jônatas had breached the principle of impartiality.[31] The proceedings, the attorneys wrote, were extravagant and had caused Nunes disproportionate economic losses.[32] They argued that "genetically improved" cows valued at $7,500 in the market had been sold at the auction "by weight" and, therefore, for a "vile price."[33] The real value of the livestock seized was more than twice the sum raised, or about a million dollars, and the value given to the vehicles was also heavily underestimated, according to the attorneys.[34] Higher courts would dismiss the appeals.[35]

The raid was widely reported by the media. "Labor Court Confiscates Assets of the [Man] Accused of Killing Union Leader" wrote *O Liberal*, referring to Nunes.[36] The operation would also cause a furor in Rondon. "It was a shock," Frank, the officer of the labor court, would recall later. "It left a mark on the townspeople. It was like the arrest of

President Lula [in 2018]: Nobody believed it could happen, and then it did."[37]

Court documents showed that despite the costly distrainment, Nunes would continue to be suspected of breaking labor laws. Only a year later, a team of inspectors from the Ministry of Labor raided several of his ranches and rescued nine people, one of them over sixty-five years old, who had been working in conditions that Brazilian law considered analogous to modern slavery. Some employees, prosecutors reported, worked up to 450 hours per month in shifts from 4:00 a.m. to 8:00 p.m., "a worktime that exceeded double the worktime established by the constitution."[38] Besides, employees were deprived of their right to paid overtime and paid vacation time.[39] "Landowner Will Be Sued for Slavery," the press wrote.[40] Nunes's defense attorneys, through appeals, would succeed in stalling these cases for years.[41] Still, many other former workers sued Nunes and won their cases before the labor courts.[42]

IN 2012, SOME 1,600 people working in slave-like conditions across Brazil were freed by federal task forces, and the data showed that Pará was the state most affected, with almost a third of all cases.[43]

Compared to the wild years of the opening up of the Amazon, when squads of anonymous laborers were brought by *gatos* or middlemen to clear the jungle, both the extent and the severity of the abuses committed against the temporary farmhands had substantially decreased. But debt bondage had never been completely eradicated, and the pervasiveness of the exploitation led to a larger question: Why did wealthy landowners like Nunes, who had amassed a fortune on the frontier, refuse to stick to the law and pay salaries that often consisted of a mere few hundred dollars? Why did compliance with the norms and even the most fundamental human rights pose such a problem in a new context in which a modern and prosperous Brazil exported tens of billions in agribusiness products across the globe?

"Individuals or corporations don't have a good or bad character," explained Judge Jônatas, who received the 2012 Human Rights Prize of the presidency of the republic, the same award Maria Joel had received

six years prior, for his quest against debt bondage.[44] "This," Judge Jônatas would note, referring to modern slavery, "is a question of profitability, an economic issue."

In a testimony to the labor justice, a former Nunes employee and
plaintiff estimated that "due to his violent character," only one or two
out of ten of Nunes's former workers ever decided to initiate legal actions against him.[45] In confidence, one of Nunes's defense attorneys
would also admit to officials that the "culture" or standard practice in
his companies was to give final paychecks or due compensation only if
judges ruled it to be necessary, which could take years.[46] Still, the judge
presiding over the case had to have a strong sense of duty—strong
enough to resist all kinds of pressure, including threats in a region
where violence was widespread.

A lawyer in Marabá named Pedro Miranda who specialized in defending laborers and farmhands would offer an insightful example to
illustrate this enduring business culture.[47] "One day, the owner of a
sawmill with fifty pending cases came to my office, and I asked him,
'Hey, why you don't just do things in the right way?'" (He was not referring to Nunes.) "He replied: 'Look, I have a workforce of about a
thousand, but only fifty of them have sued me. The others are in my
pocket." The sawmill owner, probably with the intention of intimidating the lawyer, who was representing some of his former employees,
added, "Years before, when I had a labor lawsuit, I killed either the
worker or the lawyer who represented him. But now I don't want to
kill anymore."

Miranda also represented Nunes's former woodworkers and would
have a chance to chat with the fazendeiro after the raid carried out
by Judge Jônatas. One day, unexpectedly, he showed up in Miranda's
office "accompanied by a gunman," according to the attorney. The lawyer, thinking that nothing would happen to him as long as he remained
at his office, where some other staff also worked, acquiesced to the
meeting.

"I have heard that you are the demon," Nunes bellowed, according
to his account.

"I've heard the same about you," answered the lawyer to show he wouldn't be intimidated. Months earlier, Miranda would say, some of his clients, worried about the implications for their lives if they confronted Nunes in court, had decided not to pursue litigation against him.[48]

The rancher told Miranda that he was there to inform him that one of his clients—a former employee who had sued Nunes for arrears totaling some $50,000—had accepted a settlement of $10,000. "This has been settled," he said. Miranda replied that he would not accept the deal, no matter what his client had said, unless the full amount claimed was paid. No agreement was reached.

A few weeks later, in a conversation with other lawyers, Miranda heard from colleagues that Nunes was planning to kill a worker suing him, Miranda, and Judge Jônatas.[49] According to the attorney, the information was trustworthy, because it originated with attorneys who'd worked for Nunes, so he passed it on to the judge.

After the distrainment, colleagues had suggested to Judge Jônatas that he no longer target Nunes because, by their account, he was "dangerous." Pará's justice system would also attempt to remove him from the region, Jônatas said, arguing that his protection would be "too costly." But he did not agree and responded to the apparent threat by going public about the issue, prompting a new wave of damaging notoriety against Nunes, now depicted by the media as a suspect in planning the execution of a federal judge.[50]

Without delay, the director of the Brazilian judges association traveled to Marabá to meet Judge Jônatas and urge law enforcement to provide round-the-clock security to the judge.[51] For over three months, Judge Jônatas would be escorted day and night by police.[52] Miranda, who claimed to have been followed for months, decided to get his gun license, buy a weapon, send his children away, and reduce his social life to a minimum. "I even stopped going to shopping malls out of fear," he said.

Federal officers investigated the threats and interrogated Nunes, but they never found sufficient evidence to press charges.[53] Nunes would later say of Judge Jônatas, "I didn't threaten him! He's a coward! He made up the situation to be on the front pages of newspapers."[54] His lawyers accused Judge Jônatas of bias and of using the case against the rancher "to show off" and gain publicity with the aim of advancing his career.[55]

In hindsight, Jônatas, who would leave the region years later of his own will, would say about Nunes, "I felt more threatened by a rancher than by Vale," the giant mining company he had ruled against. "Nunes is someone who lives in an underworld universe. . . . This is a paradox in a democracy. This is only possible because here, in Brazil, we live in a democracy still under construction."

22

She Is Out

NOT long after the seizure operation on Nunes's fazendas, the campaign to elect a new mayor kicked off in Rondon. With Olávio ousted and banned from running as a candidate, the front-runner was Shirley Cristina, who this time sought to win her mandate through the popular vote. One of the main projects of her pro-business platform was to spur the economy with the opening of a large mine in the municipality, where reserves of bauxite iron ore had been found.[1]

Her campaign became a magnet for donors, with almost half a million reais—or some $230,000—received. It was double the amount she had gotten in the previous election and over twice what her main rival—the candidate of the PMDB, a landowner named Edilson Oliveira Pereira—would collect.[2] Again, Shirley Cristina's family represented an important source of donations, with Josélio contributing $23,000.[3] In her asset declaration, the mayor claimed a very modest wealth of 236,000 reais, about $110,000, which included vehicles, ranches, and shares in her father's companies.[4] Her two bank accounts had a remarkably low balance—$1,900 and $1.50, respectively.[5]

Her third consecutive attempt proved successful. Shirley Cristina got 9,173 votes or about 42 percent of the valid ballots, about a thousand more than Edilson.[6] Not only was she elected, achieving the milestone of putting an end to ten years of uninterrupted victories by the PMDB in the municipal elections, but the coalition of conservative political parties that had supported her and the PSDB also elected six out of the thirteen council members. This represented a rather reassuring message from voters, signaling that she would face four years of smooth, unopposed governing. The popular vote was crucial for her, because, as she would say, the previous two years as the head of city hall "hadn't been easy."[7] During that time, Shirley Cristina had governed with the uncertainty that an appeals court could overturn the removal of Olávio and reinstate him.

On the evening of January 1, 2013, the stage was set for the Barros family to enjoy its triumphal moment. Hundreds of townspeople—former mayors, investors, fazendeiros, corporate donors of her campaign, and average citizens—gathered at a multipurpose gymnasium for the inauguration ceremony.[8] The white and green walls had been decorated with cream-colored drapes, and people sat in rows of chairs covered with white slipcovers. Plastic flowers and small palms were placed on either side of an aisle cutting through the rows of chairs and leading to a large rectangular table where the swearing-in ceremony would take place. Men and children were dressed in suits, and women wore high heels and fancy dresses. Several journalists covered the event.

The ambience was a bit kitschy, but Mayor Shirley Cristina was radiant. She wore a filmy light blue wrap over a pristine white dress, her shoulder-length dyed-blond hair covering pearl earrings paired with a matching brooch that held the wrap closed at her waist. Her customarily severe expression had vanished. She displayed a wide smile as she greeted people with her right hand, the left one holding that of her husband, visibly shorter than her and dressed in a navy blue suit. They looked like a joyful couple at their wedding ceremony.

At seventy-seven, a bespectacled, white-haired Josélio watched the scene with satisfaction. He and his wife, Sirley de Souza Barros, were seated in the first row, with the fazendeiro occupying the chair by the aisle. He wore cowboy boots, a loose pair of dark blue jeans, and a blue-and-white-striped shirt, which was opened to his torso, with the sleeves rolled up. Although his face was wrinkled, he retained the same stony countenance as half a century earlier when a black-and-white photograph of him was printed on the front pages of Espírito Santo's newspapers in regard to the murder of Major Cavalcanti.

Shirley Cristina received the ceremonial mayor's sash, a red-and-white strip of cloth, and then she recited the oath of office, swearing "to maintain, defend, fulfill, and enforce the constitution; observe the laws of the union, the state and the municipality; promote the welfare of all; and exercise my mandate inspired by patriotism, loyalty, and honor."⁹ In a statement given later to the press, Shirley Cristina would say that she intended to make Rondon "the best city in southern Pará."¹⁰

"We have renewed spirits. We now have confirmation that the people in Rondon chose us," she declared, exultant.

Just rows behind where Josélio was seated was Maria Joel. By some irony, she also had something personal to celebrate. Joelminha, her eldest daughter, had managed to break the family's curse in politics and had been elected councilwoman for the Workers' Party. Maria Joel would recall that night as one marked by opposing feelings, "a moment both very sad and joyful, because both my daughter and the people who had pursued my husband" were taking office at the same time.

At thirty years old, Joelminha, married to a man who had a small business and now the mother of several children, had decided to run for a series of reasons. "I did want to do something for the rural workers," she would say. But another goal, she admitted, was to use her position to protect her mother. "We had heard several times that the police of Pará might remove her protection. Because I was a public servant, I was able to go to talk with the governor or with the chiefs of the police to request that this not happen," she would explain. Eventually, Joelminha's political career would become greatly controversial.

Mayor Shirley Cristina didn't want the contentious history of the two families to jeopardize her mandate or even to disturb the swearing-in ceremony. Some days prior to the event, anticipating the uneasiness of the situation, she had called Joelminha for a talk in her office.

"I wanted to tell you that my father had nothing to do with the death of your father," said Josélio's daughter, according to Joelminha.

Joelminha thought for a while before answering.

"Here in city hall, I'm not Joelminha. I'm a councilwoman. And you're not Shirley Cristina; you're the mayor," she answered. Shirley

Cristina insisted—her father, she said, had no relation whatsoever with the crime.

"I thought that your invitation to talk here in your office was to discuss something related to the municipality, not to our personal lives," Joelminha recalled reiterating. "As for whether your father was or was not involved, we'll leave this for the legal system to decide."

Unsatisfied, Shirley Cristina maintained once more that Josélio was innocent. "My father didn't even know yours," she said, according to Joelminha's later recollection.

It was an odd tack for Shirley Cristina to take. Josélio had departed for a "holiday" on the eve of Dezinho's murder, and although he had been questioned afterward, he hadn't been investigated. Over a decade after the crime, only two fazendeiros—Nunes and Lourival, the owner of the Fazenda Santa Monica—were being prosecuted. So why was Mayor Shirley Cristina making this point at such a crucial time in her career? The women wouldn't discuss this further, but later, Shirley Cristina's possible intentions would become clearer.

If she was worried about opposition from Joelminha, the mayor would soon discover that her real problems were coming from another direction. The burden for Shirley Cristina, one which threatened to blemish her reputation and career, had in fact commenced when she'd won the election, because by that point, the six-foot-one green-eyed and boyish-faced Judge Gabriel had already launched a series of investigations into the funds used for her campaign.[11] It was the last of a series of actions showing the man's crusade-style approach to justice.

JUDGE GABRIEL WAS the son of a medium-size landowner from Minas Gerais and a teacher named Terezinha de Lourdes Costa Ribeiro. He was the youngest brother of a string of well-respected lawyers and judges and a senator, all of whom had made their name in Brasília through hard work and contacts that had helped them get close to the political establishment of Brazil. In the capital city, a young Gabriel studied law and became accustomed to the tidiness and order of local courts.

Soon after being named Rondon's judge in early 2009, he was

confronted with a much different situation in the frontier town. The courtroom was a mess. Thousands of cases involving "drug trafficking, adoptions, murders, and pistoleiros" and containing crime-scene reports, court transcripts, statements of suspects, and other files for which there was no digital backup lay in jumbled piles scattered throughout the building. "The documents of criminal and civil cases were piled up even in the cells used for prisoners, filled with rat and bat excrement," he would say. Pictures he took at the time showed a mass of moldy paper mixed with cardboard boxes, plastic bags, and even bike frames someone had placed in a room. "I believe that in that context, it was easy to order pistoleiros to murder people and have the cases go unpunished," Judge Gabriel declared.

The judge, eager to leave his mark and advance in his early career as state judge of Pará, decided to enact a collective, all-hands-on-deck response to the chaos. He used his contacts in Brasília to get computers and printers donated and instructed his staff to devote weeks to organizing the chaos and to logging cases into the national computerized database. Those efforts, Judge Gabriel would say, would eventually be marred by the sudden decision of Mayor Shirley Cristina to remove some of the public servants from the courthouse. That was his first clash with the Barros family.

Judge Gabriel's determination to modernize the local justice system and have rule of law prevail made him some other powerful enemies. Only six months after he moved to town, he sentenced three police officers of Rondon to prison terms after finding them guilty of extorting and torturing alleged local drug traffickers the cops had arrested inside their own home without a warrant and for no apparent reason.[12] The officers, seemingly looking for a bribe, had threatened to fabricate a case by planting fake evidence, according to Judge Gabriel's sentence.[13] The case, on which the judge ruled in September 2009, would stand out because of the actions of the defendants' colleagues. The day of the trial, anticipating a guilty verdict, fifteen officers including a captain showed up heavily armed to intimidate Judge Gabriel.[14] Fearing that a shoot-out was imminent, lawyers who happened to be at the courthouse ran away, and the judge was left with no alternative but to hide in his office for hours. He wasn't shot, but from that day on, Judge Gabriel would receive protection from two police officers he picked personally. He also bought an armored car.

Months later, Judge Gabriel would make another hard-nosed decision in a civil case involving Josélio, his wife, Sirley, and Mayor Shirley Cristina, all members of the board of a company called Fazenda Nova Delhi Agropecuaria, which had some 1.6 million reais—about $900,000—of debts with the state-controlled bank Banco da Amazônia.[15] The bank had previously initiated litigation against the debtors, and on July 7, 2010, Judge Gabriel requested that the Central Bank of Brazil search for any money in the accounts of Josélio, his wife, and their daughter to proceed with distrainment.[16] It was a kind of seizure operation like the one Judge Jônatas would carry out against Nunes, but focused on financial assets only. The Central Bank followed the order but informed the judge that the account of the Fazenda Nova Delhi Agropecuaria contained only $3.60, and there wasn't much more than that in the private accounts of the Barroses. In Josélio's two accounts, the total balance was $11.40, and in those of Shirley Cristina and her mother, Sirley, there was no money.[17]

Josélio and his wife had yet another pending civil case with Banco da Amazônia for an additional debt of 2.82 million reais, or about $1.5 million.[18] Three years earlier, when the bank had initiated litigation, Josélio had offered a ranch as payment, but Banco da Amazônia argued that the value of the land wasn't enough to settle the debt in full, so the judge had ordered that he offer another fazenda.[19] The case was apparently stalled in courts when Judge Gabriel took his post in Rondon, so he ordered Banco da Amazônia to report whether Josélio had finally settled the debt. Apparently, he hadn't.

In February 2011, Banco da Amazônia asked the judge to terminate the two civil litigations (*"extinção da execução"*). According to the bank, a renegotiation was made with the Barroses.[20] The conditions would not be published, but a bank's lawyer who worked on the case claimed that this practice was commonplace, as the bank's legal department was overwhelmed by the delinquency rate in Pará and couldn't litigate all the cases.[21] Many years later, however, Banco da

Amazônia and a lawyer of the Barros family admitted that the debts, even if renegotiated, hadn't been settled and had grown to $4 million.[22]

All those cases and situations preceding the October 2012 mayoral elections won by Shirley Cristina were a hotbed for the legal contest that was about to start between Judge Gabriel and the mayor.

Weeks prior to voting day, Judge Gabriel began to receive reports of alleged wrongdoings committed by members of Shirley Cristina's campaign, from the suspected use of public resources to benefit her campaign to the existence of a "pirate radio station" illegally broadcasting her political propaganda.[23] The discovery of the "radio station" was made when a prosecutor driving through Rondon had tuned the radio of his car to a frequency that wasn't legally assigned. (Josélio's daughter would say that she had nothing to do with it and that the propaganda wasn't being broadcast from a radio station but instead came from a Bluetooth device.[24])

Another tip indicated that Shirley Cristina might be offering public services in exchange for votes. On September 18, 2012, the judge and his staff attempted to verify this by inspecting the van of a municipal contractor who had been hired to bring sick people—especially patients requiring hemodialysis treatments—to well-equipped hospitals in Marabá.[25] Political propaganda for Shirley Cristina was found inside the vehicle, according to court documents, and the contractor was found to be transporting anyone who had been preauthorized by city hall, no matter if the person was sick or not. According to Judge Gabriel, the driver gave free rides to all those who declared they would vote for Shirley Cristina.

The mayor denied the claim, but her rivals from the PMDB party, who had been sued and ultimately ousted four years earlier, took their chance at payback, filing several lawsuits against her candidacy before electoral courts.[26] Shirley Cristina managed to win the election and take office while the case was being prosecuted. She also retaliated against Judge Gabriel, whom she now saw as her personal enemy.

From September 2012 to mid-2014, Shirley Cristina would file no fewer than six lawsuits and legal proceedings against the judge, both in state courts and with the state and federal internal affairs offices of the judiciary.[27] The mayor would accuse Judge Gabriel of, among other things, not being impartial and committing slander, requesting he be removed from the cases in which she was a defendant. In a seemingly coordinated action, the Associação funded by Josélio and presided over by Malcher—Shirley Cristina's husband—also sued the judge.[28]

Of the series of accusations made against Judge Gabriel, one had the potential to greatly damage his career. On March 6, 2013, Shirley Cristina wrote to the internal affairs office of Pará's judiciary claiming that the judge had authorized adoptions of children in Rondon "without the due enforcement of the law" and resorted to "illegality, abusiveness, arbitrariness."[29] The mayor posited in her eight-page report that Judge Gabriel had followed "some [illegal] proceedings" in the adoption of one child who was later returned to the state by the adopting parents. The violation of the law, she wrote, had ultimately resulted in psychological harm to the child, "leaving indelible marks that compromised" him.[30]

The internal office division found credibility in the story and responded without delay. A few days later, a team of inspectors led by a judge of a higher court traveled to Rondon to inspect not only the specific case of the child referred to by Shirley Cristina but all the adoptions presided over by Judge Gabriel, who fought back to defend his integrity with the help of his sister, lawyer Raquel Costa Ribeiro, and other members of his family.[31] One of their arguments was that the flood of lawsuits was a maneuver to "intimidate him . . . and push him out of" Rondon in order to prevent the judge from ruling against Shirley Cristina in the crucial case over the suspected use of public funds to benefit her campaign.[32]

Four months later, the internal affairs office concluded that Judge Gabriel had "acted within the law," had "not acted in an arbitrary way," and had indeed ruled in the case of the child mentioned by the mayor in "the best interest of the minor."[33]

But the legal battle wasn't finished yet. It continued in the electoral courts and with the investigation into whether Shirley Cristina had violated the law by granting free rides to voters. The mayor resorted to appeals courts to have Judge Gabriel removed, now claiming that he was biased because he supported politicians of the PMDB.[34] She used the legal team of city hall to sue Judge Gabriel for slander and defamation, a move of dubious legality raising more questions about her use of public funds for personal purposes.[35]

By mid-2013, the cascade of legal actions from Shirley Christina was starting to raise serious doubts in the top spheres of Pará's judiciary. Veteran judges in Belém were struggling to understand whether there were any legal foundations for all this litigation coming from Rondon or if instead the lawsuits were, as an appellate judge would say, "the perfect setup by city hall [*"armação perfeita pela Prefeitura"*] to remove him [Judge Gabriel] from the cases" that could lead to the ousting of Shirley Cristina.[36] Multi-judge panels were particularly uneasy at Shirley Cristina's use of political contacts to influence the final ruling. "I clarify to the Court," said a judge, addressing his colleagues in one of the hearings, "that the mayor was in the courthouse, accompanied by a senator, requesting to remove Judge Gabriel, and she didn't succeed. She went to the Senate, she went to the state government, and by different means she has been trying all these years [to remove him from Rondon]. It's a persecution that I believe the Court doesn't have to accept."[37] According to Judge Gabriel, the senator who was helping her was Fernando de Souza Flexa Ribeiro, the same man who had lobbied Lula's government in favor of the loggers back in 2005.[38]

Through a series of appeals, Mayor Shirley Cristina would succeed in keeping Judge Gabriel away from her electoral case—and remain in power—for almost two years. But on September 30, 2014, Brazil's top electoral court overruled a previous decision and reinstated Judge Gabriel on the case, settling once and for all the issue of whether he had any personal bias against the mayor.[39] (According to the court, he did not.) On October 15, Mayor Shirley Cristina was at a public event with teachers when the news that she had been ousted spread like wildfire through the town. Judge Gabriel had removed her from office for "abuse of political and economic power."[40] The sentence had immediate effect.

Visibly affected, she took the floor to talk to the public. "I haven't committed any electoral crime," she declared.[41] "You don't discuss the decision of a judge; you appeal. . . . We don't know whether we'll be successful, but we always put things in the hands of God." A day later, rancher Edilson, from the PMDB, was sworn in as the new mayor.

Three weeks later, on November 6, Judge Gabriel ruled against Shirley Cristina in another electoral prosecution, this time for using "civil servants, public services, and assets from Rondon do Pará for the benefit of her own candidacy."[42] The ousted mayor appealed both sentences, but the ruling for "abuse of power" was upheld by Brazil's top electoral court.[43] Because she had broken electoral law, Shirley Cristina was not only removed from her seat but also banned from holding any political office for the next eight years.[44]

To her greater dismay, she would have to face all those consequences of her downfall without the support of her father.

23

The Trial

JOSÉLIO de Barros Carneiro witnessed Shirley Cristina's troubles with the electoral justice, but he didn't live long enough to see her removed from office. On April 24, 2014, at 4:40 p.m., Josélio passed away in São Paulo at the age of seventy-nine. The cause of death, the doctors at the highly reputed Hospital Sírio-Libanês wrote, was a combination of ailments: a cerebral vascular accident, septic shock, bronchopneumonia, and acute kidney failure.[1]

The news of Josélio's death spread quickly in Baixo Guandu, where some of his brothers and sisters still lived and the third generation of the Barros family continued with the family's political tradition; José de Barros Neto, one of Josélio's nephews, had succeeded his father, Chico, as mayor.[2]

The days of rage and fury in the former Contestado area were over, but Josélio remained a mythical figure. A reporter who had investigated the Criminal Syndicate wrote his obituary and described Josélio as "one of the last members of a select group that, in the fifties, sixties, and part of the seventies, terrorized the Sweet River region and the north of Espírito Santo, imposing the trigger's law to favor its rural elites."[3] The reporter noted with a certain irony that, despite what he described as a violent past, Josélio hadn't died as a consequence of a bloody vendetta, but instead "in a bed."[4]

After a "very crowded" funeral, Josélio's body was transported by plane to Pará for further services.[5] A wake was organized in Rondon's exhibition center, the same one used for ExpoRondon, and dozens of townspeople paid their final respects. Afterward, Josélio was brought to Belém for cremation before his ashes were returned to Rondon "to be scattered on his properties," wrote a reporter.[6]

Police officers who had participated in the recapture of Adélcio Nunes Leite in the Fazenda Te-Chaga-U said that the infamous Souza paid a final tribute to "his friend" by showing up at his funeral.[7] After eighteen years in prison, Adélcio had in fact just been freed. Although the outlaw had been sentenced six times and had been found guilty of eleven homicides, his accumulated punishment reaching 229 years, Adélcio's attorneys had managed to obtain a pardon in 2013.[8] (Adélcio would nevertheless meet a tragic end; in 2020, he was assassinated by a gunman in Minas Gerais.)[9]

Senator Fernando de Souza Flexa Ribeiro was responsible for making the death of Josélio known in Brasília. He took the floor in the Senate to honor the rancher and express condolences to the family. "The cattleman," Senator Flexa Ribeiro declared, referring to Josélio, "for forty years was devoted to improving the [Guzerá] breed."[10] In the Brazilian congress, a lawmaker named Wandenkolk Gonçalves, a member of the agribusiness caucus known for his stances against landless peasants, proclaimed that "Josélio was a very important man for Pará."[11]

Josélio had transferred some of his land to his children in the 1990s, but by the time of his death, large areas remained under his ownership.[12] Strangely, he left no will and testament, according to public records.[13] This was an odd decision for a man with Josélio's profile, who was the father of six legitimate children and had amassed a considerable fortune. An inventory of assets made by his widow and daughters would show that by the time of his death, Josélio claimed ownership of six fazendas and a herd of some 2,500 cows. He and his companies, however, had also accumulated high levels of debt with banks and suppliers.[14]

It is unclear why Josélio made that decision, but, according to the publicly available information, it is reasonable to suspect that this was motivated by problems with the deeds of some of his fazendas.[15] (Some of his children, including Shirley Cristina, would get full legal control of the land only in December 2021—and after paying the state of Pará for it, because the deeds had indeed been declared fake.)[16] Whatever the case, the absence of a will would prompt a profound and long-lasting rift in his family. In 2017, José Eloy, Josélio's sole son, would initiate a series of legal proceedings against his mother for control of the assets. According to him, his mother and some of his sisters, including Shirley Cristina, were attempting to get a larger share of the possessions, mostly land, apartments, vehicles, and cattle [17]

This litigation and the documents in the court case revealed a hint of the fortune amassed by Josélio. José Eloy estimated that the value of his father's assets amounted to 93 million reais, some $28 million at the time. Josélio's widow declared that it was much lower, some 28 million reais, or $8.8 million.[18] By mid-2022, the case was still unsettled, and the children—despite the mediation of a judge—hadn't reached an amicable agreement. The legal burden only seemed to grow when a woman claiming to be a daughter of Josélio born from an extramarital relationship filed a paternity lawsuit with the aim of receiving her share of the inheritance.[19]

WHAT DID THE family of Maria Joel feel when Josélio passed away? Dezinho had many times accused the rancher of plotting his assassination, and many within the family and the Sindicato continued to nurture the idea that Josélio must have played a role in the murder, although no evidence supported this narrative. Still, Maria Joel recalled that she didn't feel anything special when she learned of Josélio's death.

By then, the widow had concluded her mandate at FETAGRI and had returned to work in Rondon. In early 2014, she had been reelected president of the Sindicato.[20] But her focus these days wasn't on the projects she had for her third term as the head of the union. Instead, her full attention was on the issue that had most determined her life—her quest for justice and memory—because, after more than thirteen years of back-and-forth, the case against Nunes reached trial on April 29, 2014, only five days after Josélio's death.

The outcome of the much-anticipated jury trial was highly unpredictable. The process now had more than 3,000 pages, but no testimony or piece of evidence had been conclusive in pointing to the King of Wood as the mastermind of the murder. The justice system had searched for a smoking gun, ordering Nunes's bank and phone records to be scrutinized to unearth any ties to Wellington, the triggerman, or any other individuals suspected of participating in the murder. Yet nothing was found, and the feeling was that the jury would render a verdict based upon the testimonies at trial.

There was a real possibility that Nunes would be acquitted and that the crime would remain forever unsolved. Wellington had never been seen again, just like the alleged two middlemen, Ygoismar and Rogério. Also, some months prior, in October 2013, rancher Lourival had been tried by a jury and had been acquitted for lack of sufficient evidence; one of his employees, Raul, suspected of being the original owner of the gun that Wellington had used to shoot Dezinho, had also been acquitted.[21]

In that environment of uncertainty, supporters of both Nunes and Maria Joel gathered in the Belém courthouse on the morning of April 29. By 7:30 a.m., members of Nunes's family and dozens of his employees lined up to enter the building, while landless peasants and friends of Maria Joel waited next to them. There was tension, but things were calm and civilized. They looked like football fans entering a stadium, each group recognizable by the inscriptions on their T-shirts—"Décio [José] Barroso [Nunes] is innocent. We trust in the justice system" and "Dezinho: Alive in the people's fight. His memory will never be forgotten by those who defend the law!"[22]

Inside the courtroom, the state attorney, Nunes's defense lawyers, and Judge Raimundo Flexa—the same man who had presided over the jury trial of Wellington and also some of the cases of Dorothy's murderers—observed official etiquette by wearing black robes. It wasn't a big courtroom, and the benches reserved for the public were quickly packed with supporters on both sides and representatives from NGOs and the state and federal governments.

From his dark wood bench, Judge Flexa instructed a female court officer, also wearing a black robe, to make an official announcement of the case's information to open the trial. She took a microphone and stood up. "Crime: Homicide. Defendant: Décio José Barroso

Nunes. Victim: José Dutra da Costa," she recited, listing Dezinho's full name.

Perfectly shaved, the fifty-seven-year-old Nunes was without hand-cuffs when he was ushered into the courtroom by an official who po-litely asked him to stand by the defendant's bench—a wooden chair placed in front of the judge—until he was given the order to sit. He wore a pair of dark blue trousers and a spotless white shirt with his sleeves rolled up to the elbows. Nunes's hair had turned salt-and-pepper, and the downturned arc of his mouth—a consequence of age-related sagging—made him appear perpetually angry. His clear blue eyes and his large aquiline nose remained the dominant features of his virile face, once chiseled and now a bit chubby. Nunes seemed some-what anxious. After all, while the King of Wood had faced many law-suits before, this was his first potential criminal conviction.

Nunes's fate would be decided by only four of the seven jurors. In Brazil, the jury isn't required to reach a unanimous decision to ren-der a verdict; a simple majority is enough to either convict or acquit the defendant. Another major difference from jury trials in America is that Brazilian jurors deliberate in what is called a secret room. Each juror votes confidentially and therefore is uninfluenced by the conclu-sions reached by the others.

Throughout the years, several attorneys had participated in his defense, but for that crucial day, Nunes chose a burly lawyer in his for-ties named Roberto Lauria.[23] The state attorney was the bespectacled and energetic Franklin Lobato Prado, who was around the same age as Lauria. The men had great professional respect for each other and occasionally met at social events.

State attorney Franklin would say that his bosses considered the trial of Nunes to be "the trial of the century." Days earlier, the office of the Brazilian presidency—still in the hands of Dilma Rousseff—had contacted the office of Pará's attorney general to make the point that they wanted the prosecutorial team led by Franklin to do a good job representing the plaintiff.[24] Despite the political pressure or the eventual prestige that a conviction could deliver for state attorney Franklin, it had not been easy for Maria Joel to get a prosecutor to handle the case. Before Franklin was ultimately given the job, four of his colleagues, providing nonspecific reasons of a private nature, had refused to represent Maria Joel in the jury trial.[25] This reluctance led to speculation over whether state attorneys feared retaliation should Nunes be sentenced and thus declined to take the case. According to Nunes's attorneys, however, the reason was simply that the case lacked sufficient evidence and prosecutors didn't want to participate in a trial that could convict a man without proving his guilt beyond reasonable doubt.

By 8:00 a.m., Judge Flexa eyed Nunes through his black-rimmed glasses and, before giving the floor to the state attorney, the first to speak, asked the defendant to confirm his name and that of his lawyer, which Nunes did. At that point, the judge ordered, "Please, sit."

The first of several witnesses called to testify, Maria Joel wore a simple black dress. Her face was pinched, and her hair was tied back in a short ponytail. The bailiff ushered her to the chair that served as the witness bench and gave her a microphone. The first thing Maria Joel did was to ask the judge to remove Nunes—seated a short distance from her—from the courtroom. "I want him to leave," the widow said timidly. His mere presence, she argued, intimidated her.

"This is a senseless act!" defense attorney Roberto Lauria replied. "Fourteen years have passed since the fact [the murder]; nothing happened that could justify [the request]," he complained.

Joelminha, present at the trial with her sisters and brother, fumed. She would tell the media that her mother's home had been watched over the previous weeks by unknown people riding motorbikes and cars. Although no threat had materialized, the family suspected that it was part of a strategy to intimidate the widow. "This is the life that we've always lived, with constant death threats," Joelminha complained.[26]

Judge Flexa acknowledged that it was Maria Joel's constitutional right to testify free from the presence of the defendant and ordered his removal. Guided by the bailiff, Nunes left the room silently, and then Maria Joel began her testimony.

Maria Joel explained that her husband, before he was killed, had devoted his life to the fight against grilagem and debt bondage as well as to promoting land reform through the occupations of fazendas. In regard to Nunes, she said Dezinho had once told her that the rancher's workers "couldn't protest, and [any] who had the courage to request a due payment was murdered." She also asserted that Dezinho was aware that the Fazenda Lacy was the product of land-grabbing, and therefore a site of potential occupation by members of the Sindicato.

State attorney Franklin wanted to make sure that the widow devoted time to "the motives that led Delsão [Nunes] to hire the murder of your husband to a pistoleiro," so he enumerated several issues and asked her to both confirm and elaborate on them: that Dezinho was aware that Nunes was in fact "the Judge" and commanded a criminal consortium of ranchers; that Dezinho had denounced cases of debt bondage and even murders among Nunes's employees; that Dezinho had been told about the suspected existence of a "clandestine cemetery," said Franklin, in the Fazenda Lacy.

"A person had informed Dezinho that Delsão [Nunes] ordered workers killed in the area and threw them in a pool," replied the widow, referring to an alleged pond infested with alligators in the Fazenda Lacy, where Nunes supposedly disposed of corpses.

In total, the state attorney and his team devoted almost an hour to Maria Joel's examination, but the testimony was disappointingly choppy and flawed. Her answers to questions like "Tell us about the threats," "Tell us about the Fazenda Lacy," and "Tell us about how you got to know that Nunes ordered the murder of Dezinho" were short and skipped relevant details like dates of events, names of informants, and firsthand context that only she—the widow of the victim and later his successor—could provide to lead the jury to conclude that Nunes had hired out the murder of Dezinho.

Maria Joel's account of the other crimes allegedly committed by the fazendeiro, not part of this case but essential as pieces of background information, were at all times preceded by phrases like "I was told by Dezinho," "Dezinho had heard from sources," or "It was said in town." She was relying upon what is known in legal jargon as hearsay evidence—secondhand information produced out of court by other sources, not by the witness testifying under oath. Given its potential untrustworthiness, the admissibility of hearsay evidence is highly restricted in many legal systems, including in the United States. In Brazil, although its use in jury trials is accepted, it is not enough on its own to convict a defendant beyond reasonable doubt.[27] Appellate courts can later nullify verdicts based upon hearsay evidence and order a retrial.

The testimony of Maria Joel presented other problems, too, like the absence of a compelling and cohesive narrative that could appeal to the emotions of the jurors. For instance, prosecutor Franklin asked her about the eight years of continued threats before Dezinho was taken away, and about the struggle she had endured afterward due to her fight for justice. She answered merely that "it was a very difficult time." The longest description she made was of the day her son Joelson had remained at home with Dezinho after peasants had come to warn the family of an imminent attack—in her words, a plan masterminded by Nunes. She devoted less than a minute to that account.

That morning, when it was so crucial for the jury to know the full account of Maria Joel's story, when her memories of the fear consuming her and her children were so important to recall, and in detail, the widow was brief and hard to follow. She would later admit that she had been too anxious and suffered from high blood pressure, marring her ability to recall particulars and provide a strong, persuasive account. This circumstance would be further exposed when Maria Joel was subjected to cross-examination by defense attorney Lauria, who focused his strategy on depicting Nunes as someone with no motive to want Dezinho killed.

"Maria Joel," said Lauria, politely and calmly, "tell me something. . . . [In the period] close to his death, how many ranches did Dezinho invade in the region?"[28]

"Four," the widow replied, noting that "Dezinho himself didn't invade" the fazendas. Instead the Sindicato organized occupations carried out voluntarily by landless families.

Lauria nodded thoughtfully, then continued. "Answer me something simple," he said. "Did Décio [Nunes] have his land invaded?"

"No."

Lawyer Lauria pressed the matter. Maria Joel had said that Dezinho had been murdered for his struggle, and in fact threats had grown after the first fazenda was occupied in Rondon in 1997. But, Lauria reasoned, if Nunes's land hadn't been invaded, how could he possibly be the main suspect?

"Maria Joel," he repeated, speaking slowly but raising his voice for emphasis, "did Nunes ever have his land invaded?"

"He did not," Maria Joel replied.

The lawyer leaned back, visibly satisfied. "Thank you very much," he said.

The defense lawyer turned and paced away to redirect his questioning, now more determined to debunk what he would argue was a case based upon the egregious lies of witnesses. He asked Maria Joel to name one by one the ranches occupied—"invaded," as he put it—by the Sindicato in the months prior to the murder. There was the Tulipa Negra, owned by Kyume Mendes Lopes and later seized by INCRA; the Santa Monica, owned by Lourival, who had been already acquitted; and the controversial Vila Gavião, owned by the murdered Hilário.

Lauria devoted some time to providing the jury with context about Hilário. He noted that unlike Nunes, Hilário's family might have had a good reason to kill Dezinho—revenge. Neto, the murderer of Hilário, had ruled out that Dezinho had participated in the execution of the rancher, but after being arrested by the police, he had managed to flee, and the crime had remained unpunished, so in Lauria's telling, it was perfectly possible that the family had taken justice into their own hands by hiring a gunman.[29]

Then with almost two hours of trial elapsed, Lauria made what would be the first of a series of mentions of Josélio and his potential involvement in the crime.

"Did they find something irregular in Josélio's ranch?" Lauria asked Maria Joel after a brief introduction of the Te-Chaga-U raid.

"Yes, they did," Maria Joel replied.

"What did they find? If you could tell me, I'd appreciate . . ."

"They found human remains," said Maria Joel.

"Human remains . . . Did Josélio have problems with the law for the charges made by Dezinho?"

"He was called to testify," she recalled.

"Was he arrested? Was his arrest ordered as a consequence of the denunciation made by Dezinho?" persisted Lauria.

"I don't remember," admitted Maria Joel.

Lauria had studied the Te-Chaga-U case and asked her to confirm that "Josélio was a large landowner in Rondon" and that he had been "a fugitive of the justice" (*foragido*) after the raid. Maria Joel confirmed this information.

"This is a very complex case; there are a lot of lies in it," Lauria proclaimed after recalling that according to the statement of a "citizen named Geovano," Olivandro—Hilário's son and the man supposedly at the wheel the day Brito was almost abducted—was seen in a bus with Wellington days prior to Dezinho's murder. (Lauria never said, however, that Geovano, the source of this information, was the brother of Nunes's father-in-law, and his statement had been given years after the execution of Dezinho, when Nunes was already being prosecuted.)[30]

To cast further doubt on the widow's testimony and, more broadly, on the narrative that Nunes was the one who had masterminded the killing of Dezinho, Lauria asked Maria Joel to name any specific evidence—other than "It was said" or "Dezinho heard from sources"—to support the accusation that the King of Wood used a pond stocked with savage reptiles to dispose of bodies on the Fazenda Lacy. Maria Joel had none. Lauria then reminded the jury that the police—Sergeant Claudio Marino Ferreira Dias, precisely—had indeed inspected the Fazenda Lacy and found, the defense attorney said with irony, "no pond with an alligator floating with an arm in its mouth."

WHEN MARIA JOEL concluded her testimony, the public feeling was that she had lost a crucial opportunity to influence the jury, and for now, the scales seemed to be weighed in Nunes's favor.

Judge Flexa ordered a short recess. From his bench, the judge gave the impression of being relaxed. He got out his phone and spent some time—almost four minutes, precisely—trying to take a selfie. "It's blurry, it's blurry," Judge Flexa complained after several attempts.[31]

Several smiling female assistants approached him to kiss his bald head, the video of the trial capturing the somewhat grotesque scene.

After the pause, the atmosphere was again suffused with gravity. The plaintiff called its second witness, and Judge Flexa instructed his staff to bring in Francisco Martins da Silva Filho. A profound silence fell over the courtroom as Francisco was ushered in by the bailiff. He wore faded jeans, black leather shoes, and a dark gray sweatshirt that hung loosely on his slender frame. It was a hot day in Belém, but only the skin of Francisco's bony hands was visible. A dark ski mask covered his head except for holes at his mouth and eyes, and he also had on a pair of glasses with tinted lenses.

Francisco reached his seat, and it became apparent that he was on edge. His legs shook with anxiety, and the fabric of the ski mask clung to him. He had a wild desire to light a cigarette. "If they had let me smoke in the courtroom," Francisco later said, "I would have consumed two packs."

The witness believed he was running a great risk by offering his testimony. Francisco claimed that he was still alive only because, right after offering his statement to Detective Resende, he had entered the Brazilian Protection of Victims and Threatened Witnesses Program (PROVITA). On November 30, 2000, just three days after Francisco made his explosive revelations, he, his wife, Maria, and their six children had been transferred by officers to Belém, where they had temporarily settled in a Catholic church. After a week in confinement, the family was brought to a village located on a small island near the state capital. All family members were asked to change their appearances—Francisco let his beard grow, and Maria and the children, all under twelve, cut their hair—and wait. A few weeks later, at night, a vehicle picked them up and took them to the airport.

"It was the first time any of us had ever boarded a plane," Francisco would say, recalling the details of those frantic days.

They flew to Brasília and afterward were driven to a city in the highlands of central Brazil. They settled in a house rented by PROVITA and were given a phone number that henceforward would become the exclusive channel for any authorized contact with the outside world. Even the most trivial issue—buying milk or medicine for the children—had to be done through PROVITA, at least theoretically. When Francisco called the officials of the program, the protected witness had to identify himself with a code name. He was prohibited from working, leaving the house, owning a cell phone, or sharing his address with relatives.

The restrictions, Francisco later said, caused his family "to live off [state] handouts," which he hated and found pitiful, and he ended up voluntarily leaving and then reentering PROVITA a few times over the years. "I worked my whole life, and my wife had never had to ask for a spoon of sugar from a neighbor," he would declare, frustrated by the stultifying restrictions imposed on protected witnesses, which, in his opinion, were even more draconian than those endured by activists like Maria Joel.

If only the protective measures had worked. Between 2000 and 2014, Francisco and his family had been relocated no fewer than ten times because either he or PROVITA officials believed that they had been exposed and were at risk. One day, for example, Francisco and his family had been hastily moved from a backwoods town when he happened to have crossed paths with a former Rondon neighbor at a local market.

Later, Francisco would say that he had refused Nunes's alleged bribes. One of the attempts to buy his silence, according to Francisco, had taken place years prior to his testimony, when a group of lawyers met Francisco in a hotel room in Brasília and offered him a million reais, some $500,000.[32] The latest proposition had come on the eve of the trial, when, Francisco would later claim, his half sister had been told by Nunes's second wife that she would pay Francisco about $100,000.[33]

"Take the money and forget about this issue," the half sister, who lived in Rondon, had suggested to Francisco over the phone.

"I won't take it," Francisco had replied.

FRANCISCO HAD BARELY had time to sit when defense attorney Lauria, aware that the sight of the balaclava could have a powerful impact on the jurors, quickly objected to what the protected witness was wearing. "Why is he dressed this way?" he asked, agitated.

He and state attorney Franklin immediately clashed over this issue, the latter arguing that Francisco was really afraid. "After all, his brother was allegedly murdered by the defendant," Franklin said, referring to Nunes, whom Francisco claimed had ordered the execution of Pedro.

The judge allowed Francisco to testify in disguise and with Nunes out of the courtroom. But he warned Francisco that he was under oath and had to tell "the truth" or he might be charged with perjury and immediately taken into custody. "Yes, sir," Francisco replied.

State attorney Franklin began his examination of the witness by reading the statement Francisco had given to Resende back in 2000. He asked Francisco to confirm or correct several details to provide important context to the jury about who he was and why he was there to testify—for instance, that his brother had been a pistoleiro for Nunes; that Pedro had lived in and had an office in Rondon; that Francisco, for "three or four months," had worked with Pedro offering services like investigating and collecting arrears. This last issue explained why Francisco, although he stated he had never participated in any crimes, knew many details about Pedro and his deals with Nunes, including the existence of the criminal consortium of fazendeiros and the original plan to murder Dezinho.

Before the state attorney had devoted two minutes to the questioning, and as soon as he asked Francisco to explain "what your brother told you about the deaths that happened in the place known as Presa de Porco," Lauria objected. In his opinion, those were "facts unrelated" to the case and the question wasn't relevant. State attorney Franklin replied that this line of questioning was fundamental to understanding who Nunes truly was, including the nature of his relationship with Pedro and the motive behind Dezinho's murder. Within seconds, the argument abruptly escalated.

"What are you afraid of?!" Franklin yelled at the defense attorney. One of his assistants approached him and whispered something into his ear. Then he took a deep breath and addressed the judge. Franklin argued that Lauria "knew that Pedro was the pistoleiro of Delsão

[Nunes]" and therefore feared the implications of Francisco's testimony.

"I'm afraid of lies!" retorted Lauria.

Judge Flexa, struggling to make his voice heard, considered the issue for a moment. He then asked the prosecutor whether there was a link between the questions he planned to put to the witness and Nunes's alleged involvement in Dezinho's murder.

"Yes," state attorney Franklin replied.

"Then," the judge said, "ask your question."

24

A Certain Sense of Justice

PEDRO'S life, according to Francisco, went off the rails after he undertook an initial "service" for Nunes that, considering the businessman's well-known reluctance to settle his debts with employees, wasn't without irony: Nunes hired Pedro to pursue a client in arrears.

It was the second half of the 1990s, and the delinquent was Vicente Paulo Favoreto, a shady timber entrepreneur who had once employed both Pedro and Francisco. One day Favoreto vanished, and his sawmill in Rondon was suddenly closed, leaving behind tens of thousands of dollars of unsettled bills with local employees and suppliers. One of them was the King of Wood, to whom Favoreto owed $70,000 in timber.

Chatter indicated that Favoreto had escaped to Rio. It was said, though, that he had died in a car crash soon after fleeing. But Pedro didn't believe the story. He suspected that Favoreto might have faked his death to evade angry lenders. In fact, it wasn't the first time that Favoreto had apparently disappeared without first fulfilling his obligations; banks had previously filed lawsuits against him for borrowing money he didn't pay back.[1]

Aiming to earn both a juicy commission and Nunes's favor, Pedro proposed a deal to the businessman: he would go to Rio to check the information and eventually collect the arrears, and in exchange, Nunes would give him a cut. Nunes found the story credible and accepted, asking Pedro to also collect payment from a furniture factory in Minas Gerais that had bought timber on credit.

On March 23, 1998, *Folha de S. Paulo* reported the deaths of two men in a multiple-car collision that took place on a highway connecting Rio de Janeiro and São Paulo. One of the victims, according to the paper, was Favoreto, who was forty-four. Pedro spent several weeks investigating in Rio and confirmed that the death had indeed occurred—or so he told Nunes. Afterward, he collected the debt from the furniture company in Minas Gerais, earning Nunes's trust to the point of becoming a regular contractor.[2]

That morning in the courtroom, Francisco claimed that up until the day Pedro went after Favoreto, his brother's first assignment for the King of Wood, he had never heard anything about Nunes being involved in felonies. "He was an exemplary boss," Francisco declared when questioned about his years as an employee of the businessman.

It was only when Pedro got more involved with Nunes that Francisco realized his former boss had a dark side. The witness then testified about three bloody crimes his brother had said he committed for or in collaboration with Nunes. Francisco had previously mentioned them to Detective Resende, but now he provided further details.

The first was the execution of a pair of Nunes's former workers who had stolen valuable equipment from his sawmill before fleeing to Maranhão and hiding in Presa de Porco. Pedro and another gunman working for the rancher—a Rondon-based policeman, in Francisco's account—had hunted the thieves down, executing one of them with a shot in the head and the other with a bullet in the chest. "Pedro showed me the photographs. . . . He said he had also shown the photographs to Nunes as proof of the murder. Then my brother ripped up the evidence in front of me," declared Francisco.

The murders at Presa de Porco, he proclaimed, were the first executions Pedro had ever committed.[3] A third hired murder was that of Bodão, a driver and an overseer of lumberjacks. Bodão was allegedly on good terms with Nunes until, one day, according to Francisco, the boss ordered Pedro to kill him, a crime the gunman had said was intended to "burn the archives."

But the most damning account for Nunes's image would be that of the homicide of Piaui, a former worker who Francisco said had threatened the entrepreneur with litigation over unpaid salaries. Francisco explained to the court how he got to know about it. He said that his wife, Maria, had been at Pedro's home with her sister-in-law, Regina, when she spotted Nunes in a black pickup truck parked in front of the house. The businessman didn't leave the vehicle, but Maria recognized his face through the lowered window.

"Pedro then took two hoods, one black and another pumpkin-colored, went out, got into the vehicle, and left," Francisco said, recalling what his wife had told him.

Later that day, around lunchtime, Nunes drove Pedro back home. Maria reported to her husband that when Pedro entered the house, he was "very nervous and his boots were stained with blood." Regina immediately questioned him, but he replied that it wasn't important and went into his bedroom. The two women wondered about Pedro's odd reaction. By that time, it wasn't a secret in the family that Pedro might be involved in criminal activity.

At dusk Nunes showed up a third time, according to Francisco. Before Pedro and Nunes disappeared again in the pickup, Maria saw a pair of old tires in the bed of the truck. She shared this information with Francisco, who asked Pedro about it later that evening. Pedro assured him that everything was all right, but when Francisco announced that he would be going to Dom Eliseu, Pedro immediately became worried. "Don't go tonight. A murder took place on that road. You may get into trouble," he warned.

Francisco set off the next day and stopped at a bar by the highway to get information about the murder. A bartender confirmed that the night before, a man had been killed and his body had been burned along with some tires. "It seems that the one who did it is the owner of a sawmill in Rondon," the source declared, according to Francisco. Back in Rondon, Francisco asked his brother again about the issue.

"I'll tell you the truth, but don't share the story with anyone, even your wife," Pedro said.

According to Francisco, Pedro told him that he and Nunes had gone to Piauí's home right after Maria had spotted the fazendeiro in the pickup. Pedro had entered the house while Nunes waited in the vehicle until Piaui was brought out. Then they had taken him to an isolated area near a fazenda and the rancher had executed him. In the afternoon, when Nunes returned with the old tires, they had burned Piaui in a bush close to the BR-222. (A police report naming Pedro and Nunes as suspects stated that a burned male body had been found in the BR-222, but the deterioration of the corpse didn't allow them to identify the victim, so he was interred in a mass graveyard where homeless people were buried. Ultimately, no charges had been pressed against Nunes; Francisco would later say that the family of Piaui had received money to leave town.)[4]

At that point of the testimony, Francisco made the connection between those crimes and the killing of Dezinho. The union leader, he explained to the jury, had obtained information about some of these executions and reported them to the authorities. In Francisco's account, that issue had prompted Nunes to hire Pedro to assassinate Dezinho. He attempted the job twice, said the witness, revealing that Dezinho had reported Pedro's threats to the police.

A crucial—and never fully uncovered—part of the story involving Pedro, Nunes, and Francisco was the reason the gunman had failed to kill Dezinho. Francisco claimed to have had a fundamental role in preventing the crime from eventually happening, but he offered two different accounts. In the statement he gave to Detective Resende, Francisco had said that he was "a friend" of Dezinho and "had met him some four months prior" in the occupation of the Fazenda Tulipa Negra. To the jury, Francisco would argue that he had never described Dezinho as "a friend" to the officers.

Maria Joel's sister Eva posited that Dezinho may have agreed for Francisco to join the occupation of the Tulipa Negra in order to control Pedro through his brother. "Dezinho always said to me that one had to keep the enemies close," Eva would explain. "He probably thought: *I'll take Francisco to the occupation, and Pedro will feed me information from the fazendeiros.*"

Francisco's argument ran like this: the fact that Pedro, only some months prior to Dezinho's execution, had been assigned the job by Nunes proved that the King of Wood had been the instigator of the homicide. Francisco would also tell the jury that Pedro had actually been murdered for failing to kill Dezinho. He had previously said so to Detective Resende, and that statement had prompted the officer to open an investigation into Pedro's case. But now he offered details of how his brother had been killed, and this part of his testimony would be the most riveting—and a firsthand account supporting his assertion that several fazendeiros in Rondon had coalesced to form a criminal consortium.

Pedro was shot in a bar around 11:00 p.m. on November 3. At the time of the attack, committed by several men traveling in a car, Pedro was accompanied by a policeman named Luiz Carlos Aviz de Oliveira, apparently a friend.[5]

Francisco was already sleeping when Lucas, another brother, showed up with the news that Pedro had just been shot. Both men raced to the bar, but Pedro wasn't there anymore. Eyewitnesses told them that he had been taken to St. Joseph Hospital, which had been founded by the Lopes family. "At that time," Francisco declared to the jury, "I suspected that if he had managed to survive, Pedro would not leave that hospital alive unless we could get him out."

Francisco and Lucas rushed to the hospital, intending to take their brother to Marabá. Dashing into the one-story building, Francisco and Lucas found Pedro alive, lying on a gurney in the emergency room, his clothes soaked in blood. According to Francisco's testimony, Dr. Antônio Lopes de Angelo was beside him. An hour had passed since Pedro had been shot in the hand, the chest, and the abdomen, but according to Francisco, Pedro hadn't received any treatment yet. Promptly, Francisco approached him.

"Are you able to get out of here?" he asked. Pedro nodded.

Francisco asked Dr. Antônio whether he could lend him an oxygen cylinder so he could transport Pedro to Marabá.

"We're here to take care of him," said Dr. Antônio, according to Francisco.

Francisco, deeply suspicious, said angrily, "Fuck! He's been here thirty minutes, and you didn't fucking do anything at all!"

"We were waiting for the family to come!" replied Dr. Antônio.

"You know him quite well, Doctor, you know him very well!" yelled Francisco, signaling to the doctor that he was fully aware who Pedro was and what his links were with Nunes and the gang and that Dr. Antonio didn't need to wait to treat him. (A brother of Dr. Antônio, José Lopes de Angelo, would admit to the police that "he knew Pedro personally because he always treated the relatives that Pedro brought to the hospital for consultations."[6] Francisco would claim that Regina, Pedro's wife, had previously given birth at St. Joseph, and that Nunes had covered the expenses.)

"We'll take care of him!" insisted Dr. Antônio, ordering a nurse to go get some medicine.

Francisco addressed Pedro again; it would be the last conversation they would ever have. "My brother. Who did this to you?"

"My brother," Pedro responded. "You're smart."

According to Francisco, Dr. Antônio told him that Pedro needed an urgent blood transfusion and asked him and Lucas to get their blood type tested. Only Francisco's was compatible. While blood was being drawn, Dr. Antônio approached him, resting his hand on Francisco's shoulder. "Unfortunately, your brother has left us," he announced.

Luiz, the officer who had accompanied Pedro when the attack had been carried out, was waiting outside the hospital. He would admit to the police that, right after the attack and prior to going to the hospital, he had gone to Nunes's house to inform him about the crime.[7] Luiz, a suspected coconspirator, was also spotted by eyewitnesses talking at the entrance of the hospital with three people inside the car allegedly used by the men who had shot Pedro.[8]

After Pedro passed away, something happened that made Francisco even more suspicious. He testified that Dr. Antônio, right after informing him about Pedro's death, asked him some odd questions. "He wanted to know whether I knew about my brother's criminal life and who Pedro worked for," Francisco told state attorney Franklin. There was a profound silence in the courthouse, and the jury followed this part of the testimony with rapt attention, as if they were watching the climax of a soap opera. "For sure I denied it! I said to Dr. Antônio, 'I know that my brother made a lot of mistakes, but who he did them for,

I don't know." I wasn't stupid. If I had told him the truth, I wouldn't be here today with you!"

Joaquim Carlos Costa, a doctor at St. Joseph, would tell the police that Pedro had died while he was being taken to the operating room.[9] Both he and Dr. Antônio would admit they were in the hospital that night and had met Francisco, but they denied any negligence or participation in Pedro's death.[10] Their statements were taken by Resende only after the detective, who had originally been sent to Rondon to investigate Dezinho's murder, found something murky in the fact that Rondon's police hadn't opened an investigation into Pedro's killing and decided to open a probe himself. Resende had believed the murders of Pedro and Dezinho were connected.

Looking for leads, Resende had also interrogated other men suspected of participating in Pedro's murder, eventually pressing charges against three men: Luiz; a councilman named Francisco Veloso de Freitas; and the overseer of the Fazenda Primavera II, Paulo Ferraz Rodrigues.[11] On December 27, 2000, Resende had also requested an autopsy to establish the cause of Pedro's death. Astonishingly, five years passed before that autopsy was performed. By then, the deteriorated state of the remains made it impossible to establish the cause of death.[12] Ultimately, this issue would become crucial to the eventual shelving of the case. Without an official report stating that Pedro had died by gunshots, prosecutors found no evidence to press charges against Luiz, Veloso, and Ferraz.[13]

But much earlier than any of these events—actually only hours after Pedro's death—Francisco had made a decision that would be of great importance to Dezinho's case. He went straight to Pedro's office and forced open a cupboard. He found a handful of audiotapes, documents, radio transmitters, and dozens of photographs, many of them of clandestine graveyards used to dispose of bodies. Days later, Francisco asked his mother-in-law to be his mule and transport the evidence—hidden under her clothes and stuck to her body with plastic film—to Dom Eliseu.

Francisco claimed he didn't know in advance about the existence of the secret archive, although he said he was aware that Pedro occasionally recorded phone conversations. There's a real possibility, though, that Francisco knew beforehand and went to the office with the plan to enact a bloodless vendetta against the people he accused of being part of the criminal consortium. Whatever the case, the tapes, although they didn't reveal any conclusive evidence of the alleged existence of the murderous gang, were pivotal in bringing Detective Resende to the conclusion that Francisco's statement was reliable and that Nunes might have hired the murder of Dezinho. (The conversations in the tapes also seemed to indicate that Favoreto had in fact succeeded in faking his death, that he later met Pedro in a coastal city of Brazil, and that he planned to escape to the United States using a fake ID.)[14]

IT WAS 12:30 p.m. when Francisco concluded his testimony and was removed from the courtroom. In total, his deposition had lasted about two hours.

Although Francisco's testimony was not unflawed and many questions remained—for instance, why he and his family had looked the other way as far as Pedro's criminal activities were concerned if they were not emmeshed in them—defense attorney Lauria had failed to expose any contradictions during cross-examination.

Lauria had warned the jury that Francisco was delivering "a big farce," but never did Francisco grope for his words or recant or deviate from his previous statement to Resende, despite the great number of details Francisco had provided and the fact that over a decade had passed since he had made his revelations. When Lauria had formulated questions and had sometimes misquoted Francisco's previous statement during the cross-examination, the witness had displayed an impressive recollection and had reacted almost smugly, correcting the lawyer on the passage, the sentence, or the word Lauria had attempted to attribute to him. The underlying feeling was that Francisco, despite his early anxiety, had been in full control of the narrative; that he knew too much, much more than he could ever admit under oath; and that he also knew how to use that information. "*Quem com ferro fere, com ferro sera ferido,"* Francisco had said to Lauria, quoting the Portuguese translation of the biblical proverb that says, "He who lives by the sword shall die by the sword."

Many questions remained unanswered about who Francisco was, the most evident being whether he had been a pistoleiro, an assumption that he would rebut. At that point in the trial, however, the issue of Francisco's background seemed largely irrelevant to the verdict. The emotional impact of the man in the ski mask and the cohesive nature of his account—with Pedro depicted as the key man in a crime with many layers and players that had originally been expected to remain unsolved—had made Francisco's testimony central to the outcome of the trial. Whatever else would be said in that courtroom henceforth, it would orbit around his deposition. Therefore, if Nunes wanted to dodge a conviction, he would have to deliver a powerful account that could counter Francisco's accusations. The rancher would soon have his chance, but before that, other important witnesses were called to testify.

After a recess ordered by Judge Flexa, state attorney Franklin informed the court that he would pass on examining his third and last witness, the former president of the Sindicato, José de Soares Brito. But Lauria said he wanted to hear from Brito, probably to explore any rancor he might feel after his arguments with Dezinho and Maria Joel, so the activist was brought to the bench.

Brito, who had suffered a stroke some years prior that had affected his ability to recall dates, had no intention of betraying his erstwhile partner. "We didn't have enmity; we had differences," he said, referring to Dezinho. He also confirmed that the Sindicato had been informed by former employees that Nunes had an alligator-infested pond and that, days before his murder, Dezinho was scared because he had learned of "a plot to assassinate him" that was being planned by the King of Wood. Still, Brito acknowledged that the information was hearsay.

Again, the name of Josélio was mentioned in the courtroom. Defense attorney Lauria put several questions to Brito about the rancher, who, in his argument, may have played a role in Dezinho's murder. Brito was questioned about the escaped workers and the Te-Chaga-U raid, and he recalled that "human remains were indeed found" by the police in the fazenda. Asked whether he believed ranchers had formed a criminal gang to confront the Sindicato with blood and bribes, Brito answered affirmatively and said, "The first name [of this criminal group] was that of Josélio, because I had endured two attacks, and Josélio had been involved in both." He was referring to the 1991 arson of his home and the 1996 kidnapping attempt.

Brito concluded his statement, and Nunes was finally allowed to return to the courtroom where his fate was being decided.

The defense's first witness to testify was a police officer who was on duty the day of Pedro's murder. In his brief statement, he said that the night Pedro was shot, eyewitnesses had told him that a car had hurriedly left the bar where the crime was committed. The second witness was a smallholder and councilman from Dom Eliseu named Alberto Nogueira dos Santos. He was what is called in legal jargon a character witness—that is, a person who, despite having no information on the crime itself, has known the defendant for a long time and can provide background information about his life and how he interacted with the community. He described Nunes as a dedicated businessman who helped his neighbors and employed hundreds of breadwinners.

More than six hours had passed by the time Nunes had his chance to talk and tell his version of the story. Nunes was given a microphone and was asked by Judge Flexa to confirm his full name and address. The judge also asked him some introductory questions.

"Did you know the victim?" Judge Flexa asked, referring to Dezinho.[15]

"I didn't know him," Nunes said.

"If you didn't know him, you never had contact with the victim," commented the judge.

"Absolutely not," Nunes replied.

"What is your involvement in this crime?"

"None, Your Excellency. Absolutely none," stated Nunes, pleading not guilty.

The judge wanted the defendant to explain the essence of his relationship with Pedro. Nunes had previously admitted to the police that

Pedro had performed some "services" for him, notably the collection of debts, and that he had paid Pedro's expenses during business trips he made to Rio de Janeiro and Maranhão, but now it was time to make clear, and in detail, that there was nothing illegal about it. Nunes confirmed both the information he had provided in the previous statement and the fact that Pedro had pursued Favoreto. "Pedro called me from Rio saying that Favoreto had died," he recalled. Then Judge Flexa asked him whether that was all the business he and Pedro had done together.

"Then," he replied, speaking a bit hesitantly, as if weighing his words carefully, "he made some further debt collections for me . . . some other things, Your Excellency."[16] A long silence ensued. Nunes never explained to the jury what "other things" Pedro had done for him, choosing instead to equivocate on the issue.

Judge Flexa shifted topics, asking Nunes whether he had any pending labor lawsuits. The businessman explained that he did have some cases pending, but in his opinion, such litigation "was common" for a man like him, "with so many workers."

"In any of those legal actions was it established that you employed people in slave-like conditions?" the judge inquired.

"No, no, no. . . . I was accused of it, but it wasn't found [to be] slave labor," replied the defendant.

State attorney Franklin was then given the floor. He asked Nunes about the accusations made by Francisco—the murders at Presa de Porco and those of Piauí, Bodão, Pedro, and finally, Dezinho. Nunes was adamant that all of it was a bunch of lies. "No" or "not true," he answered as the prosecutor enumerated each execution, failing to tell his own version of events or provide counterarguments that could cast doubts on Francisco's account. He offered no viable explanation about what could have prompted the witness to tell all those lies—if they were truly lies—about him.

The argument Nunes followed to support the claim he was innocent was that none of his fazendas had been occupied by the Sindicato, and therefore he had no reason to order Dezinho's murder. He even asserted that he didn't know the activist at all, a statement that was untenable and was picked apart by a member of Franklin's team.

"You didn't know him closely, you mean?" the attorney inquired.

The rancher replied that he was so busy working that, up to Dezinho's death, he had never heard about the activist and his work at the Sindicato. However, in June 1999, about a year prior to the murder, Nunes had taken legal action against Dezinho and the Sindicato after the media had reported that occupations were being organized against the Fazenda Lacy. That same year, the spread had also been inspected by Sergeant Marino because Dezinho and FETAGRI had reported the existence of illegal deforestation and slave labor in the Fazenda Lacy.

Afterward, state attorney Franklin brought up Josélio. He told Nunes that the defense witnesses believed the story that he was "the Judge" and had succeeded the former Contestado man in Rondon. "At the beginning, the leader of the [criminal] group was Josélio," Franklin said, "but afterward, you rose socially and became the leader of the group."

"That never happened," replied Nunes laconically.

The defendant stated that he'd only been an entrepreneur—a pioneer who had moved to the Amazon to prosper. "I arrived at Rondon at the end of the 1970s, attracted by the propaganda of the federal government that there was 'land for people, for people without land.' I always saw that place as ideal for me. I always liked to work hard. I came [to Rondon] when I was twenty-two and knew no one beforehand. I found there, in Vila Rondon, what I needed: a lot of work. I started in logging; I opened a company," Nunes said, now responding to his own lawyer's softball questions. "The business was very good, first of all for my perseverance. . . . I started to buy land to be able to harvest the wood and started having two kinds of businesses at the same time: timber and cattle ranching. It all worked very well, and today I have a modern breed of cattle—very good, excellent genetics."

After the self-laudatory remarks, Nunes concluded his hour-long deposition by depicting himself as a victim in a miscarriage of justice. "I had absolutely no reason to kill Dezinho. . . . For about fourteen years now I've been receiving attacks from the media, the newspapers, the televisions. . . . I never defended myself [publicly]. Why did I never defend myself? Because I had many obligations, many things, a lot of people who relied upon me," he went on, referring to his hundreds of employees. "I couldn't cause instabilities in my business. . . . And so today, I'm here for the court to recognize my innocence."

WITH NUNES'S TESTIMONY concluded, prosecutor Franklin and defense attorney Lauria presented their closing arguments. It was their last chance to influence the seven jurors.

Franklin was the first to talk. Having memorized the names of the jurors and their faces, he addressed them individually, insisting that Francisco had delivered a "robust and firm testimony."

"What just happened here," Franklin proclaimed, his voice rising, "was unprecedented in my twenty-year career. . . . For a witness to come here, to have the courage to come and testify before everyone, before the public, that it was him, Delsão [Nunes], who ordered the murder of Pedro, Bodão. . . This is unprecedented."

Anticipating the defense's strategy to present Nunes's companies as essential for breadwinners in Rondon, the state attorney, almost yelling and gesturing wildly, asked the jury, "[Is] the murder of Dezinho going to be unpunished because he is a businessman?! . . . I won't allow it! I'll fight until the end; I promised the widow to have justice served. I won't accept [Nunes being acquitted] because he is rich, because he is a businessman, because [his prison sentence] could bring Rondon do Pará to ruin!"

Defense attorney Lauria remained more sober. He pressed the jurors to think hard before "wiping out both the life of Décio [Nunes] and a great share of Rondon's economy." He described Francisco's testimony as "crazy" and insisted on Nunes's alleged lack of motive, citing other landholders as the ones who had a direct interest in the death of the activist. Again, Josélio's name was mentioned in the courtroom.

"Josélio hated Dezinho because he was put under investigation based on Dezinho's charges," asserted Lauria. "But why accuse Décio [of the homicide]? It was a strategic choice. At that time, nobody wanted to get justice for the death of Dezinho; at that time, they wanted to make the death of Dezinho an instrument to fight for land," argued the defense attorney. "Why pound on Josélio, who was already in a clear decline? They chose Décio [Nunes] so as to pound on the strongest one. . . . If Décio was indicted, arrested, and sentenced, all the land invasions in the regions would be legitimized," he concluded.[17]

Judge Flexa announced it was time for deliberation and instructed the jury to go to the secret room. He and his assistants left the bench, and tension filled the courtroom. The judge returned quickly, subsequently ordering everyone to stand. Holding a microphone, he quoted Martin Luther King Jr. ("Injustice anywhere is a threat to justice everywhere") before finally reading the verdict rendered by the jury. It was 9:20 p.m. on April 29, 2014, and after almost twelve hours of trial, darkness had fallen over Belém and the muddy waters of the Pará River.

In the first row of chairs reserved for the public, Maria Joel awaited the outcome transfixed. She was in the middle of a human chain formed by her and her four children. The widow had her arms crossed and her thin lips pinched. The four children—wearing black T-shirts emblazoned with the phrase "My father, my dear! Eternal longing . . ."—had strained faces. Their severe gazes were fixed on Judge Flexa.

"The jury found Décio José Barroso Nunes guilty of being a coconspirator and mastermind of the crime that resulted in the death of José Dutra da Costa," read the judge, sentencing Nunes to twelve years in prison. However, due to what he described as good behavior and a cooperative attitude—"he has been appearing in court since the preliminary hearings," he said—Judge Flexa ruled that Nunes could appeal the sentence at liberty. Attorney Lauria immediately announced that he would appeal.

The doors of the courtroom were opened, and the media rushed toward Maria Joel, who was surrounded by friends and supporters.

"My heart is filled with joy," Maria Joel told reporters, her voice slightly shaky with the strong emotions.[18] "I said I would never stop pursuing justice."

Accompanied by her children, the widow left the courthouse. She had tears rolling down her face but didn't make a sound during the walk to the car where officers were waiting to escort her back to the hotel. The family was filled with a mix of satisfaction and sadness, a sense of deep relief and yet of profound injustice. Nunes had been sentenced after an almost-fourteen-year legal quest, but he would still manage to dodge prison, at least for now.

That night, Maria Joel, now alone in her room, devoted some words to her dead husband. "Dezinho, wherever you are, I did this for you. Justice has been served. You're still present in my life, in my heart."

The next day, Maria Joel flew to Brasília. She planned to stay away from Rondon for a while. But the courageous woman had no intention of surrendering.

Epilogue

MARIA Joel's quest for justice didn't end that April 2014 when Nunes was convicted. As of this writing, in June 2022, her story is still unfolding.

After the trial, defense attorney Lauria appealed, arguing that the verdict rendered by the jury wasn't supported by evidence presented in court—in other words, that Nunes had been convicted solely on hearsay evidence. On October 25, 2016, a court in Pará sustained the appeal and gave Nunes the right to a retrial. "Nothing concrete was proved in relation to the accusation that he was the instigator of Dezinho's murder. There were only mere suspicions," wrote a judge.[1] A second jury trial would be held almost three years later, but not before new dramatic twists in the legal saga.

In July 2019, the media published an audio recording of a phone conversation in which Judge Flexa was apparently negotiating a bribe of 60,000 reais (some $15,000) with a former mayor who had been ousted for buying votes and aimed to be reinstated.[2] The content of the recording instilled serious doubts about the integrity of a man who had presided over hundreds of jury trials in his career, including some of the most notorious land-related murders in the Amazon, like that of Sister Dorothy.[3] The attorney general of Pará requested an immediate investigation of the issue by Brazil's National Justice Council, the highest organ conducting disciplinary proceedings against judges.[4]

An investigation was opened, and Judge Flexa's choice for a defense lawyer raised questions about his impartiality in Nunes's case. To defend him against these serious allegations, the judge picked none other than attorney Lauria. By then, Lauria was no longer defending Nunes in the case of Dezinho's murder, but the coincidence certainly led many to wonder about the relationship between Lauria and Judge Flexa at the time of Nunes's jury trial and whether those links had somehow influenced the judge's ruling.

Maria Joel and her lawyers had never hidden their suspicions regarding Flexa's impartiality and sense of duty. In jury trials, Brazilian judges have no authority over whether a defendant is found guilty or not, but they determine crucial issues once an individual is convicted, such as the period of imprisonment and whether the sentence must be served immediately. Flexa had in fact given Nunes a rather light sentence—twelve years—and he had virtually suspended it when he'd ruled that the rancher could appeal at liberty. Those two decisions greatly disappointed Maria Joel and her legal team, who believed that, for ordering the hit on Dezinho, Nunes should have been given at least the same punishment as Wellington, who had been sentenced to twenty-nine years. The fact that Maria Joel lived under state protection was also a powerful reason for Flexa to rule that the King of Wood serve his sentence right after the trial. Batista had appealed the two decisions made by the judge, but he was unsuccessful.[5]

The plaintiffs' suspicions about Flexa didn't end there. On April 23, 2019, weeks before the alleged bribing scandal broke, the first retrial of Nunes had been scheduled in Belém, with Flexa presiding over the judgment, but Maria Joel and her lawyer Batista had decided to refuse to participate. They argued that the trial was a sham, intended only to acquit Nunes.

That trial did begin, but just two hours later it was suspended. After the state prosecutor left the courtroom in protest over some controversial decisions by Judge Flexa—such as limiting the time allowed to hear the recorded testimony provided by Francisco during the previous trial—the process was put on hold. Before leaving the courtroom, the state prosecutor, irritated, addressed Flexa from his bench and called the trial a *palhaçada*—a travesty of justice.[6]

Then the media published the audio of the alleged bribery conversation, and the authorities launched an investigation into Judge Flexa. It lasted two years, and the judge and his attorney did everything they could to shelve the case. Lauria argued, among other things, that the voice on the audio recording was not actually Flexa's, though it did sound a lot like him. The defense attorney claimed that the audio leaked to the media had somehow been digitally fabricated by someone for an ulterior motive he never explained.[7]

Responding to this allegation, prosecutors asked Judge Flexa to submit to a voice test. However, once the test was scheduled at the headquarters of the scientific police of Pará, the judge didn't show up.[8] The internal affairs office presiding over the investigation then used a recording from one of Flexa's jury trials to compare the sound of his voice with that of the audio. It matched beyond any reasonable doubt.[9]

On October 13, 2021, a multi-judge state court panel decided unanimously to force Flexa into retirement.[10] As of this writing, Lauria is still appealing that decision at the National Justice Council.[11] A ruling to reinstate Flexa seems highly unlikely.

ON AUGUST 13, 2019, four months after the first retrial of Nunes was suspended because the state prosecutor had left the courtroom, a new trial finally began in Belém. This time, the trial was presided over by Judge Angela Alice Alves Tuma; Maria Joel and Batista decided to participate.

Although state attorneys often shift posts, prosecutor Franklin again represented the plaintiff. Franklin was key in convincing Francisco to testify again, in the retrial. Nunes, slightly fatter and visibly older, had picked a new defense attorney, a tall and bulky man named Antônio Maria Freitas Leite Jr. He'd made that choice carefully—Freitas was the attorney who had succeeded in having Lourival acquitted by the jury.

Though the ambience in the courtroom was more tense than during the 2014 trial, Maria Joel was more persuasive in establishing connections between the murder of Dezinho and the role Nunes might have played. Maybe because of this, defense attorney Freitas, soon after the widow began her testimony, abused her verbally, approaching Maria Joel and screaming menacingly, "You're a liar!" and "We're going to crush this deposition!" The lawyer even wrestled with Maria Joel's lawyers and the courtroom guards as he attempted to approach the widow, probably aiming to intimidate her.[12] Judge Alves, shocked by the lawyer's behavior, suspended the trial for some minutes. Once it resumed, Freitas apologized, but it was too late to repair the damage made to the defendant's already bad reputation. The trial only went downhill for Nunes from that point on.

Francisco, despite what he claimed were renewed attempts to buy his silence, offered a powerful and detailed deposition about Pedro and his criminal links with the King of Wood. Once again testifying with his identity hidden completely, Francisco nevertheless exuded self-confidence. He had no trouble confronting Nunes when his former boss called him a liar. "Pedro told me it was you [who committed the crimes]," said Francisco to Nunes, who was seated a short distance from the witness.

Francisco maintained that if he ever returned to Rondon, he would be a dead man. He was very critical of the treatment he had received from the state after becoming a protected witness and lambasted the police of Pará for never fully investigating the murders of both Dezinho and Pedro, ignoring one of the main leads he had provided—the photographs of spots where bodies had been buried.

Before the jury, Nunes would stick to the account that he was not involved in any crimes. The cattleman reiterated that he didn't know Dezinho and therefore had no interest in his death. He depicted himself as the victim of a legal hell and suggested that Maria Joel was responsible for it. "She uses the death of Dezinho and advances herself in the world by accusing an innocent man like me. She knows it."[13]

The jurors didn't believe him. On August 14, 2019, after sixteen hours of arguments over two days, the jury found Nunes guilty of masterminding Dezinho's murder. He was sentenced to twelve years in prison, but Judge Alves again allowed him to appeal at liberty. The very same night that he was convicted of murder for a second time, Nunes was seen in a pickup truck parked near Franklin's house, according to the prosecutor's account. Franklin believed the action was a veiled threat against him and his family. Weeks later, Franklin went public about the issue and requested his removal from the case.[14]

As of this writing, Nunes is still fighting in the courts to dodge jail time. His attorneys have requested yet another retrial, an appeal that courts have so far dismissed.[15] Many in Maria Joel's entourage believe that by filing any number of appeals and exploiting the flaws in the legal system, Nunes, now in his mid-sixties and claiming to have cancer, might succeed in avoiding ever serving his sentence. His case is not uncommon in Brazil, where money plays a crucial role in enabling people who have been convicted to stay out of prison. As a *Bloomberg* reporter covering the country's business elite would write in a book on Brazilian millionaires, "As long as you can afford to appeal, and your lawyer can submit arcane requests to bog down the process, convictions don't mean much because your sentence is usually suspended during the appeal process."[16] Impunity continues to be the norm in rural Brazil. Data from the CPT show that from 1985 to 2018, about 92 percent of all land- and resource-related murders logged in Brazil— some 1,790 in total, most of them in the Amazon—resulted in no arrest or trial whatsoever.[17]

In addition to the criminal conviction for Dezinho's murder, Nunes faces other legal troubles. According to public records, he and his companies owe millions of dollars in federal and state taxes and dues, while the case about grilagem in the Fazenda Lacy is still unfolding in federal courts.[18] In 2020, Nunes was fined $600,000 for environmental offenses, and a year later he settled a labor lawsuit for 600,000 reais ($120,000).[19]

His controversial record does not seem to have jeopardized the expansion of the business empire he cultivated. In 2020, the meat plant formerly belonging to Nunes, which is now registered under the name R. C. Moreira Costa, his second wife, received authorization from the federal authorities to export beef to Hong Kong, Brazil's second largest meat market.[20] The meat plant has been the object of recent investigations by the Ministry of Agriculture for undisclosed misconduct, and yet the state of Pará has granted the company tax benefits.[21]

Many in Rondon do Pará continue to believe that no indisputable evidence proves he was the mastermind behind Dezinho's murder. It is fair to say that despite the conviction, many unanswered questions remain about the case, partly because Wellington, the hitman, never revealed who had hired out the crime. Since he left the prison on Christmas 2007, Wellington has never been found, and some speculate that he was killed.

However, Wellington's brother Rogério, suspected with Ygoismar (the man on the motorcycle) of being a middleman in the crime, was found. A fugitive since November 2000, Rogério was arrested in July 2020 by law enforcement in Minas Gerais, where he had been using a cell phone number registered under his real name and ID—a mistake that led investigators to him. During the pretrial inquiries, Rogério denied having participated in the plot and claimed to be ignorant of his brother's fate, refusing, for now, to provide information.[22] Jailed in Pará until his trial is held, likely in 2023, Rogério might plead guilty, cooperate, and solve the case once and for all.[23] Or perhaps not, and Dezinho's case will prove again that, in the Amazon, subcontracting a murder to gunmen is an effective strategy to provide impunity for the masterminds behind the crimes.

WHEN I FIRST thought about writing a book on the Amazon, my initial focus was investigating the causes of deforestation. I wanted to understand the reasons behind the forest's destruction, and my plan was for the jungle to have a central place in the narrative.

Maria Joel's case made me change my mind. I realized that her saga, with all its twists and chilling details, epitomized what I consider today to be the central element fostering deforestation, human degradation, and organized criminality in the world's largest rainforest: impunity. My intention with this book is to shed light on the mechanisms that, in most cases, succeed in undermining the rule of law in the Brazilian Amazon. In my view, impunity is ultimately the consequence of actions by controversial individuals, more than of general flaws in the legal system. As the case of Maria Joel shows, when the people accused are wealthy and powerful, policemen and judges often treat them with extreme leniency.

This has real consequences for the Amazon and for the people who fight to protect it. Since I began to investigate this book in 2017, about 180 land and environmental activists have been murdered in Brazil.[24] The number of homicides and the degree of brutality is appalling. Some of the latest victims were a couple and their teenage daughter known for their efforts to protect river turtles in Pará.[25] Another victim was a nine-year-old boy, son of a farmworkers' leader in Pernambuco state.[26] He was hiding under the bed with his mother after seven mobsters invaded their tumbledown home and shot his father, injuring him in the arm. The boy was then pulled out from under the bed and massacred. The mother survived and witnessed the killing.

On a global scale, this horror targeting land and environmental defenders is sharply on the rise. During the period from 2017 to 2022, over 800 people have been slain worldwide. Countries with tropical forests and rich in natural resources like Colombia, the Philippines, the Democratic Republic of Congo, and Cambodia are magnets for violence. The year 2020 was the deadliest for campaigners, at least on record, with 227 lethal attacks, or "an average of more than four people a week," according to Global Witness. Agribusiness, logging, mining, and poaching are the main sectors driving this violence.[27]

The stories of land and environmental defenders are of critical importance, and yet they are underreported. Attempting to tell such stories to the world is a dangerous endeavor. As I was putting the final touches on this book, the federal police confirmed the murders of British journalist Dom Phillips and Indigenous advocate Bruno Pereira. They had disappeared on June 5, 2022, while investigating the efforts at conservation by Indigenous groups in the Javari Valley, a region in the far west of the Brazilian Amazon. Phillips and Pereira were apparently killed by members of illegal fishing gangs. In the Javari Valley, Indigenous organizations have been crucial in exposing the lawlessness caused by the actions of illegal loggers, poachers, and gold miners, especially after President Bolsonaro won his seat and emboldened the offenders plundering the jungle.

The fate of the Amazon rainforest rests on the shoulders of environmental campaigners, Indigenous leaders, land activists, and courageous and determined officials (judges, prosecutors, and state and federal investigators). Every day, they serve on the front lines of social conflict and are key players in the race against climate change.

Still, while writing this book, I thought a lot about the personal price those individuals pay for protecting the jungle and exposing the abuses. I wondered whether, in the end, Maria Joel was a winner or a loser in this story. One of the last times I met with her, I asked her about her view of the role martyrs play in social movements and how her work had contributed to making Rondon a more just place for poor people. "They took Dezinho," Maria Joel replied. "But the seed planted never stops growing and eventually bears fruit."

This view isn't shared by all of the members of her family. Dezinho's children, even if they agree that their father's and mother's struggles resulted in hundreds of poor families getting a plot, do not see—and do not even want to see—their story as epic. Often, when I asked them about those achievements and how their family had been crucial in exposing the lawlessness, they replied with accounts of pain and trauma. They see themselves as having won nothing. After the murder of Dezinho they had decided to challenge the case's apparent outcome (impunity and oblivion), but it was only because, as a family, they had reached the conclusion that if they didn't do something about the situation, it would be much harder to overcome the loss. As Joélima, the middle daughter, told me, they got involved in the fight for justice and memory because, had they not done so, it would have been "as if Dad was murdered twice."

One of the last times I met the family in Rondon I asked Joelminha whether she shared her mother's view that, despite the loss, they had achieved something. We were in Maria Joel's house, and the police officer in charge of protecting the widow was in another room. Despite the noise of an old fan circulating the hot air, we could hear the sounds outside the walled residence—boisterous children playing on the street and the hum of conversations of ordinary people living ordinary lives not shrouded by death and relentless fear.

Joelminha looked at her mother and thought for a moment before replying. "My father knew that he would die, and, after his death, the occupations of fazendas would be transformed into settlements," she told me, tears rolling silently down her cheeks. "But our story isn't a beautiful or triumphant one, because my father was taken away from us."

By then, Joelminha was the object of fierce criticism even by members of the Sindicato. In mid-2016, she had left the Workers' Party to join the conservative Social Christian Party (PSC). Her political shift, an unpleasant surprise for many close to the family, became even more shocking when it was revealed that she would be a candidate for vice-mayor on the ticket with Arnaldo Ferreira Rocha. Arnaldo was a former mid-ranking official in the city hall who had been picked as the mayoral candidate for the PSDB—the party of none other than Shirley Cristina.

Arnaldo was so unknown to the PSDB electorate that townspeople whispered that he was a puppet—a way for Josélio's daughter, banished from holding any political office but commanding the PSDB in Rondon do Pará, had found to remain in power. Indeed, some decisions Arnaldo made later bolstered that thinking, like the choice of Shirley Cristina's husband for a key post in the government. These facts made it even more difficult to understand why Joelminha had decided to participate in the PSC-PSDB coalition, although she maintained that the agreement had been made by the top ranks of the two parties and had not directly involved her and Josélio's daughter. Shirley Cristina had long perceived Joelminha as a rising star in politics and as someone useful for her own political interests; that was probably why she had approached Joelminha back in 2012, on the eve of taking office, to try to persuade her that Josélio had never participated in Dezinho's murder.

Whatever the case, the political alliance worked, and in the 2016 municipal elections Arnaldo was elected mayor and Joelminha vice-mayor. Still, if she had thought that such success might be a jumping-off point for her political career, she was wrong. Her four years in power were anything but smooth. In addition to confronting some obvious fierce criticism from her own group, she would claim that multiple attempts were made to inveigle her into corruption. She also argued that she was routinely sidelined in key decisions at city hall, and that when she spoke out, she began fearing for her life. "It was rumored that I would be killed in a fake traffic accident while driving my motorbike," she told me. (Some sources in Rondon downplayed the risks and criticized the family for continually overstating the dangers surrounding them.)

Ultimately, Joelminha would pay a high political price. In 2018, she failed to be elected state legislator, and two years later, when the new mayoral elections were held in Rondon, she didn't even get the votes to be elected councilwoman. (Shirley Cristina also ran for state legislator and fought in courts to have the right to campaign despite the eight-year ban; in the end, she lost the legal battle, and regardless, she hadn't received enough votes to earn a seat. As of this writing, she is again a candidate, now running for federal lawmaker in the 2022 general elections.)[28]

Why had Joelminha decided to shift her political allegiance? Maria Joel would admit that the choice was "very difficult to accept," although she made clear that she had never intended to dictate her children's future. "Dezinho and I have a story, but they are free to make their own decisions," she said. I pressed Joelminha on the issue, asking her some tough questions. She denied having ever taken a bribe. Ultimately, she admitted that she had decided to join that political alliance because it was a way to get into power and to use her political clout to help her family.

"If today my mother receives protection from the state, it is thanks to my former work as vice-mayor and to my enduring political influence," Joelminha explained, referring to her lobbying with senators to prevent Maria Joel from losing her police escort. Joelminha had shifted parties the same year that Nunes got his sentence annulled and was given the right to a retrial, a decision that severely exacerbated the family's sense of injustice.

It's easy for someone from the outside to make a moral judgment on

Joelminha's controversial decision. Throughout my research, many interviewees would accuse Joelminha of betraying both Dezinho's ideals and her mother's story of struggle. But I wondered whether that was true, or whether that questionable choice was simply a consequence of a deep sense of vulnerability, with clear roots in a perverse system unable to guarantee justice for victims and punishment for offenders. After all, this was a woman who, since her early youth, had dreamed of nothing more than a "normal life."

I tried to put myself in Joelminha's shoes. Using a public office to obtain benefits for oneself, even if it is something as understandable as the protection of a threatened mother, is wrong. There were many contradictions in her decision, but I wondered what my choice would have been had I been confronted with the same dilemma. I asked myself: if my father had been killed, would I have stayed in that lawless town of receding jungle? Would I have remained where people—often hemmed in by their narrow-mindedness and economic dependence— vilified my siblings and me just because we wanted certain criminals to be punished? Would I have had the same courage to refuse to sing the song of propaganda and to keep questioning whether land-grabbers and illegal loggers really deserved any credit for creating an "economic prosperity" sustained by environmental catastrophe and human rights abuses?

While her father was alive, Joelminha was probably the one in the family who most disapproved of Dezinho's staunch determination to build the movement and push the agenda of the landless families, and that criticism didn't end with his assassination. Over time, Joelminha's opinions about her father's career would evolve, and I could see that she was proud of him, but she never hid from me that she had always wanted something different for her family. That didn't mean that Joelminha was betraying her father's memory, much less wishing to forget him. "I think about him every day," she told me.

During one of our conversations, Joelminha told me a story that epitomized how difficult it was for her to deal with her past, especially due to pent-up resentment. One day, after Nunes had been convicted the first time, Joelminha and her eldest son entered a café. Seated at a table in the back was the rancher. This wasn't a common encounter as Nunes lived on his estate, and the families attended very different kinds of social events. When the fazendeiro saw her, he recognized her and lowered his eyes. But she couldn't control herself at the sudden realization that that man, despite the conviction, was enjoying freedom. She looked at her young son and told him, "You see that man? He's the one who killed your grandfather."

Seconds later, she realized her mistake when the boy, roused by his mother's anger, approached the table and fixed his menacing gaze on Nunes.

"What are you doing?!" Joelminha said, pulling him away.

"Mom, if I turn eighteen and this man still isn't jailed or dead, I'll kill him myself."

From that point forward, Joelminha would tell me, she would always try her best to control her anger. But it was hard, she admitted, and I believed her. The state still needs to prove to her—and to the dozens of families who every year lose a beloved one because he or she was trying to save the forest or to fight appalling social inequalities—that breaking the law has real consequences.

There is no real democracy without the rule of law, and the rule of law is still largely lacking in the Amazon. The rainforest's economic frontiers were opened up half a century ago by a military regime that, at the time, promised to spark progress in the forest but, in fact, undertook a state policy that laid the foundations for contemporary social and environmental problems in the region.

AS OF THIS writing, democracy in Brazil seems to be heading toward a crossroads. Opinion polls for the 2022 presidential election show that Lula is the front-runner and might return to power, frustrating President Bolsonaro's attempts at reelection. But even if Lula seems to have a firm lead, the real question remains whether Bolsonaro will accept an eventual defeat or whether he will use his clout over large pockets of far-right voters and the army to cause instability or even a coup d'état. His attacks on the Supreme Court and his deliberate attempts to erode the institutions present a far from reassuring message. Since day one of his presidency, Bolsonaro has been sowing the seeds of chaos.

Lula was never an environmentalist, but he has promised that if he wins, his government will tackle the drivers of destruction. There's room for hope. In his previous two terms, IBAMA was given more power and funds to fight offenders, and federal prosecutors gained more independence to jail criminals. Empowering Indigenous groups with key positions in the federal government—for instance, in FUNAI—and expanding the number of protected Indigenous territories should be a priority, especially to prevent large areas in the western and central Amazon from becoming new frontiers for cattle and grilagem. Many responsible Brazilian agribusiness leaders and fazendeiros agree with the idea that deforestation is no longer needed to expand the production of agro-commodities.

One crucial question is that of the international community's role in protecting the forest and people's rights. China has surpassed the United States and Europe as Brazil's top trading partner. In 2021, trade between China and Brazil amounted to a historic $135 billion, and soy and meat currently constitute nearly a third of all of Brazil's exports.[29] Besides, Brazil runs a spectacular annual trade surplus of some $40 billion. This has made China a pivotal economic partner for Brazil, one essential to its economic stability. China is also a country that poses no questions about deforestation, human rights, or land-related murders, at least publicly. This is not to say that China is ultimately responsible for the jungle's destruction; many other countries buy Brazilian agricultural products unconditionally. But Beijing undoubtedly holds unparalleled sway over Brazil and could play a decisive role in the Amazon's future. Meanwhile, its official silence only weakens the efforts of the international community to pressure Brazil on the issue.

In the frontier towns of Pará, the message I heard from fazendeiros was rather discouraging. "China has freed us from Europe," Agamenon da Silva Menezes told me. He was the leader of an association of big landowners suspected of organizing riots, deforestation, and grilagem along the BR-163—the "soy highway." Near his office in the town of Novo Progresso stood some of the most extraordinary Indigenous reservations of the state, where the Kayapó tribe continued to drive away trespassers from their towering jungle and risk their lives for it.

Not too far from there, subsistence peasants had set their plots and produced crops without the use of pesticides. They supplemented their income by harvesting açai, cupuaçu, cacao, banana, and many other fruits. They had won their land—and gained their dignity—after enduring years of threats by land-grabbers and after losing some colleagues in the battle.

If I am honest with myself, I have to admit I would probably not have been able to respond to those hardships with the courage and fortitude demonstrated by so many others. One day I shared that realization—my innermost feelings—with an extraordinary woman named Maria Joel Dias da Costa. She looked me in the eyes, with empathy and pride, and then said, "You should care about the jungle, but never forget about us."

Acknowledgments

My first thanks go to G.S., O.B., and C.S. I feel incredibly lucky to have crossed paths with them. This is family.

Many people in Rondon do Pará helped me in researching this story. I could fill several pages with their names, but most have chosen not to be identified. I'm grateful to all of them.

Maria Joel and her family submitted to repeated interviews and never refused to answer questions. Brito was of crucial importance in understanding the rise of Dezinho at the Sindicato. Lawyers Carlos Guedes and José Batista Gonçalves Afonso played a decisive role in helping me understand the story of Dezinho and Maria Joel. Francisco devoted many hours to reply to my tough questions about him and his brother Pedro.

I'm thankful to Judge Gabriel Costa Ribeiro, Judge Jônatas dos Santos Andrade, Detective Antônio Carlos Corrêa de Faria, and state prosecutor Franklin Lobato for telling me their stories and devoting time to explain their tenacious work to enforce the rule of law. A special thanks to Judge David Guilherme de Paiva Albano and all the other officials at the state justice of Pará and Espírito Santo who granted me access to court files. Also thanks to Camila Pitanga, Salete Hallack, Father Ricardo Rezende, and Wagner Moura for replying to my queries.

The work of many people at archives and libraries made it possible for me to access confidential files that validated events and details of this story. At the CPT office in Marabá, I owe a lot to Batista, Iranete, Andreia, and Clara; in Vitória, to historian Tiago de Matos Alves from the Arquivo Público do Estado do Espírito Santo; in Pará, to Valdecir at the Arquivo Público do Estado do Pará. In Brasília and Rio de Janeiro, I was very lucky to receive help from committed public servants preserving Brazil's memory and history in the extraordinary Arquivo Nacional and Biblioteca Nacional.

Several people were critical in helping me to reach sources. Renata M., E. Rodrigues and Matheus F. were of great importance.

To the extraordinary Veronica Goldstein, my agent—this book is the result of her confidence in the project. And thanks to Peter Hubbard, Molly Gendell, Andrew Jacobs, Kell Wilson, Laura Brady, Emily Snyder, Yeon Kim, Aryana Hendrawan, and all the people at HarperCollins/Mariner who made this book better. Any errors are the author's alone.

To Hilary L. Gunning for her incredibly important work during the writing and editing of this book. She's determined, patient, and remarkably talented. And *grazie* to the great G.S.

I'm grateful to my friends and great reporters Jenny Barchfield and Melissa Chan, who devoted their precious time to reading part of the manuscript and provided input. Anthropologist Iara Ferraz, who has for years studied the Gavião and defended their rights to their land, also provided input to improve this book. Thanks to my friends Pedro Moreno, Pablo Barrientos, Emiliano Guanella, and his lovely wife Juliana Nunes for their support in Rio, SP, and elsewhere.

Scholars and journalists have previously explored social and economic aspects of Pará that were important for writing this book. I'm grateful for the work of the missionary and historian Jean Hébette and the reporters and photographers of *A Província do Pará, O Liberal, Folha do Norte, Folha Vespertina, A Gazeta, Jornal do Brasil,* and *Diário do Pará* who covered some of the early chilling episodes of the fight for land in Vila Rondon. The complex issue of land-grabbing or grilagem was—and remains—a central one in the Amazon, and I was lucky to benefit from the groundbreaking work of scholars like Girolamo Treccani, one of the top experts in Brazil on the issue, and of Jeremy M. Campbell, a professor of anthropology at George Mason

University, who has written a great book on grilagem titled *Conjuring Property: Speculation and Environmental Futures in the Brazilian Amazon*. The work of geographer Susanna Hecht, historians John Hemming and Barbara Weinstein, and reporter Sue Branford was also very important. Last but not least, a great thanks to Danicley de Aguiar, from Greenpeace Brazil, who first made me aware of Maria Joel's story.

A Note on Sources

This book is a work of nonfiction, and no dialogue or event is invented. The unfolding of events is also presented in the real chronological order.

Very few works exist on the history of Vila Rondon. Most studies are prepared as master's and doctoral dissertations. Therefore, the backbone of my research stems from two categories of sources: the testimony and accounts of some two hundred interviews, mostly performed by me in Rondon do Pará and other areas of Brazil, and about 100,000 pages of documents, most of them unpublished, drawn from Brazilian state and federal courts, notary public offices, press clippings, and private and public archives and libraries located in five Brazilian states, Italy, and the United States.

Oral sources—especially the living memories of pioneers who were willing to talk to me—were crucial for re-creating some passages, events, and places in the most detailed and faithful way possible, which I believed was important to tell this story. Whenever possible, I gave voice to eyewitnesses, but hearsay accounts from secondary sources, like the relatives of dead victims, were also important. Officials were equally central to writing this book, and I interviewed dozens of former ministers, current and former state and federal lawmakers, law enforcement agents, federal and state judges, and prosecutors. Most interviewees were willing to talk on the record, but others asked to remain completely anonymous.

Individual memory is fallible, especially when sources are recollecting events from several decades prior, and the debate over land ownership and frontier expansion in the Amazon is polarized and controversial. For instance, many used the word *fazendeiros* or large landholders as a euphemism for a world of people who are half rural businessmen and half criminals. (Of course when I refer to a person in this book as a fazendeiro, that by no means indicates that he is an outlaw or someone who has ever committed a crime. Most Brazilian fazendeiros are law-abiding citizens and successful and ethical businessmen.) Therefore, I have undertaken extensive cross-checking and have tested the memories of my most important sources in order to write an account as close to the facts as possible. Maria Joel, Brito, and Francisco, for example, were interviewed multiple times. When possible, I also corroborated oral accounts with documentary evidence. As a result, this book contains a large number of endnotes.

Brazil and its constitutional and legal framework allowing journalists to access public information and documents deserve credit, and the evidentiary lattice of this work would have been much different otherwise. Through Brazil's freedom of information act (Lei de Acesso à Informação. or LAI), I filed dozens of requests for access to documents from the municipal, state, and federal governments, and, in most cases, I received the information I requested. LAI helped validate many stories I was told orally in Rondon do Pará. Some of the most sensitive material of this book, however, was already available to the public in the extraordinary Arquivo Nacional, especially hundreds of confidential dossiers produced by several information agencies during the military dictatorship (1964–1985).

Accessing court documents proved slightly more difficult and time-consuming. I had to hire lawyers to get access to some court cases archived in Rondon do Pará's house of justice, and the files of the "chainsaw case" were not found in the archives when I was granted access by the president of Pará's justice system. But in general terms, and despite some bureaucratic snags, I felt that the right to information prevailed in most cases, and civil servants and judges granted me unrestricted access to information crucial to writing the book.

According to written official records, Josélio de Barros Carneiro admitted to his participation in five murders: three in Espírito Santo and two in Rondon do Pará.[1] Several reliable accounts attribute many more crimes to him, though. This myriad of accusations made it difficult for me to verify what was true and what was perhaps an exaggeration or part of a defamation campaign, as Josélio often claimed. Therefore, to tell the most controversial parts of his personal story, I have relied almost entirely upon written records from official sources, such as the documents of some twenty civil and criminal cases. They validated, in most cases, what I had heard about Josélio from sources in both Pará and Espírito Santo.

Understanding the backgrounds of Josélio and Nunes was fundamental, and I have also relied on oral sources sympathetic to them. One of Josélio's brothers, José Francisco de Barros, was interviewed twice. His information proved fundamental to understanding Josélio's early days and the story of his family in the complex and volatile Espírito Santo of the 1950s and 1960s. However, none of Josélio's daughters or his son cooperated with the research for this book. I was particularly interested in interviewing Shirley Cristina, who was Rondon's mayor in addition to Josélio's most trusted daughter, because she had her own story to tell. But she never responded to my requests, although she has given many interviews to the media over the years. Only one of Josélio's six children ever replied to my emails, phone calls, or letters— but only to say that she'd never say a word to me, and that if I persisted in requesting an interview, she would sue me.

I also contacted Décio José Barroso Nunes through one of his brothers and, more important, through his lawyers, but I failed to get an interview with him. The two attorneys who represented him in the 2014 and 2019 jury trials for the murder of Dezinho, Roberto Lauria and Antônio Maria Freitas Leite Jr., respectively, also declined to be interviewed.

Still, this book is largely built on information from Josélio and Nunes. Both fazendeiros have been involved in civil and criminal litigation for several decades. Their version of the facts was in tens of thousands of court files, as well as in occasional interviews with the press, so I have incorporated their words in the text and duly annotated the sources. Some of Nunes's defense attorneys in civil cases also granted me interviews.

Anything that appears in the text between quotation marks has been told to me during interviews or is sourced in police files, court documents (including transcripts and informant statements), diplomatic cables, and files from the Brazilian congress and senate, the Indian Protection Service (SPI), the National Indian Foundation (FUNAI), the Brazilian Institute of Environment and Renewable Natural Resources (IBAMA), the Land Institute of Pará state (ITERPA) and the National Institute for Colonization and Agrarian Reform (INCRA). I've also drawn upon a wide range of documents from institutions such as the International Committee of the Red Cross (ICRC); Centro Missionario—Diocesi di Piacenza Bobbio; the Pastoral Land Commission (CPT); and Armazém da Memória. They are all mentioned in the notes.

When I attribute some thoughts to the sources, it is because those sources expressed them to me or to an interviewer like a police officer, a judge, or an attorney, or I read about it in official documents. At times, the thoughts of some characters were conveyed to a third person I have later interviewed. This is the situation, for example, with Maria Joel (Dezinho) and with Francisco (Pedro).

Notes

PREFACE

1. Rondon do Pará has 8,246 square kilometers, according to the Brazilian Institute of Geography and Statistics (IBGE).

2. In Brazil, the main source for data on land-related murders is the Pastoral Land Commission, which the author cites extensively in the book. However, for a global comparison, the author has chosen here the data provided by Global Witness on annual reports from 2002 to 2021.

3. For a global overview of the conflicts and the roots of violence against campaigners, see Arnim Scheidel et al., "Environmental Conflicts and Defenders: A Global Overview," *Global Environmental Change* 63 (2020): 102104. DOI: 10.1016/j.gloenvcha.2020.102104. Also important to note is the data from the Environmental Justice Atlas database (www.ejatlas.org).

4. Data from Seth Garfield, *In Search of the Amazon: Brazil, the United States, and the Nature of a Region* (Durham, NC: Duke University Press, 2013), 8.

5. "Making Sense of Amazon Deforestation Patterns," undated, Earth Observatory, NASA https://earthobservatory.nasa.gov/images/145888/making-sense-of-amazon-deforestation-patterns.

6. Jochen Schöngart et al., "Age and Growth Patterns of Brazil Nut Trees (*Bertholletia* excelsa Bonpl.) in Amazonia, Brazil." *Biotropica* 47, no. 5 (2015): 550–58. DOI: 10.1111/btp.12243.

7. On the pet jaguar story, pictures showed to the author by pioneers.

8. Referring to the small town of Buriticupú, near the border between Pará and Maranhão.

9. Referring to the city of Novo Progresso, in Pará. On the bullets, see Jesse Hyde, "The Lawless Frontier at the Heart of the Burning Amazon," *Rolling Stone*, September 17, 2019. On the attacks against helicopters and vehicles, see "Manifestantes invadem sede do Ibama no Pará," *G1*, April 5, 2011, and *Folha de S. Paulo*, July 7, 2011. For a larger story of the city, see Mauricio Torres, Juan Doblas, and Daniela Fernandes Alarcon, "'Dono é quem desmata':

Conexões entre grilagem e desmatamento no sudoeste paraense," Instituto Agronômico da Amazônia, 2016, https://www.socioambiental.org/sites/blog .socioambiental.org/files/nsa/arquivos/dono_e_quem_desmata_conexoes _entre_gril1.pdf; and Jeremy M. Campbell, *Conjuring Property: Speculation and Environmental Futures in the Brazilian Amazon* (Seattle: University of Washington Press, 2015).

10. Quote is from Professor Clifford Welch in Global Witness, *Deadly Environment* (London: Global Witness, 2014).

11. Barbara Demick, *Eat the Buddha: Life and Death in a Tibetan Town* (New York: Random House, 2020). Quote is from the Author's Note.

12. Carl Bernstein, *Chasing History. A Kid in the Newsroom* (New York: Henry Holt, 2022). The quote is from chapter 20. His colleague Bob Woodward offered a slightly different but equally valid definition for journalism: "the best obtainable version of the truth."

1: THE ESCAPE

1. Gil's account based on statements provided to the police on June 28, 1995, and July 5, 1995, court case 0000272-32.1997.8.14.0046, Pará state.

2. On the meeting with Manuel and the bonfire, Gil's statement to the police. Description of the place using police photographs filed in court case 0000272-32.1997.8.14.0046. Also based on author's interviews with Luiz Cavalcante and Sueny Feitosa.

3. On his previous fears, Gil's statement to the police on June 28, 1995, and July 5, 1995, court case 0000272-32.1997.8.14.0046. He would say to law enforcement that it was whispered that "when Josélio didn't like a laborer or didn't want to pay them, his hitmen killed them."

4. The ranch's description is based on author's interview with former laborers and police photographs filed in court case 0000272-32.1997.8.14.0046.

5. Diana Jean Schemo, "Amazon Is Burning Again, as Furiously as Ever," *New York Times*, October 12, 1995.

6. Gil's statements to the police on June 28, 1995.

7. Gil's statement to the police on July 5, 1995, in court case 0000272-32.1997.8.14.0046 and police inquiry 95-DRPA-018/95.

8. Ceará's story and description is based on the information provided by Gil to the police and police report 95-DRPA-018/95, court case 0000272-32.1997.8.14.0046.

9. John Hemming, *Tree of Rivers: The Story of the Amazon* (London: Thames & Hudson, 2008), quote from chapter 10.

10. The recruiter's full name was Francisco das Chagas Abreu. Here I refer to his statement to the police on July 27, 1995, and *Correio do Tocantins. O Jornal de Carajás*, July 21–27, 1995. Josélio confirmed in his statement dated August 18, 1995, that Chico recruited laborers for his ranches.

11. In his statement to the police on July 27, 1995, Chico admitted he recruited

workers in hotels in Paragominas to work as laborers in Josélio's land and paid their bills.

12. On the size of the fazenda, see Escritura Pública, Livro 113, Folha 106, Cartório Condurú, dated May 20, 1987, and contained in civil court case 0010474-32.2017.8.14.0046, Rondon do Pará. About the meaning of Te-Chaga-U, Juan Francisco Recalde, "Vocablos designativos de relações e contatos sociais nas linguas tupi e guarani," *Revista do Arquivo Municipal de São Paulo*, September 1937.

13. Gil's statement on June 28, 1995, court case 0000272-32.1997.8.14.0046.

14. Josélio's physical description based on confidential file ACE 3582/83, August 28, 1975, and eyewitnesses.

15. Gil's statement on June 28, 1995, court case 0000272-32.1997.8.14.0046 and police inquiry 95-DRPA-018/95.

16. Gil's statement on June 28, 1995, court case 0000272-32.1997.8.14.0046.

17. Gil's statement on June 28, 1995, court case 0000272-32.1997.8.14.0046.

18. Gil's statement to the police on June 28, 1995, and July 5, 1995, court case 0000272-32.1997.8.14.0046.

19. Gil's statement to the police on July 5, 1995, in court case 0000272-32.1997.8.14.0046 and police inquiry 95-DRPA-018/95.

20. Luiz Cavalcante and Sueny Feitosa, interviews with the author.

21. Gil's statement to the police on July 5, 1995, in court case 0000272-32.1997.8.14.0046 and police inquiry 95-DRPA-018/95.

22. Gil's statement to the police on July 5, 1995, in court case 0000272-32.1997.8.14.0046 and police inquiry 95-DRPA-018/95.

23. Gil's statement to the police on June 28, 1995, in court case 0000272-32.1997.8.14.0046.

24. The name of Sueny Feitosa Cavalcante is cited in police statements as "Sueli," but in interviews with the author, the source said it had been misspelled.

25. Luiz and Sueny, interviews with the author.

26. Sueny in statement of July 5, 1995, and Luiz of June 28, 1995, and July 5, 1995, court case 0000272-32.1997.8.14.0046.

27. Quote by Sueny in statement of July 5, 1995, court case 0000272-32.1997.8.14.0046.

28. Luiz's statement of July 5, 1995, court case 0000272-32.1997.8.14.0046.

29. *Diário do Pará*, April 27, 1990, and May 17, 1990.

30. Interviews with the author and Luiz's statement on 05/07/1995.

31. Sueny's interviews with the author.

32. Luiz's interviews with the author.

33. Quote is from author's interview with Sueny. Her claim seems to be confirmed by the statement of Amaro José de Sousa, a Te-Chaga-U worker who said that he had health problems and received no help. Statement provided by Amaro to the police on July 14, 1995, court case 0000272-32.1997.8.14.0046.

34. "Sindicalista é jurado de morte," *O Trabalhador Rural*, CONTAG, June 1993.

35. According to the author's interviews with Rita Serra, a FETAGRI leader, Dezinho had told her that "a friend who works on the ranch tells me that people enter that fazenda but never leave." On Dezinho's suspicions regarding Josélia Leontina, author's interviews with Maria Joel, Brito, Zé Geraldo, Rita Serra, and Sergio Galiza.

36. The exact date of departure wasn't recalled by Sueny and Luiz in interviews with the author, but the information is consistent with accounts provided—to either the author or the police—by Brito, Gil, Zé Geraldo, and Rita Serra.

37. "Report on the Situation of Human Rights in Brazil," the Inter-American Commission on Human Rights, OEA/Ser.L/V/II.97, Document 29, revision 1, September 29, 1997.

38. Sueny and Luiz in interviews with the author.

39. *A Província do Pará*, July 17, 1991.

40. *A Província do Pará*, July 17, 1991.

41. Author's interview with Zé Geraldo.

42. Author's interview with Zé Geraldo.

43. For a historical background on Brazil's forced labor, see International Labor Organization, *Fighting Forced Labour: The Example of Brazil* (Geneva: ILO, 2009).

44. For the working conditions in the rubber areas or *colocações*, see Garfield, *In Search of the Amazon*, and the classic book on the topic: Barbara Weinstein, *The Amazon Rubber Boom, 1850–1920* (Stanford, CA: Stanford University Press, 1983).

45. On the exchange of goods for work during the colony, see Weinstein, *The Amazon Rubber Boom, 1850–1920*, chapter 1, "Tappers and Traders." For the importance of rubber made out of natural latex, see also review by Patricia Perkins, "*The Amazon Rubber Book 1850–1920* by Barbara Weinstein," *Technology and Culture* 26, no. 4 (1985): 865–67.

46. Susanna Hecht and Alexander Cockburn, *The Fate of the Forest: Developers, Destroyers, and Defenders of the Amazon* (Chicago: University of Chicago Press, 2010), 72.

47. For a history of abolition in the Americas, see Laird W. Bergad, *The Comparative Histories of Slavery in Brazil, Cuba, and the United States* (Cambridge and New York: Cambridge University Press, 2007). See also chapter 4 of Thomas E. Skidmore, *Brazil: Five Centuries of Change* (New York: Oxford University Press, 2010).

48. "Slavery, Debt Bondage and Forced Prostitution in Brazil: ASI Reports," Anti-Slavery International press release, April 30, 1992.

49. Data on number of cases is from "Contra exploração, desapropiação," *O Globo*, January 16, 1996.

50. The source for the 80 percent data is "FHC cria um grupo para combater trabalho escravo," *Folha de S. Paulo*, June 28, 1995. On land-related violence, see 1994 and 1995 annual reports of the Pastoral Land Commission (CPT).

51. International Labor Organization, *Fighting Forced Labour: The Example of Brazil* (Geneva: ILO, 2009). On the confidential report, "OIT acusa o Brasil de nao punir trabalho escravo," *O Globo*, November 17, 1995.

52. On Mrs. Barros Lopes being a shareholder, see information in *Processo NUP—28.650/003657/91*, SUDAM.

53. State legislator Zé Geraldo's report of the mission, July 3, 1995, in court case 0000272-32.1997.8.14.0046 and police inquiry 95-DRPA-018/95.

54. *A Província do Pará*, May 25, 1995.

55. This issue will be explained later in the book, but the statement is sustained by information on court cases numbers 0000443-15.1995.814.0039 and 0000667-78.1995.8.14.0039, Pará state.

56. Detective Moraes, interview with the author.

57. State legislator Zé Geraldo's report of the mission, July 3, 1995, in court case 0000272-32.1997.8.14.0046 and police inquiry 95-DRPA-018/95.

58. On the ski mask, information provided by Rita Serra, who was with Gil in the van. Also, Termo de Assentada, Detective Moraes, June 28, 1995, court case 0000272-32.1997.8.14.0046.

59. Description provided to the author by eyewitnesses Rita Serra, Sergio Galiza, Zé Geraldo, and Detective Moraes. Also, based on photographs taken that day by law enforcement and contained in court case 0000272-32.1997.8.14.0046.

60. The contents are described using the forensic information (reports Laudo de Ossada and Laudo de Exame) in court case 0000272-32.1997.8.14.0046.

61. *Relatório de missão determinada pela portaria n 763/95-DGPC*, Joao Nazareno Nascimento Moraes, court case 0000272-32.1997.8.14.0046.

62. On the illicit arms trade, see James L. Cavallaro and Anna C. Monteiro, *Fighting Violence with Violence: Human Rights Abuse and Criminality in Rio de Janeiro*, Human Rights Watch Short Report (1996).

63. This issue is detailed later in the book, but the statements were given by Brito and Dezinho to officer Raimundo Wilson Souza Rêgo on August 28, 1996.

64. State legislator Zé Geraldo's report of the mission, July 3, 1995, in court case 0000272-32.1997.8.14.0046 and police inquiry 95-DRPA-018/95.

65. State legislator Zé Geraldo's report of the mission, July 3, 1995, in court case 0000272-32.1997.8.14.0046 and police inquiry 95-DRPA-018/95.

66. State legislator Zé Geraldo's report of the mission, July 3, 1995, in court case 0000272-32.1997.8.14.0046 and police inquiry 95-DRPA-018/95.

67. Detective Moraes, interview with the author.

68. Police report dated June 29, 1995, court case 0000272-32.1997.8.14.0046.

69. On the content of the human remains found by the police, see Oficio n.001/IPL 018/Depol Rondon/PA, July 3, 1995, court case 0000272-32.1997.8.14.0046.
70. Judge Ana Lúcia Bentes Lynch, July 5, 1995, in court case 0000272-32.1997.8.14.0046 and police inquiry 95-DRPA-018/95.
71. Judge Lynch's rule on July 5, 1995.
72. *O Liberal*, editions from June 29, 1995, to July 7, 1995; *A Provincia do Pará*, editions from June 29, 1995, to July 1, 1995.
73. *Folha de S. Paulo*, June 30, 1995.
74. *Diário do Pará*, July 30, 1995.
75. "Secretario pede a apuração de mortes," *O Liberal*, July 4, 1995.
76. On the announcement of an antislavery task force, see *Folha de S. Paulo*, July 28, 1995.
77. *Folha de S. Paulo*, July 28, 1995.
78. Fax dated on June 29, 1995, sent to *O Liberal*, court case 0000272-32.1997.8.14.0046.
79. Afonso Dias Soares's statement to the police on July 13, 1995, in court case 0000272-32.1997.8.14.0046, Pará.
80. *Relatório de Missão*, Miguel Tomaz Neto, November 8, 1997, court case 0000272-32.1997.8.14.0046.
81. Afonso Dias Soares's statement to the police on July 13, 1995, court case 0000272-32.1997.8.14.0046.
82. Josélio de Barros's statement to the police in Maceió, on July 17, 1995, court case 0000272-32.1997.8.14.0046.
83. Josélio de Barros Carneiro's statement on August 18, 1995.
84. *Correio do Tocantins, O Jornal de Carajas*, July 21–27, 1995.
85. Eliana Vilaça de Lima, motion filed on August 25, 1995, court case 0000272-32.1997.8.14.0046.
86. Petition dated July 11, 1995, contained in court case 0000272-32.1997.8.14.0046.
87. Statement of Armando Novaes Viana, July 14, 1995, court case 0000272-32.1997.8.14.0046.
88. Eliana Vilaça de Lima, motion filed on August 25, 1995, court case 0000272-32.1997.8.14.0046.
89. State attorney Aline Moreira Barata, September 26, 1995.
90. Oficio n.001/IPL 018/Depol Rondon/PA, July 3, 1995, court case 0000272-32.1997.8.14.0046.
91. Police report 97-SUS, February 18, 1997, filed in court case 0000272-32.1997.8.14.0046.
92. Police report 97-SUS, February 18, 1997, filed in court case 0000272-32.1997.8.14.0046.
93. Judge Paulo Cesar Pedreira Amorim, May 27, 1999, court case 0000272-32.1997.8.14.0046.

94. Motion of district attorney Raimundo Antonio Silva Aires, March 22, 1999, court case 0000272-32.1997.8.14.0046.

95. Rita Serra, interviews with the author.

2: THE CRIMINAL SYNDICATE

1. On Josélio's age and physical description, ID documents and photographs contained in court case 1.277/67, Vitória, Espírito Santo.

2. "Nova Almeida: distrito da Serra já foi município e 'resume' história do Brasil," *A Gazeta*, October 26, 2019, and "História municipio Serra," IBGE Cidades.

3. Statement given by Josélio de Barros on March 2, 1970, court case 1.277/67, Vitória, Espírito Santo.

4. Physical descriptions from photographs published from newspaper *A Gazeta* from March 19 to March 29, 1967.

5. *A Gazeta*, March 29, 1967.

6. Family description given by one of Major Cavalcanti's daughters, Rossana Ferreira da Silva Mattos, to journalist Paulo César Dutra in "Morte do Major Orlando Desencadeou uma Série de Execuções," *Folha Diária*, February 24, 2017.

7. Statements in court case 1.277/67, Vitória, Espírito Santo, and *A Gazeta*, March 29, 1967.

8. Statement given by Fausto Ferreira Santos on May 28, 1967.

9. *A Gazeta*, February 28, 1967.

10. See, for instance, feature stories published by magazine *Revista Alterosa* in January 1958 and the series on the Contestado by the newspaper *Última Hora* published daily from November 1 to November 13, 1958.

11. Quote from "Morte do Coronel Binbim ativou a guerra entre os pistoleiros," *Jornal do Brasil*, June 12, 1968.

12. For a general picture of political-related violence, see Walace Tarcisio Pontes, *Conflito agrário e esvaziamento populacional: A disputa do Contestado pelo Espírito Santo e Minas Gerais (1930-1970)*, dissertation, Universidade Federal do Espírito Santo, Vitória, 2007; also Hélio dos Santos Pessoa, *O Negociador de Vidas na Saga do Rio Doce. II. As Lutas* (Belo Horizonte: Plurarts, 2001); and Antonio Carlos Corrêa de Faria, *O Sangue do Barão: As Histórias de Nossa História* (Belo Horizonte, self-published, 2007).

For a nationwide context of political polarization in the 1960s previous to the coup, see Thomas E. Skidmore, *The Politics of Military Rule in Brazil, 1964–85* (New York: Oxford University Press, 1988).

13. *A Gazeta*, February 28, 1967.

14. Journalists of *A Gazeta* who went into the field wrote that twelve shots killed him; on June 12, 1969, *Jornal do Brasil* said that he was hit by fifteen bullets.

15. Statement given by Fausto Ferreira Santos on May 28, 1967, and *A Gazeta*, February 28, 1967.

16. Statement given by Fausto Ferreira Santos on May 28, 1967, and *A Gazeta*, February 28, 1967.

17. *A Gazeta*, February 28, 1967.

18. *A Gazeta*, March 29, 1967. About the jeep's rushed departure, see *A Gazeta*, February 28, 1967.

19. About the time of arrival, see police report by major and detective José Tavares, quoted in Corrêa de Faria, *O Sangue do Barão: As Histórias de Nossa História*, 217–18.

20. *Jornal do Brasil*, June 12, 1969.

21. *A Gazeta*, March 22, 1967.

22. The details of the escape are from the statement given by Fausto Ferreira Santos on May 28, 1967.

23. For details on the escape plan, see police report by major and detective José Tavares, quoted in Corrêa de Faria, *O Sangue do Barão: As Histórias de Nossa História*.

24. *A Gazeta*, March 29, 1967.

25. Statements of the three outlaws, contained in court case 1.277/67, Vitória, Espírito Santo, and *A Gazeta*, March 29, 1967.

26. *A Gazeta*, March 13, 1967.

27. *A Gazeta*, March 21, 1967.

28. *A Gazeta*, March 21 and 22, 1967.

29. *A Gazeta*, March 29, 1967.

30. *A Gazeta*, March 29, 1967.

31. *A Gazeta*, March 29, 1967.

32. *A Gazeta*, March 29, 1967.

33. In a signed statement addressed to the court in Espírito Santo, Josélio admitted to the murders of two alleged hitmen who were about to kill his father in Baixo Guandu's train station. The signed statement and the motion filed by his defense attorney, Mario Rocha, is in the files of court case 1.277/67, Vitória, Espírito Santo.

34. Statement given by Josélio de Barros on March 2, 1970, court case 1.277/67, Vitória, Espírito Santo.

35. State attorney Romualdo Cola's address to the judge on December 14, 1971, court case 1.277/67, Vitória, Espírito Santo.

36. On the officers and the ten crimes, see *Correio Braziliense*, December 4, 1967.

37. Ministério do Exercito, CMA 80 RM, ACE 3582/83, August 28, 1975.

38. Author's interviews with Chico.

39. Author's interviews with Chico.

40. Warren Dean, *With Broadax and Firebrand: The Destruction of the Brazilian Atlantic Forest* (Berkeley: University of California Press, 1995), 6.

41. Author's interviews with Chico.

42. Jório de Barros Carneiro, FGV CPDOC, accessible at http://fgv.br/cpdoc /acervo/dicionarios/verbete-biografico/jorio-de-barros-carneiro.

43. Statements by several teachers and the school's directors, contained in court case 0000272-32.1997.8.14.0046.
44. Chico's interview with the author.
45. Chico's interview with the author.
46. On Barros's election, see *80 Anos Baixo Guandu. 1935–2015,* undated. Magazine published by the municipal council of Baixo Guandu. On the consequences for the Barroses, author's interview with Chico.
47. For an account of the suicide and the Vargas era, see Richard Bourne, *Getulio Vargas of Brazil, 1883–1954: Sphinx of the Pampas* (Knight, 1974); Robert M. Levine, *Father of the Poor? Vargas and His Era* (New York: Cambridge University Press, 1998); and Lira Neto, *Getúlio (1945–1954): Da volta pela consagração popular ao suicídio* (Companhia das Letras, 2014). Also chapter 9 of Lyman L. Johnson, ed., *Death, Dismemberment, and Memory: Body Politics in Latin America* (Albuquerque: University of New Mexico Press, 2004).
48. Vitória Amylton de Almeida, *Carlos Lindenberg: um estadista e seu tempo* (Arquivo Público do Estado do Espírito Santo, 2010), 426
49. "Delay in Finding Killers Irks Rio," *New York Times,* August 8, 1954.
50. Sam Pope Brewer, "Vargas Commits Suicide After Ouster by Military; Violence Flares in Brazil," *New York Times,* August 2, 1954.
51. Referring here to the Revolta de Jacareacanga. Quote is from Lilia M. Schwarcz and Heloisa M. Starling, *Brazil: A Biography* (New York: Farrar, Straus and Giroux, 2018).
52. The hijack and rebellion were known as Revolta de Aragarças, an isolated spot in Central Brazil where the insurgents organized. Quote is from "Brazil Airmen Rebel and Seize 5 Planes," *New York Times,* December 4, 1959. See also "Em 1959, militares sequestraram avião com passageiros e tentaram derrubar JK," *Agência Senado,* December 2, 2019.
53. See, for instance, historian Antonio Barbosa, from the University of Brasília. "Morte de Getúlio, em 1954, adiou o golpe em 10 anos, diz historiador." *Senado Notícias,* August 22, 2014.
54. *Última Hora,* November 5, 1958.
55. Tarcisio Pontes, *Conflito agrário e esvaziamento populacional,* 45.
56. Tarcisio Pontes, *Conflito agrário e esvaziamento populacional,* 16.
57. Tarcisio Pontes, *Conflito agrário e esvaziamento populacional,* 52.
58. Tarcisio Pontes, *Conflito agrário e esvaziamento populacional,* 52.
59. "Morte do Coronel Binbim ativou a guerra entre os pistoleiros," *Jornal do Brasil,* June 12, 1968.
60. *Século,* November 21, 2021.
61. Although a work of fiction, Adilson Vilaça's novel *Cotaxé* (Vitória: SEJUC/SPDC/ISNJN, 1997) is often mentioned as a reference to learn about the little-known story of the Union of Jehovah and its founder, Udelino Alves de Matos.

62. Thomas E. Skidmore, *Brazil: Five Centuries of Change* (New York: Oxford University Press, 2010), 21–23.

63. Leslie Bethell, ed., *The Cambridge History of Latin America, Vol. 2: Colonial Latin America* (New York: Cambridge University Press, 1984), 594.

64. Gabriel Bittencourt, *História Geral e Econômica do Espírito Santo: do engenho colonial ao complexo fabril-portuário* (Vitória: Multiplicidade, 2006), 11.

65. José Teixeira de Oliveira, *História do Estado do Espírito Santo* (Vitória: Arquivo Público do Estado do Espírito Santo, 2008), 185.

66. Teixeira de Oliveira, *História do Estado do Espírito Santo*, 517–18.

67. Francieli Aparecida Marinato, *Índios imperiais: os botocudos, os militares e a colonização do Rio Doce (Espírito Santo, 1824–1845)*, dissertation, Universidade Federal do Espírito Santo, Vitória, 2007.

68. Teixeira de Oliveira, *História do Estado do Espírito Santo*; Bittencourt, *História Geral e Econômica do Espírito Santo*, 162. For a story of Baixo Guandu's foundation, see Manoel Milagres Ferreira, *Histórico do Município do Baixo Guandu* (Baixo Guandu: Prefeitura Municipal, 1958).

69. H. S. Espindola and M. T. B. Vilarino, "'O tempo é a minha testemunha': só as pedras estavam aqui, todo o resto é imigrante," in M. Gerhardt, E. S. Nodar, and S. P. Moretto, eds., *História ambiental e migrações: diálogos* [online] (São Leopoldo: Oikos; Editora UFFS, 2017), 225.

70. Julia Louisa Keyes, *Nossa vida no Brasil. Imigração norte-americana no Espírito Santo 1867–1870* (Vitória: Arquivo Público do Estado do Espírito Santo, 2013). For data on number of migrants, see 28 and footnotes.

71. Cyrus B. Dawsey and James M. Dawsey, eds., *The Confederados: Old South Immigrants in Brazil* (Tuscaloosa: University of Alabama Press, 1995).

72. Eugene C. Harter, *The Lost Colony of the Confederacy* (Jackson: University Press of Mississippi, 1985), 63.

73. Dawsey and Dawsey, *The Confederados*, quote from chapter 3.

74. "Vargas Proposes Coffee Export Tax," *New York Times*, July 25, 1951. And *Brazilian Bulletin*, October 15, 1962.

75. Description based on photographs published by the press of the time. See, for instance, *Última Hora*, May 11, 1958.

76. Author's interview with Hélio dos Santos Pessoa, the son of an influential politician in Aimorés and the author of *O Negociador de Vidas na Saga do Rio Doce As Lutas*.

77. *Correio da Manha*, April 24, 1964.

78. *Jornal do Brasil*, June 12, 1968, 14.

79. Minas Gerais's legislator Sebastião Cypriano do Nascimento, nicknamed Toto, was elected in 1954, 1958, and 1962. *Dados Estatísticos: eleições federais e estaduais*, from the Tribunal Superior Eleitoral (TSE). He would admit to journalists that he had led a group of gunmen.

80. On the 500 gunmen, see *Última Hora*, October 31, 1958. On the 8,000 deaths and gunmen on election day, see magazine *Século*, November 21, 2001.

81. *Jornal do Brasil,* December 6, 1968.
82. Author's interviews with Chico.
83. Author's interviews with Chico.
84. On the funeral, see *Século,* November 21, 2001; on the homage, see *Correio da Manha,* April 24, 1964.
85. To understand the consequences of Bimbim's death, see the work of journalists Rogério Medeiros and Paulo César Dutra, both interviewed by the author. For example, "O início do crime organizado no Estado e a onda de crimes." *Eshoje,* February 24, 2017.
86. *Jornal do Brasil,* December 6, 1968.
87. Josélio's signed statement in court case 1.277/67, Vitória, Espírito Santo. Also Hélio dos Santos Pessoa, *O Negociador de Vidas na Saga do Rio Doce. II. As Lutas.*
88. Josélio's written statement on March 2, 1970, court case 1.277/67, Espírito Santo.
89. Josélio's written statement on March 2, 1970, court case 1.277/67, Espírito Santo. The resemblance of Josélio and the victim was confirmed by the author with photographs of the time.
90. *Jornal do Brasil,* December 6, 1968, and author's interviews with Josélio's younger brother, José Francisco de Barros.
91. Author's interviews with Chico.
92. Author's interviews with Chico.
93. *A Gazeta,* April 9, 1967.
94. *A Gazeta,* April 9, 1967.
95. *A Gazeta,* April 9, 1967.
96. Letter sent by Josélio to a politician named Celso Borges on October 4, 1966, court case 1.277/67, Vitória, Espírito Santo.
97. Letter sent by Josélio to a politician named Celso Borges on October 4, 1966, court case 1.277/67, Vitória, Espírito Santo.
98. No transcription of Josélio's statement before the jury was found by the author in the court files, so the author quoted a later signed deposition of Josélio to present his views to judges: statement given by Josélio on March 2, 1970, court case 1.277/67, Vitória, Espírito Santo.
99. *A Gazeta,* June 21 and 29, 1968.
100. *Acordão,* Revisão criminal n 1.383, Vitória, September 24, 1970.
101. Decision by *desembargador* Crystallino de Abreu Castro, Tribunal de Justiça di Espírito Santo, on March 7, 1972.
102. *O Jornal,* January 26, 1966.
103. *Jornal do Brasil,* January 26, 1966.
104. Murder of officer José Scardua in February 1973. See *O Estado de S. Paulo* April 2, 1973, and *A Gazeta* 04 and February 6, 1973. Josélio would declare himself to be Scardua's friend in his written statement to the court, March 2, 1970, court case 1.277/67, TJES.

105. Author's interviews with Chico.

106. Josélio's letter to the court in Espírito Santo, December 10, 1973.

107. Josélio's written statement on March 2, 1970, court case 1.277/67, Vitória, Espírito Santo.

108. Susanna Hecht and Alexander Cockburn, *The Fate of the Forest: Developers, Destroyers, and Defenders of the Amazon* (Chicago: University of Chicago Press, 2010), 120.

109. The size of the Amazon biome differs from the political boundaries (Legal Amazon) or the Amazon Basin. Here I refer to the data provided by the minister of transportation, Mário David Andreazza, quoted in *Jornal do Brasil*, November 29–30, 1970. For a study on population in the Amazon, see Stephen G. Perz, "Population Growth and Net Migration in the Brazilian Legal Amazon, 1970–1996," in *Deforestation and Land Use in the Amazon*, Charles H. Wood and Roberto Porro, eds. (Gainesville: University Press of Florida, 2002).

110. Statement of Mário David Andreazza quoted in *Jornal do Brasil*, November 29–30, 1970.

111. Hecht and Cockburn, *The Fate of the Forest*, 112–14.

112. Hecht and Cockburn, *The Fate of the Forest*, preface.

3: TERROR ON THE NUT ROAD

1. NASA's "July 20, 1969: One Giant Leap for Mankind," released on July 20, 2019.

2. For an overview of the Gaviões' history, see Expedito Arnaud, *Os índios Gaviões do Oeste. Pacificação e Integração*, Museu Goeldi (Publicações Avulsas no. 28), 1975; and the work of Brazilian anthropologist Iara Ferraz, like *De 'Gaviões' a comunidade 'Parkatêjê': uma reflexão sobre processos de reorganização social*, dissertation, UFRJ, 1998.

3. The fleeing of colonists is based on reports by *Folha do Norte* and *Jornal do Brasil* on July 29, 30, and 31, 1969. Also, Lamartine Ribeiro de Oliveira, *Relatório da Missão de observacao sobre occorrências verificadas nas circunvizinhanças da rodovia PA-70—entre o grupo Kyikatêjê—"Gavião"—e as frentes de expansão da sociedade nacional*, Fundação Nacional do Indio, September 18, 1969.

4. Lamartine Ribeiro de Oliveira, *Relatório da Missão*. Also based on reports by Brazilian newspapers *Folha do Norte* and *Jornal do Brasil*.

5. "Morte no Tocantins," *Itatocan*, January 1953, n. 31, 15.

6. "O povo Gavião," *Itatocan*, year II, July–August 1954, n. 10.

7. Author's interview with pioneer Rita Belem.

8. Quote from author's interview with Rita Belem in Rondon do Pará, formerly Vila Rondon. On the purpose of the road, the source is the official report of the project written by DER-PA and provided by DER-PA to the author, undated.

9. Based on Ribeiro de Oliveira, *Relatório da Missão.*
10. Rita Belem's statement to the author. In reports of FUNAI officials, it is claimed that it was Cotrim who informed the authorities.
11. Travel log of Antonio Cotrim Soares, personal archive, and author's interview with Cotrim.
12. Lykke E. Andersen et al., *The Dynamics of Deforestation and Economic Growth in the Brazilian Amazon* (New York: Cambridge University Press, 2002), 15.
13. Based on Ribeiro de Oliveira, *Relatório da Missão.* Also on eyewitnesses like journalist José Ribamar Fonseca, interviewed by the author and present at the DER camp the day of the events.
14. The growing tension in the area, from as early as 1968, was described by Cotrim in the report Protocolo 1748, 28/11/1968, 2ª Inspetoria Regional—Pará, sent to Joao Oscar Henriques.
15. "Funai nao responde se muda índios mediante indenização," *Jornal do Brasil,* August. 1, 1969.
16. Ribeiro de Oliveira, *Relatório da Missão.* Also, *Folha do Norte,* July 22 and 23, 1969.
17. Ribeiro de Oliveira, *Relatório da Missão.* Also, *Folha do Norte,* July 22 and 23, 1969.
18. Ribeiro de Oliveira, *Relatório da Missão.*
19. Details and account from Ribeiro de Oliveira, *Relatório da Missão.*
20. Details provided by *Folha do Norte,* July 29 and 30, 1969, and August 1, 1969.
21. Ribeiro de Oliveira, *Relatório da Missão.*
22. Description based on Ribeiro de Oliveira, *Relatório da Missão,* and photographs published by *Folha do Norte.*
23. Description based on Ribeiro de Oliveira, *Relatório da Missão.*
24. The sound was described by a reporter of *Folha do Norte* in a piece published on August 1, 1969.
25. Ribeiro de Oliveira, *Relatório da Missão.*
26. Ribeiro de Oliveira, *Relatório da Missão.*
27. *Folha do Norte,* August 1, 1969.
28. On the local fauna, see the comprehensive *Plano de Controle Ambiental,* BR-222/PA, Marabá-Dom Eliseu, SETRAN, June 1999.
29. *Folha do Norte,* August 1, 1969.
30. *Folha do Norte,* August 1, 1969.
31. Ribeiro de Oliveira, *Relatório da Missão.*
32. Author's interview with Rita Belem.
33. Rubens Valente, *Os fuzis e as flechas: História de sangue e resistência indígena na ditadura* (São Paulo: Companhia das Letras, 2017). Background information and quote from chapter 1.
34. Work contract signed on August 30, 1968, Fundação Nacional do Índio. Source: personal archive of Cotrim.

35. Although the author interviewed Sydney Possuelo, the quotation is from chapter 9 of Scott Wallace, *The Unconquered: In Search of the Amazon's Last Uncontacted Tribes* (New York: Crown, 2011).

36. Jonas Gregorio de Souza et al., "Pre-Columbian Earth-Builders Settled Along the Entire Southern Rim of the Amazon," *Nature Communications* 9, no. 1 (2018): 1125.

37. Charles R. Clement et al., "The Domestication of Amazonia Before European Conquest." *Proceedings of the Royal Society B : Biological Sciences* 282, no. 1812 (2015): 20150813.

38. Charles R. Clement et al., "The Domestication of Amazonia Before European Conquest."

39. Charles R. Clement et al., "The Domestication of Amazonia Before European Conquest."

40. Charles R. Clement et al., "The Domestication of Amazonia Before European Conquest."

41. Report written by Cotrim and sent to FUNAI, 2a Inspectoria Regional, Pará, on November 28, 1968.

42. Travel log of Cotrim, personal archive.

43. "Relatório. Assunto: Epidemia gripal que assolou comunidade indígena e suas consequencias," communication of Cotrim with Dr. Paulo Monteiro, personal archive.

44. John Hemming, *Tree of Rivers: The Story of the Amazon* (London: Thames & Hudson, 2008). Quote from chapter 1.

45. For a short story of the Gaviões' interaction with mainstream society, see Expedito Arnaud, "O comportamento dos índios Gaviões de Oeste face a Sociedade Nacional," *Boletim do Museu Paraense Emilio Goeldi. Série Antropologia* 1, no. 1 (1984): 5–66.

46. Travel log of Cotrim, personal archive.

47. Author Candice Millard, who wrote a book about Roosevelt's trip to the Amazon, argued that "Rondon's success in the Amazon had depended on this dictum. It was the only reason the Indians had ever dared to trust him." Candice Millard, *The River of Doubt: Theodore Roosevelt's Darkest Journey* (New York: Broadway Books, 2005). Quote from chapter 9.

48. Travel log of Cotrim, personal archive, and the author's interview with Cotrim.

49. On the size of gardens, see Arnaud, *Os índios Gaviões de Oeste pacificação e integração*, 43.

50. Antonio Cotrim's personal, unpublished chronicle of the contact.

51. Cotrim's interview with the author.

52. Decreto no. 63.515, October 31, 1968.

53. *Jornal do Brasil*, July 31, 1969.

54. Cotrim's interview with the author.

55. "Os Gaviões acossados pela civilização, atacam colonos nas margens da PA-70," *Jornal O Maraba*, November 3, 1968.
56. *Folha do Norte*, July 29, 1969.
57. *Folha do Norte*, July 22, 23, and 25, 1969.
58. Cotrim went back to the jungle on July 30, 1969, as reported by *Jornal do Brasil*, July 31, 1969, and *Folha do Norte*, July 31, 1969.
59. *Folha do Norte*, August 9, 1969, and *Jornal do Brasil*, July 9, 1969.
60. Quotes from Antonio Cotrim's personal unpublished chronicle of the contact.
61. *Folha do Norte*, August 9, 1969.
62. *Jornal do Brasil*, August 10–11, 1969.
63. *Jornal do Brasil*, August 10–11, 1969.
64. "Conferência de paz deve acabar hoje com a guerra entre brancos e índios," *Jornal do Brasil*, August 2, 1969.
65. See *Folha Vespertina*, August 4–9, 1969.
66. The mayor was identified as Simplício Miranda and the event apparently took place in 1956. Source is Cotrim.
67. The meetings between FUNAI officers and the governor of Pará was reported by *Folha Vespertina*, August 8, 1969. *Jornal do Brasil* said on August 12, 1969, that Governor José Sarney, future president of Brazil, had meetings with federal officers and FUNAI to abort the mission. Quote is from Paul Montgomery, "Killing of Indians Charged in Brazil," *New York Times*, March 21, 1968.
68. Shelton H. Davis, *Victims of the Miracle: Development and the Indians of Brazil* (Cambridge University Press, 1977).
69. Montgomery, "Killing of Indians Charged in Brazil."
70. Montgomery, "Killing of Indians Charged in Brazil."
71. *"Lost" report exposes Brazilian Indian genocide*, April 25, 2013, Survival International (https://www.survivalinternational.org/news/9191).
72. "Documento recuperado após décadas aponta crimes contra índios." *GloboNews*, April 27, 2013.
73. Elena Guimarães, *Relatório Figueiredo: entre tempos, narrativas e memórias*, Rio de Janeiro, 2015. Dissertation, Universidade Federal do Estado do Rio de Janeiro.
74. Guimarães, *Relatório Figueiredo: entre tempos, narrativas e memórias*.
75. *Folha do Norte*, August 14, 1969.
76. *O Estado de S. Paulo*, August 2, 1969.
77. Antonio Cotrim's personal unpublished chronicle of the contact.
78. *Memo 011/COORD. GT/86*, FUNAI, March 21, 1986. For an overview of the reservation, see *Terras Indigenas no Brasil*, ISA, https://terrasindigenas.org.br/pt-br/terras-indigenas/3750, accessed June 24, 2022.
79. *Jornal do Brasil*, August 10–11, 1969.

80. The exact date of the removal wasn't published by the press. Contacted by the author, FUNAI claimed having no report or details on the process of removal of the Indigenous group.

81. On the pictures, source is Father Giuseppe Castelli, interviewed by the author and whose camera was taken by the army.

82. Author's communication with FUNAI, May 2019.

83. *Jornal do Brasil*, May 20, 1972.

84. Scott Wallace, *The Unconquered*. Quote from chapter 9.

85. Bo Akerren, Sjouke Bakker, and Rolf Habersang, *Report of the ICRC Medical Mission to the Brazilian Amazon Region* (May–August 1970). Geneva, October 1970, 47–48.

86. Akerren, Bakker, and Habersang, *Report of the ICRC Medical Mission to the Brazilian Amazon Region (May–August 1970)*, 47–48.

87. Camilo M. Vianna, "Vila Rondon." *Projeto Rondon*, Belém, October 13, 1970.

88. Quote from *Diário de Notícias*, February 22, 1970.

89. Jean Hébette, *Area de fronteira e conflito. O leste do médio Tocantins* (Belém: NAEA, 1983).

90. Interview with the author.

91. Sue Branford and Oriel Glock, *The Last Frontier: Fighting Over Land in the Amazon* (London: Zed Books, 1985), 55.

92. *Intentional Forest Fires and Covid-19*, fact sheet of the virtual briefing "What Happens When Forest Fires and Covid-19 Collide," hosted by the Columbia Earth Institute on June 17, 2020. https://docs.google.com/document/d/1fE1zylmGqm5QY6SId4iqCqQ6RRCyS-w6DvoxMXPVMGk/edit.

93. Hecht and Cockburn, *The Fate of the Forest*, 38. An excellent work on the use of fire by Indigenous tribes in the southern edge of the Amazon is the work of the anthropologist James R. Welch. For instance, James R. Welch, "Xavante Ritual Hunting: Anthropogenic Fire, Reciprocity, and Collective Landscape Management in the Brazilian Cerrado," *Human Ecology* 42, no. 1 (2014): 47–59.

94. In Guimarães, *Relatório Figueiredo*, a FUNAI official noted: "The devastation of the forest was dramatic."

95. In interviews with the author, other pioneers, like Lourival de Souza and his wife, Dona Rosa, confirmed the existence of "clandestine nightly burials" in early Vila Rondon.

96. Thomas E. Skidmore, *The Politics of Military Rule in Brazil, 1964–1985* (New York: Oxford University Press, 1988), 138.

97. Hecht and Cockburn, *The Fate of the Forest*.

98. On the international loans, see Davis, *Victims of the Miracle*.

99. Skidmore, *The Politics of Military Rule in Brazil, 1964–1985*, 9.

100. *Jornal do Brasil*, October 9, 1970.

101. Branford and Glock, *The Last Frontier*, chapters 2 and 3.

102. Branford and Glock, *The Last Frontier*, chapters 2 and 3.

103. According to Minister of Transportation Mário David Andreazza, Brazil's federal, state, and municipal roads totaled 62,000 kilometers in 1973. *Jornal do Brasil*, January 26, 1977.

104. Data from Jeremy M. Campbell, *Conjuring Property: Speculation and Environmental Futures in the Brazilian Amazon* (Seattle: University of Washington Press, 2015), chapter 1.

105. *O Jornal*, April 16, 1970.

106. Alfonso Henriques, "The Awakening Amazon Giant," in Athens Center of Ekistics, vol 34, no. 202 (1972).

107. Campbell, *Conjuring Property*, 32–33.

108. Greg Grandin, *The End of the Myth: From the Frontier to the Border Wall in the Mind of America* (New York: Metropolitan Books, 2019).

109. Seth Garfield, *Indigenous Struggle at the Heart of Brazil: State Policy, Frontier Expansion, and Xavante Indians, 1937–1988* (Durham, NC: Duke University Press, 2001), 156–57.

110. "Amazônia: Chega de lendas, vamos faturar!" Official propaganda published in newspapers, for instance, in *Jornal do Brasil*, December 7, 1970. "Integrar para não entregar" and "Amazônia: Desafio Que Unidos Vamos Vencer" were also common slogans.

4: THE CHAINSAW MURDER

1. Interview with the author.

2. Castelli's interviews with the author.

3. Pioneer interviewed by the author, who preferred to remain anonymous.

4. For instance, João Moreira, said to be a member of a gang who came from a tiny village in the contested area in Minas Gerais called Carlos Chagas, had moved to Vila Rondon with his relatives and had brought, according to historians, "some money and the weapons of the Far West pioneers. . . . In a short time, they managed to control the whole valley south [of] the road. . . . Everyone who moved into the area did it with their approval." Another feared man was Veríssimo Correa Gusmão, called Vivi, an "arrogant" land-grabber who was said to operate with total impunity. "He is the owner of almost the whole Vila Rondon: speculates, practices loan-sharking. . . . He lives threatening the authorities who oppose to him," wrote a group of scholars who visited the area. Jean Hébette, *Area de fronteira e conflito: O lesto do médio Tocantins* (Belém: NAEA, 1983).

5. Hecht and Cockburn, *The Fate of the Forest*, 123.

6. On the American families, the author refers to the story of John Weaver Davis and its early conflicts, told to the author by John Weaver Davis Jr. At the time, it was reported in *O Estado de S. Paulo*, November 4, 1971, and *Folha de S. Paulo*, May 24, 1975. On the early use of pistoleiros in Vila Rondon, see *O Estado de S. Paulo*, August 18, 1973, *Diário do Paraná*, August 26, 1973, and *Jornal do Brasil*, August 18, 1973, and August 23, 1973. On debt

bondage, see *Jornal do Brasil*, February 25 and 28, 1972, April 11, 1973, and July 29, 1972.

7. *O Estado de S. Paulo*, December 22, 1974.

8. *A Província do Pará*, August 17, 1973.

9. Pedro dos Santos's account was published by *A Província do Pará*, August 17, 1973. Estimates about his ranch and the labor from Lamartine, *Relatório da Missão*.

10. *A Província do Pará* published stories on the case on August 17, 21, and 23, 1973.

11. Data from *Diário do Paraná*, May 19, 1972.

12. *Diário do Paraná*, August 26, 1973.

13. Ersilio Fausto Fiorentini, *Con il Cuore in Brasile. Missioni diocesane piancentine: quarant'anni in terra brasiliana (1967–2007)* (Edizioni Berti, 2007), 266–68.

14. Informação 006/74, January 8, 1974, PM/PA, Ministério da Aeronautica.

15. The memoirs of Father Fontanella, *Pellegrini Sulle Strade del Mondo* (self-published, nd).

16. Diário do Congresso Nacional. Seçao I. Ano XXVIII, n. 98, September 5, 1973.

17. *A Província do Pará* published stories on Vila Rondon on August 17, 21, and 23, 1973.

18. Milton Honório Pinheiro, a twenty-one-year-old peasant, declared that his father had been slaughtered after their family farm was set alight by gunmen operating on Alves's orders. After the murder, the widow had decided to remain on the plot, but not for long. "We continued to be harassed by the fazendeiro, and my mother decided to sell her land," explained Pinhéiro, who criticized the police for never catching the killer or solving the case. *A Província do Pará*, August 21, 1973.

19. *A Província do Pará*, August 21, 1973.

20. *A Província do Pará*, August 21, 1973.

21. *A Província do Pará*, August 21, 1973.

22. On when Josélio moved to the Amazon, see Informação 341/E2/75 (ACE.3582/83), confidential report by the Ministry of the Army, August 28, 1975.

23. *A Província do Pará*, September 3, 1975.

24. João Hébette and José Alberto Colares, "Small-Farmer Protest in the Greater Carajás Programme," in David Goodman and Anthony Hall, eds., *The Future of Amazonia: Destruction or Sustainable Development* (New York, Macmillan, 1990).

25. *A Província do Pará*, September 3, 1975.

26. Author's interviews with Conceição dos Reis.

27. Author's interviews with Conceição dos Reis.

28. Letter published in *A Província do Pará*, October 9, 1975.

29. *A Província do Pará*, September 10, 1975.

30. *A Província do Pará*, September 10, 1975.
31. *A Província do Pará*, September 10, 1975.
32. Quote from state legislator Antônio Teixeira published by *A Província do Pará*, September 9, 1975.
33. State legislator Gerson Peres's speech in Para's state legislature on September 9, 1975.
34. *A Província do Pará*, September 11, 1975.
35. *A Província do Pará*, September 12, 1975.
36. For a comparative study on the social role of hitmen in the Amazon, see Ed Carlos de Sousa Guimarães, *A violência desnuda: justiça penal e pistolagem no Pará*, dissertation, Universidade Federal do Pará, 2010; and Violeta Refkalefsky Loueiro and Ed Carlos Guimarães, "Reflexões sobre a pistolagem e aviolência na Amazônia," *Revista Direito GV* 3, no. 1 (2007): 221–46.
37. "Fazendeiros guardam orelhas como troféu," *Jornal dos Trabalhadores Sem Terra* 6, no. 59 (1987).
38. On Alves's previous criminal record, see Informações 238671/ABE/75, December 22, 1975, and Informação 0386/71/ABE/76, March 11, 1976.
39. On Canuto, see *A Província do Pará*, September 3, 1975. On Canuto's criminal record, see Informação 0351/08/ABE/83, Serviço Nacional de Informaçoes, May 3, 1983.
40. "He makes illegal land transactions," read a confidential police file on Antônio Fernando Machado da Cunha. Source: Informação 044402/ABE/75, in ACE 3561/83, March 10, 1975.
41. On the support of Marechal Joaquim Justino Alves Bastos for Antônio Fernando, see letter addressed February 6, 1975, by the high-ranking military officer, filed in confidential file ACE3582/83. On Alves Bastos's role in the 1964 coup, see biography of Justino Alves Bastos in Fundação Getúlio Vargas CPDOC.
42. *A Província do Pará*, September 3, 1975.
43. Josélio declared to the police that he had settled in Paragominas in 1973. Report of detective Adonias Marques dos Santos, December of 1975, in classified file ACE 3582/83.
44. Relatório, Ace 2582/83, December 1975.
45. Confidential report by the Ministry of the Army, number 34/E2/75, August 28, 1975.
46. Confidential report by the Ministry of the Army, number 34/E2/75, August 28, 1975. The policeman was Carlos Alberto Rufino.
47. On Josélio's contacts, Informação 34/E2/75, Ministry of the Army, ACE 3582/83, August 28, 1975.
48. Report of detective Adonias Marques dos Santos, December 1975, in classified file ACE 3582/83.
49. Version of facts provided by detective Adonias Marques dos Santos, in his Relatório of December 1975, in classified file ACE 3582/83.

50. Report of detective Adonias Marques dos Santos, December 1975, in classified file ACE 3582/83.

51. Report of detective Adonias Marques dos Santos, December 1975, in classified file ACE 3582/83.

52. Report of detective Adonias Marques dos Santos, December 1975, in classified file ACE 3582/83.

53. Account by Canuto to the police, quoted in report of detective Adonias Marques dos Santos, December 1975, in classified file ACE 3582/83.

54. See, for instance, *Brasil de Fato*, December 1–7, 2011.

55. Afonso Dias Soares, statement to the police on July 13, 1995, contained in court case 0000272-32.1997.8.14.0046.

56. Author's interviews with Conceiçao dos Reis.

57. Goodman and Hall, eds., *The Future of Amazonia*, 292.

58. Goodman and Hall, eds., *The Future of Amazonia*, 292.

59. Account by Josélio to the police, quoted in report of detective Adonias Marques dos Santos, December 1975, in classified file ACE 3582/83.

60. Account by Josélio to the police, quoted in report of detective Adonias Marques dos Santos, December 1975, in classified file ACE 3582/83.

61. Account by Canuto to the police, quoted in report of detective Adonias Marques dos Santos, December 1975, in classified file ACE 3582/83.

62. Account by Canuto to the police, quoted in report of detective Adonias Marques dos Santos, December 1975, in classified file ACE 3582/83.

63. The dates of when Josélio presented himself before the notary public are found in his record files as a jailed man, in the court files found in case 1.277/67, Vitória, Espírito Santo.

64. Relatório by detective Adonias Marques dos Santos, December 1975, in classified file ACE 3582/83.

65. *A Província do Pará*, September 12, 1975.

66. *A Província do Pará*, October 1, 1975.

67. Informação 0386/71/ABE/76, March 11, 1976.

68. *O Diário do Pará*, December 30, 1975.

69. Informação 009/117/Ago/SIN/82, Serviço Nacional de Informações, April 30, 1982.

70. Informação 34/E2/75, Ministry of the Army, confidential report, August 28, 1975.

71. Informação 009/117/Ago/SIN/82, Servicio Nacional de Informaçoes, April 30, 1982.

72. On his alleged participation on the arms disappearance, see Informação 009/117/Ago/SIN/82, Seriçio Nacional de Informações, April 30, 1982.

73. On the relocation, see "Lincoln empossado na PF," *Correio Braziliense*, March 11, 1976.

74. Letter addressed by Epitácio Ramalho Alves to the president of Brazil, Ernesto Geisel, August 10, 1976.

75. Ministry of Justice file 065799, August 18, 1976, available at the Arquivo Nacional.
76. Letter addressed by eighteen smallholders, contained in confidential file ACE 3583/83.
77. Letter addressed by eighteen smallholders, contained in confidential file ACE 3583/83.
78. Informação 1232/116/ABE/76, Serviço Nacional de Informações, August 12, 1976.
79. During those years other settlers along the PA-70 contacted the Brazilian president to report land-related crimes near Vila Rondon. See, for instance, Informação 0274/117/ABE/81, Serviço Nacional de Informações, May 19, 1981.
80. Sue Brandford and Jan Rocha, *Cutting the Wire: The Story of the Landless Movement in Brazil* (London: Latin America Bureau, 2002). Quote from sociologist José de Souza Martins is from p. 6.
81. On Fontanella, see Informação 006/74, January 8, 1974, PM/PA, Ministério da Aeronautica. Also, Encaminhamento 007/75-SI/SR/DPF/PA, March 3, 1975.
82. The authorities offered a series of financial incentives to companies setting up ranches in the region before the end of 1974. These included "a 50 percent reduction in their corporate income taxes earned in other parts of Brazil if they reinvested taxable monies in the Amazon," a supplementary ten-year exemption from income tax for the new enterprise based in the rainforest, the levying of duties on imported machinery for farming, and a permissive attitude toward environmental and labor practices. See K. Bertha Becker, *Amazônia* (São Paulo: Ática, 1990); *As Amazônias de Bertha K. Becker* (Rio de Janeiro: Editora Garamond, 2015), 22–24. For data on tax incentives and the companies who benefited, see Anthony Hall, *Developing Amazonia: Deforestation and Social Conflict in Brazil's Carajas Programme* (Manchester, UK: Manchester University Press, 1991), 21–24; S. G. Bunker, *Underdeveloping the Amazon: Extraction, Unequal Exchange, and the Failure of the Modern State* (Chicago: University of Chicago Press, 1985), chapter 3; and F. H. Cardoso and G. Müller, *Amazônia: expansão do capitalismo* (Rio de Janeiro: Centro Edelstein de Pesquisas Sociais, 2008). Quote from Shelton H. Davis, *Victims of the Miracle: Development and the Indians of Brazil* (New York: Cambridge University Press, 1977), 37. The 1.7-million-acre ranch was the Suiá Missu, whose size appears in Davis, *Victims of the Miracle*, 115.
83. https://www.cptnacional.org.br/quem-somos/-historico.
84. In southern Pará, the rural workers' union in Paragominas was infiltrated for years. The source is the author's interview with Father Castelli and several confidential files of the secret services—for instance, Informação 1590/750/ABE/76, October 21, 1976.
85. According to his own account, Fontanella went to Marabá to inform Bishop

Estêvão Cardoso de Avelar, a well-connected reformist. Father Fontanella's memoirs, *Pellegrini Sulle Strade Del Mondo* (self-published, nd).

86. Informe 001/77/AG-SI/CR, January 31, 1977. INCRA.

87. Josélio was "detained for ten months" but later was "simply freed without further consequences," according to Jean Hébette, *Area de fronteira e conflito: O lesto do médio Tocantins* (Belém: NAEA, 1983), 25. The same outcome is mentioned in Hébette and Colares, "Small-Farmer Protest in the Greater Carajás Programme."

88. Since 2018, and up until the delivery of the final manuscript in mid-2022, the author searched for the court records containing the police reports, interviews, and eventual court transcriptions. He visited the central archive in Belém and Marabá and successfully requested the president of the justice system of Pará to order comprehensive research in the archives of courthouses located in several municipalities of the state. The author also asked the military authorities for any information on the case. The court records of the double homicide weren't found.

Following the lead of the Lincoln case, made public in documents found in the Arquivo Nacional, the author filed a freedom of information act request that the federal police grant access to the case. The petition was refused several times under the argument that personal information, even if it involved dubious public officials, was classified for a hundred years.

89. Deed contained in processo 2002/60712, ITERPA, Pará.

90. Escritura de compra e venda, República Federativa do Brasil, Cartório de Paragominas, Fls 39/42. Deed contained in processo 2002/60712, ITERPA, Pará.

91. Informe 001/77/AG-SI/CR, January 31, 1977. INCRA.

92. *O Liberal*, December 14, 1976.

93. Informe 132/S102-A3-CIE, Ministerio do Exército, April 8, 1976.

94. *Il Nuovo Giornale*, no. 47, December 22, 1976.

95. Father Fontanella self-published his memoirs in *Pellegrini Sulle Strade Del Mondo*.

96. On Josélio's alleged role in the priest's deportation, the author interviewed multiple sources in Rondon do Pará. A local scholar claimed that he had once interviewed Josélio, and the cattleman had apparently said, referring to Fontanella, "Yes, we managed to get that terrorist out of Rondon!" While the case against the Italian priest for allegedly faking his resident permit was being built by the authorities, Josélio's friend Lincoln had occupied the post of superintendent of the federal police in Brasília.

97. The author has investigated the story of the Davis family accessing declassified files in both Brazil and the United States. He also interviewed John Weaver Davis Jr. For an English-language account of the case, see *Washington Post*, December 26, 1987. Also a cable from the American embassy in Brasília to State Department, December 13, 1976, in the National Archives.

98. Relatório, secretaria de estado de segurança pública, Governo do Estado do Pará, July 5, 1976.
99. On the claims of the bereaved family, interviews with John Weaver Davis Jr. On the police investigation, Inquérito policial 50/76, SR/PA, Superintendência Regional do Pará, Polícia Federal.
100. "Squatters Ambush Rancher and Sons," Associated Press, July 16, 1976. Jonathan Kandell, "Violence in the Amazon: Brazil Echoes U.S. West," *New York Times*, July 20, 1976.

5: THE BOOMTOWN

1. Quotes, descriptions, and accounts in this chapter are from the author's interviews with Maria Joel and her family.
2. Data on the number of sawmills from Maria Lúcia do Amor Divino Xavier, *O declínio das indústrias madeireiras no município de Rondon do Pará*, dissertation, Universidade Federal do Pará, 2003, and interviews of the author with Rondon's timber industry researcher Maria Lúcia do Amor Divino Xavier.
3. Quote from author's interview with pioneer José Coutinho de Queiroz.
4. Author's interviews with sawmill owner Odeb Moreira.
5. Documentary on Rondon's cinema directed by Ricardo Tavares D'Almeida: *Cine Ideal: Memórias de um cinema de rua de Rondon do Pará*, 2016.
6. Statement by Gildeu Miranda, Rondon's first mayor, in interview with the author. Miranda claims to have made himself the first census of the town. Data consistent with *Censo Demografico 1991, resultados preliminares*, Brazilian Institute of Geography and Statistics, Rio de Janeiro, 1992.
7. For an account of Rondon's early health problems, see F. C. Dos Santos Silva, *História, Trabalhos Pastorais e Sociais da Paróquia Nossa Senhora Aparecida, 1968 a 1995*, dissertation, Universidade Federal do Pará, 2003.
8. Data compiled by the author from daily editions of *Diário do Pará* from June 1984 to September 1985. For a general overview of violence during those years, see *Situaçao sócio-economica na regiao sul do Pará, Rondon do Pará*, ACE 068291 88, October 24, 1987.
9. The rancher's murder was published by *Diário do Pará* on September 11, 1984; the peasants' killing and amputation by *Diário do Pará* on December 4, 1984; the rape and subsequent killing by *Diário do Pará* on May 24, 1985.
10. *Diário do Pará*, October 28, 1984.
11. *Diário do Pará*, October 7, 1984.
12. Interviews of the author with Maria Eva dos Santos Dias.
13. Author's interviews with sawmill owner Odeb Moreira.
14. Thomas E. Skidmore, *The Politics of Military Rule in Brazil, 1964–1985* (New York: Oxford University Press, 1988), 230.
15. *Inflation, consumer prices (annual %)—Brazil*, World Bank. https://api .worldbank.org/v2/en/indicator/FP.CPI.TOTL.ZG?downloadformat=excel.

16. *Economic Cycle Dating Committee*. Fundação Getúlio Vargas and Instituto Brasileiro de Economia, Rio de Janeiro, May 27, 2009.

17. Fernando de Holanda Barbosa, "Economic Development: The Brazilian Experience," in Akio Hosono and Neantro Saavedra-Rivano, eds., *Development Strategies in East Asia and Latin America* (London: Palgrave Macmillan, 1998), 69–87.

18. Aristides Monteiro Neto, César Nunes de Castro, and Carlos Antonio Brandao, órganizers, *Desenvolvimento regional no Brasil: políticas, estratégias e perspectivas* (Rio de Janeiro: Ipea, 2017). Details for Pará can be found in chapter 4, Valdeci Monteiro dos Santos, "A economia do sudeste paraense: evidências das transformações estruturais." Also, Lykke E. Andersen et al., *The Dynamics of Deforestation and Economic Growth in the Brazilian Amazon* (New York: Cambridge University Press, 2002), 17.

19. On the LSE scholar, Anthony L. Hall, *Developing Amazonia: Deforestation and Social Conflict in Brazil's Carajas Programme* (Manchester: Manchester University Press, 1989), 41. Quote is from Cesar Cals, minister of mines and energy, in Warren Hoge, "Big Amazon Project Unfolds," *New York Times*, November 18, 1981.

20. Hall, *Developing Amazonia*.

21. Alan Riding, "Mining for Profits in the Jungles of Brazil," *New York Times*, May 19, 1985.

22. Hall, *Developing Amazonia*, 41.

23. Pere Petit Peñarrocha, *Chão de promessas: elites políticas e transformações econômicas no estado do Pará pos-1964* (Paka-Tatu, 2003), 264.

24. Anthropologist David Cleary notes that the discovery of the mine is "encrusted by legend." David Cleary, *Anatomy of the Amazon Gold Rush* (London: Macmillan, 1990). For a history of the mine, Antonio Carlos Prado, "A alucinante corrida do ouro de Serra Pelada," *Istoé*, July 19, 2019, https://istoe.com.br/a-alucinante-corrida-do-ouro-de-serra-pelada/, and *Folha de S. Paulo* July 8, 2019. Also the documentary by Victor Lopes, *Serra Pelada: Gold Rush*, or in Portuguese, *Serra Pelada—A Lenda da Montanha de Ouro*, 2013.

25. *Time*, September 1, 1980.

26. About the number of gold seekers, see *New York Times*, August 23, 2004. About the crater size, see *New York Times Magazine*, June 7, 1987. On Bolsonaro's father, see Simone Preissler Iglesias and Bruce Douglas, "Amazon Gold and Army Suspicion Fuel Bolsonaro's Rainforest Rage," *Bloomberg*, August 28, 2019.

27. Author's interview with I., who remains anonymous at the source's request.

28. Relatório da Comissão Interministerial, Criada pelo decreto no. 99.386, de 12 de julho de 1990, sobre o garimpo de Serra Pelada, in ACE 39490/84, Arquivo Nacional.

29. David Cleary, *Anatomy of the Amazon Gold Rush*, 172.

30. Prado, "A alucinante corrida do ouro de Serra Pelada."
31. Susan Bright and Hedy van Erp, *Photography Decoded: Look, Think, Ask* (London: Ilex Press, 2019).
32. On murderous land slippages, see *Jornal de Jacundá*, October 16–30, 1986. On female teenagers being smuggled, see *O Marabá*, no. 733, December 31, 1985. Also interview with journalist José Ribamar Fonseca, who claimed to be one of the early Brazilian journalists in visiting the mine.
33. *Time*, September 1, 1980.
34. *Time*, September 1, 1980.

6: EARLY CHALLENGES

1. Dezinho's background and early life account based on author's interviews with his brother Valdemir Dutra da Costa and Maria Joel.
2. Data on population and number of college degrees from *Censo Demográfico: Maranhão* (Rio de Janeiro: IBGE, 1982).
3. On the detention and judgment of French priest Xavier Gilles de Maupeou d'Ableiges, see *Jornal do Brasil*, August 29 and October 8, 1970.
4. The lack of capital or access to loans was mentioned to the author by multiple migrants in Rondon do Pará as a reason for failing to become smallholders. For smallholders' struggles to access credit in Brazil, see the study about the northeast region, which shares common trends with the Amazon of those years. G. Howe and D. Goodman, *Strategies for the Alleviation of Rural Poverty in Latin America and the Caribbean: Smallholders and Structural Change in the Brazilian Economy* (Inter-American Institute for Cooperation on Agriculture, 1992).
5. F. C. Dos Santos Silva, *História, Trabalhos Pastorais e Sociais da Paróquia Nossa Senhora Aparecida, 1968 a 1995*, dissertation, Universidade Federal do Pará, 2003, 35.
6. Data on the arc, in general agreement among sources, come from *FAO Meeting on Public Policies Affecting Forest Fires*, FAO Forestry Paper 138, October 28–30, 1998. Food and Agriculture Organization of the United Nations (FAO), 97.
7. Quote from John Hemming, *Tree of Rivers: The Story of the Amazon* (London: Thames & Hudson, 2009). The study on the impact of chainsaws in Pará can be found in *Amazônia: Meio Ambiente e Desenvolvimento Agrícola* (Brasília: Embrapa-SPI, 1998), 161–85.
8. On the number of fires, Susanna Hecht and Alexander Cockburn, *The Fate of the Forest: Developers, Destroyers, and Defenders of the Amazon* (Chicago: University of Chicago Press, 2010), 44. On deforestation rates, see S. G. Perz, "Population Growth and Net Migration in the Brazilian Legal Amazon, 1970–1996," in *Deforestation and Land Use in the Amazon*, Charles H. Wood and Roberto Porro, eds. (Gainesville: University Press of Florida, 2002).

9. *Intentional Forest Fires and Covid-19*, fact sheet of the virtual briefing "What Happens When Forest Fires and Covid-19 Collide," hosted by the Columbia Earth Institute on June 17, 2020.

10. The speeches of the minister of transport, Mario David Andreazza, represent a good example of the thinking. See, for instance, the speeches reproduced by *Jornal do Brasil*, November 29–30, 1970, and *Jornal do Brasil*, January 26, 1977.

11. Data from Gilbert King, "Where the Buffalo No Longer Roamed." *Smithsonian Magazine*. July 17, 2012. Quote is from Roxanne Dunbar-Ortiz and Dina Gilio-Whitaker, *"All the Real Indians Died Off": And 20 Other Myths About Native Americans* (Boston: Beacon Press, 2016), 142.

12. Steven R. Weisman, "Reagan Warns Japanese on Trade," *New York Times*, October 28, 1989.

13. Author's interview with former nurses in Rondon who wish to remain anonymous.

7: CRICKETS AND CATTLE

1. Data on number of political parties from U.S. Department of State, *Country Reports on Human Rights Practices for 1988*, February 1989. On the number of voters, see *O Globo*, November 15, 1988.

2. "O Oligarca perfeito," *Revista Veja*, April 4, 2009. On Sarney's background, see Marlise Simons, "Man in the News; The Reluctant Successor," *New York Times*, April 23, 1985.

3. Simons, "Man in the News; The Reluctant Successor."

4. *Country Reports on Human Rights Practices for 1988*, February 1989.

5. Lilia M. Schwarcz and Heloisa M. Starling, *Brazil: A Biography* (New York: Farrar, Straus & Giroux, 2018), chapter 18.

6. *Constitution of the Federative Republic of Brazil*, 3rd ed., Biblioteca Digital da Câmara dos Deputados, 2010.

7. *The Brazilian Indians*, Survival International, 2021.

8. *Falling Short: Donor Funding for Indigenous Peoples and Local Communities to Secure Tenure Rights and Manage Forests in Tropical Countries (2011–2020)*, Rainforest Foundation Norway. For a recent study on the crucial role of Indigenous peoples in protecting the environment, see *Forest Governance by Indigenous and Tribal People: An Opportunity for Climate Action in Latin America and the Caribbean*; FAO and FILAC, 2021.

9. Alan Riding, "Brazil Opposition Parties Win Important Cities in Elections," *New York Times*, November 17, 1988.

10. Information about Lula's biography based on Richard Bourne, *Lula do Brasil: A História Real* (São Paulo: Geração Editorial, 2009); John D. French, *Lula and His Politics of Cunning: From Metalworker to President of Brazil* (Durham: University of North Carolina Press, 2020); multiple articles in the Brazilian media; and information available at Instituto Lula.

11. Age is from Lula's own account of the accident. "Como Lula perdeu o dedo?" January 7, 2022. https://lula.com.br/como-lula-perdeu-o-dedo/.
12. "Bodas de chumbo," *Época*, edition 251, March 10, 2003, and "O nada e o ódio," *El Pais*, March 7, 2017. See also French, *Lula and His Politics of Cunning*, 164.
13. French, *Lula and His Politics of Cunning*.
14. On Volkswagen, "Exclusive: Volkswagen spied on Lula, other Brazilian workers in 1980s." Reuters, September 5, 2014.
15. On Walesa, "Making Trouble on Human Rights," *New York Times*, November 30, 1981. Quote is from James Brooke, "Workers' Advocate, a Front-Runner in Brazil, Shows He Thinks Like a Boss," *New York Times*, April 30, 1989.
16. "Between 1966 and 1976 Brazil's annual GDP growth averaged an impressive 9.2%, yet income inequality rose sharply. From 1960 to 1977 inequality increased from 0.50 to 0.62 on the Gini coefficient scale." M. Carter, ed., *Challenging Social Inequality: The Landless Rural Workers Movement (MST) and Agrarian Reform in Brazil*, Centre for Brazilian Studies, University of Oxford, 2007.
17. On inflation rates, data from World Bank.
18. Brooke, "Workers' Advocate, a Front-Runner in Brazil, Shows He Thinks Like a Boss."
19. José de Souza Martins, "A reprodução do capital na frente pioneira e o renascimento da escravidão no Brasil," *Tempo Social* 6, no.1–2 (1994): 1–25.
20. *Trabalho Escravo no Brasil do Século XXI*, ILO report, 2006.
21. Chained workers were discovered by police in Fazenda São Judas Tadeu, near Paragominas. This was reported by *Jornal do Brasil*, August 13, 1988.
22. *Diário do Pará*, July 10, 1990, and *Correio Braziliense*, February 7, 1997. The use of pigs by the Italian Mafia to eat corpses has been reported in several media and books. Here, I quote Arcangelo Badolati, *Mamma 'ndrangheta. La storia delle cosche cosentine dalla fantomatica Garduña alle stragi moderne*, 2nd ed. (Pellegrini, 2020), chapter 22.
23. Relatório "Visita a Carvoarias no Pará," Brasília, Congressman Orlando Fantazzini, December 6, 2001.
24. Relatório "Visita a Carvoarias no Pará," Brasília, Congressman Orlando Fantazzini, December 6, 2001.
25. Hecht and Cockburn, *The Fate of the Forest: Developers, Destroyers, and Defenders of the Amazon*, 107.
26. Jeremy M. Campbell, *Conjuring Property: Speculation and Environmental Futures in the Brazilian Amazon* (Seattle: University of Washington Press, 2015).
27. See Campbell, *Conjuring Property*, 61 and subsequent for context. Quote on 1850 land policy is from *Brazil Land Governance Assessment*, World Bank Report 89239-BR. June 2014, 28.

28. Sue Branford and Jan Rocha, *Cutting the Wire: The Story of the Landless Movement in Brazil* (London: Latin America Bureau, 2002). Quote from historian P. Monbeig is from p. 55.

29. On the size of the fraud in Rondon, see Informação 0451/01/ABE/75, Serviço Nacional de Informações, March 11, 1975. On the size of London, some 607 square miles or 388,480 acres, see www.cityoflondon.gov.uk.

30. Philip M. Fearnside, "Spatial Concentration of Deforestation in the Brazilian Amazon," *Ambio* 15, no. 2 (1986); 74–81.

31. Fernando H. Cardoso and Geraldo Müller, *Amazônia: expansão do capitalismo* (Rio de Janeiro: Centro Edelstein, 2008), 130.

32. *O Liberal*, April 3, 1989.

33. Stephen G. Bunker, *Underdeveloping the Amazon: Extraction, Unequal Exchange, and the Failure of the Modern State* (Chicago: University of Chicago Press, 1990).

34. Campbell, *Conjuring Property*. Regarding the comparison with the United States, Professor Campbell writes: "When Frederick Jackson Turner declared the US frontier 'closed' in 1893, he was using a conceptual category—the frontier—that had itself been a product of a unifying state vision and a material process of state building. Before homesteading, railroad profiteering, or land-jobbing was allowed in the American West, congressionally mandated surveyors sketched a grid of the country and brought it back to centralized land offices that were beholden to Washington. Little more was needed to establish the legal priority of the state's claim on resources. Though genocidal violence, forced removal, and rampant land speculation were the engines of US conquest of Native American territories, the discursive grid for expropriation was established with the humble cadastre, an instrument to which all future claims and disputes would refer. If Turner could imagine the frontier as closed, it was because there was a log, a map, and a registry that said so: the state's unity backstops the frontier's unity as an economic and demographic fact."

35. Author's interviews with Professor Girolamo Domenico Treccani. Based also on confidential reports of the Brazilian police and the author's examination of documents and reports from INCRA, ITERPA, and documental evidence gathered in the cartório of Rondon do Pará.

36. Comissão Camponesa da Verdade, *Relatório Final: Violações de direitos no campo 1946 a 1988*, Brasília, December 2014, 183.

37. Campbell, *Conjuring Property*, 63.

38. *Folha de S. Paulo*, August 6, 2000. According to the CIA World Factbook, Cuba has 110,860 square kilometers, and Honduras, 112,090.

39. Felipe Lobo, "O latifundiário fantasma," (o)*eco*, September 10, 2007. https://oeco.org.br/reportagens/2060-oeco_24013/.

40. *O Livro Branco da Grilagem de Terras no Brasil*, Ministério da Política Fundiária e do Desenvolvimento Agrário, 1999, 57.

41. Informações 044402/ABE/75, ACE 3561/83, and 0356/08/ABE/83, Serviço Nacional de Informações, found in Arquivo Nacional.
42. Informações 044402/ABE/75, Serviço Nacional de Informações, March 10, 1975.
43. Quote is from Informação 0356/08/ABE/83, May 4, 1983.
44. *O Tablóide*, October 6–13, 1975.
45. Informe 001/77/AG-SI/CR, January 31, 1977. INCRA.
46. Informação 0316/119/ABE/78, Serviço Nacional de Informações, February 23, 1978.
47. ITERPA's Relatório de Análise de Documento 188 and ITERPA's president letter, dated February 28, 1979, found in processo 05264/78.
48. Instituto Nacional de Criminalística, Laudo 9763, February 9, 1973, and contained in Land Institute of Pará (ITERPA) processos 2002/96.387, 2002.60712, 2008.121168, and 2008.458543. Also, report of ITERPA official Yara Maria Ribeiro Chaves, January 2, 1979.
49. See Land Institute of Pará (ITERPA) processo 2002/96.387 and *Diário Oficial do Pará*, 880 da Repíblica, no. 23.975, March 15, 1979.
50. Informação 0316/119/ABE/76, Serviço Nacional de Informações, February 23, 1978.
51. In 1978, Josélio asked the authorities to issue a certificate stating that he was the legal owner of the land, probably to access tax incentives or cheap loans to make improvements in his fazenda, but officials rejected his demand on the basis that the deed he had provided was fake. Josélio's written request to ITERPA was made on December 20, 1978, and is found in ITERPA *processos* 05264/78 and 2008/121168.
52. Certidão 0037, ITERPA, issued to Josélio on March 5, 1987.
53. On Hugo's police records, see Informação 667-S/102-A7-CIE.
54. The author contacted Pará's land agency, ITERPA, for information on faked deeds for Josélio's ranches and was informed that the Fazenda Serra Morena (processo 2002/96387) was confirmed as being based on a forged deed. On state subsidies enjoyed by the Te-Chaga-U, information provided by the Ministry of Regional Development to the author through the freedom of information act, petition no. 59011.000007/2021-88. Processo 28.650/003657/91, Fazenda Te-Chaga-U Agrocpecuaria S.A., SUDAM.
55. The Te-Chaga-U Agropecuaria SA received $764,000 from February 1989 to November 1994, according to information provided by the Ministry of Regional Development to the author through the freedom of information act, petition 59011.000007/2021-88. Processo 28.650/003657/91.
56. Jean Hébette, *Area de fronteira e conflito: O lesto do médio Tocantins* (Belém: NAEA, 1983).
57. The candidate for vice-mayor was Jairo Marçal de Moura, and his murder was reported in *Alguns casos de violência cometidos em Rondon do Pará, pelo fazendeiro Josélio de Barros Carneiro contra trabalhadores*, unsigned,

undated document, CPT, Marabá. The author confirmed that Marçal had been murdered in 1984 and requested access to the case from the court in Rondon do Pará, but the request was never answered by the judge.

58. The settler was Antonio Roldão, a smallholder who lived on a farm adjacent to one of Josélio's estates, according to Brito's interviews with the author. He said that he knew Roldão personally. Some details of the story were also reported by the advocacy group Justiça Global in Sandra Carvalho, *Direitos humanos no Brasil 2003 Relatório anual do Centro de Justiça Global* (Rio de Janeiro: Justiça Global, 2004), 58.

59. *Diário do Pará*, June 4, 1984.

60. On the death of Fonteles, see police confidential report ACE6154/87.

61. On the family claims, see *O Estado de S. Paulo*, May 14, 1993. For an account of Fonteles's murder, see Comissão Camponesa da Verdade, *Relatório Final: Violações de direitos no campo 1946 a 1988*.

8: NO LONGER MEEK

1. *O Globo*, November 16, 1988, and *Folha de S. Paulo*, November 16, 1988.

2. *Folha de S. Paulo*, November 16, 1988.

3. Author's interview with former Workers' Party leader and mayor candidate Maria de Jesus Silva Moura.

4. Ata do I Congresso dos Trabalhadores Rurais de Rondon do Pará, March 17–18, 1993.

5. Documentary "Dezinho: Vida, Sonho, e Luta," Federação dos Trabalhadores na Agricultura do Estado do Pará (FETAGRI) and Comissão Pastoral da Terra (CPT), 2006.

6. Interview with Carlos Guedes, friend of Dezinho and lawyer of the CPT.

7. Branford and Rocha, *Cutting the Wire: The Story of the Landless Movement in Brazil*, 254.

8. John L. Hammond, "The MST and the Media: Competing Images of the Brazilian Landless Farmworkers' Movement," *Latin American Politics and Society* 46, no. 4 (2004): 61–90.

9. Branford and Rocha, *Cutting the Wire*. Data on 90 claims that "about 100,000 families" lived in MST settlements by 1999.

10. Wilder Robles and Henry Veltmeyer, *The Politics of Agrarian Reform in Brazil: The Landless Rural Workers Movement* (London: Palgrave Macmillan, 2015), 97.

11. Miguel Carter, "The Landless Rural Workers Movement and Democracy in Brazil," *Latin American Research Review* 45, Special Issue: Living in Actually Existing Democracies (2010): 186–217.

12. Carter, "The Landless Rural Workers Movement and Democracy in Brazil."

13. *Jornal do Brasil*, 08/12/1985.

14. Written statement by Laércio José Segatti, president of the Associação, on March 19, 1997, found in case number 046.1997.2.000029-0, Justice of Pará.

15. A written statement by the president of the Associação Agropecuária Rondonense in 1995 and presented to the justice by Josélio's lawyer in the Te-Chaga-U case indicated that the rancher was president from 1987 to 1988 and from 1993 to 1994.
16. Author's interview with Odeb Moreira.
17. See *A Gazeta do Comércio*, August 2004, and *Evolução*, July 12–18, 1999.
18. The names of all sources interviewed by the author giving details or opinions on Barroso Nunes have been omitted for their anonymity and safety.
19. *Correio do Tocantins*, December 5–9, 2000.
20. Certidão, 10/03/1994, no. 378, fls. 078, Livro 2-B, Registro Geral, Cartório do Único Ofício de Rondon do Pará.
21. Nunes declared to the justice that the combined value of the Fazenda Lacy, including machinery, houses, the land, and the timber, was $107 million. See 566–80 of the federal court case 0006677-95.2012.4.01.3901/PA, TRF1. Also based on the police report on the Fazenda Lacy by Sergeant Claudio Marino Dias, November 25, 1999.
22. On the size of the ranch, see *O Liberal*, March 21, 1995.
23. Information provided by Josefa de Oliveira Avila, wife of rancher José Hilário, in her statement to the police on February 21, 1995, contained in court case 0000700-89.2000.814.0401, Pará criminal justice.
24. File AGE/ANO 77679/94, Arquivo Nacional.
25. Oficio AQA/804/93, CONTAG, July 22, 1993.
26. Of. 005/93, Sindicato dos Trabalhadores Rurais de Rondon do Pará, signed by Dezinho, August 3, 1993.
27. State legislator Nonato Guimaraes communication of June 16, 1993, and author's interview with Guimaraes.
28. Oficio AQA/804/93, CONTAG, July 22, 1993.
29. Oficio AQA/804/93, CONTAG, July 22, 1993.
30. Of. 005/93, Sindicato dos Trabalhadores Rurais de Rondon do Pará, August 3, 1993.
31. Family account and statement by Dezinho on file 005/93, STR, August 3, 1993.
32. Eugene Linden, "The Road to Disaster," *Time*, October 6, 2000.
33. Oficio 005/93, 03/08/1993, Sindicato do Trabalhadores Rurais de Rondon do Pará.
34. Author interview with Carlos Guedes.
35. Josélio communication with ITERPA on October 19, 1993. ITERPA case 2009/194655.
36. José Dutra da Costa's communication with FETAGRI and judge Rosana Lúcia Canelas Bastos, March 21, 1994.
37. Relatório sobre a problemática da posse na área denominada Fazenda São Jorge, localizada na Vila Gavião. Sindicato do Trabalhadores de Rondon do Pará, March 1995.

38. In March 10, 1994, Hilário sold Nunes a piece of land in Rondon, according to Escritura Pública de Compra e Venda, livro 027, folhas 057 e 058, Cartório do Único Ofício de Rondon do Pará. Quote from Rita Serra in interview with the author.

39. On Josélio's friendship with Souza, whose real name was Adélcio Nunes Leite, I quote Josélio in his statement to the police on August 18, 1995, court case 0000272-32.1997.8.14.0046, Pará.

9: HUNTING SOUZA

1. Unless otherwise stated, the source for the account of facts, quotes, and data in this chapter is the author's multiple interviews with Detective Antônio Carlos Corrêa de Faria, his personal blog (cyberpolicia.com.br), and his memoir: Antônio Carlos Corrêa de Faria, *O Sangue do Barão: As Histórias de Nossa História* (Belo Horizone, self-published, 2007).

2. On the number of homicides, see *Hoje em Dia*, May 3, 1994. On the quote, see *Estado de Minas Gerais*, July 17, 2020.

3. See, for instance, *Diário Mercantil*, June 13, 1974, and *Diário do Rio Doce*, April 29, 1981.

4. On the support of the military, see *Estado de Minas Gerais*, July 17, 2020.

5. *O Globo*, December 28, 1991.

6. *Diário Mercantil*, September 13, 1974.

7. *Estado de Minas*, May 3, 1997.

8. *Estado de Minas*, July 17, 2020.

9. Globo Reporter, 1990. https://www.youtube.com/watch?v=bVyqnbfXvDA.

10. *Estado de Minas*, July 17, 2020.

11. *O Globo*, July 20, 1990.

12. *O Globo*, July 25, 1990. On karate and Mello, see James Brooke, "Brazil Labor Chief Gets Place in Runoff," *New York Times*, November 20, 1989.

13. *Estado de Minas*, May 20, 1990.

14. *Estado de Minas*, June 12, 1993.

15. *Estado de Minas*, June 12, 1993.

16. *Estado de Minas*, June 12, 1993.

17. Author's interview with Paulo Orlando Rodrigues de Mattos, retired detective of the Minas Gerais police. He's the person who originally received the tip about Faria.

18. *O Estado de Minas*, December 29, 1993.

19. *O Estado de Minas*, December 29, 1993.

20. Correa de Faria, *O Sangue do Barão: As Histórias de Nossa História*, 214.

21. Photographs provided to the author by officer Faria; one of the photographs of Josélio as described can be found in the file court case 0000272-32.1997.8.14.0046.

22. *Estado de Minas*, editions of June 15 through 25, 1994.

23. *Estado de Minas*, editions of June 15 through 25, 1994.

24. Adélcio Nunes Leite interrogation, August 7, 1995, court case 0000272-32.1997.8.14.0046, Pará.
25. Adélcio Nunes Leite interrogation, August 7, 1995, court case 0000272-32.1997.8.14.0046, Pará.
26. *Estado de Minas*, editions of June 15 through 25, 1994.
27. Adélcio Nunes Leite interrogation, August 7, 1995, court case 0000272-32.1997.8.14.0046, Pará.
28. *Estado de Minas*, September 29 and 30, 1994.
29. "Indulto a condenado a 229 anos," *O Tempo*, September 11, 2013.
30. *Estado de Minas*, September 29 and 30, 1994.
31. Account provided to the author by retired officer Paulo Orlando Rodrigues de Mattos; he's also a distant relative of the Cordeiros.
32. On the police suspicion of Josélio, court cases numbers (CNJ) 0000443-15.1995.8.14.0039 and 0000667-78.1995.8.14.0039, Pará state. The media also wrote he was a suspect. *O Liberal*, May 25, 1995, June 13, 1995, and *A Província do Pará*, May 25, 1995.
33. Judge Amorim ordered Josélio's arrest on March 19, 1997, and granted him an habeas corpus the day later, according to documents in court case number 046.1997.2.000029-0, Justice of Pará.
34. Author's interview with Maria Joel, Rita Serra, and Zé Geraldo. Zé Geraldo's suspicions about Mrs. Barros were also reported by *Diário do Pará*, June 30, 1995.
35. On May 22, 1995, state legislator Zé Geraldo and members of FETAGRI met with the attorney general of Pará to request Josélia Leontina's transfer, according to a press release from Zé Geraldo's office dated July 3, 1995.
36. "Certidão" from Rosilda Pacheco e Silva, ministério público do Estado do Pará, February 6, 2002, court case AREsp 518863/PA (2014/0116742-0).
37. "Certidão" from Rosilda Pacheco e Silva, ministério público do Estado do Pará, February 6, 2002, court case AREsp 518863/PA (2014/0116742-0).
38. See Relatório sobre a participação do Deputado na diligência em Rondon do Pará, Zé Geraldo, July 3, 1995. Quote is from Geraldo's interviews with the author, 2018 and 2019.
39. Court case AREsp 518863/PA (2014/0116742-0).
40. Sentence by judge Guísela Haase de Miranda, December 13, 2002, court case AREsp 518863/PA (2014/0116742-0).
41. Recurso Extraordinário com Agravo 869.180, Pará, May 10, 2015.
42. The internal affairs office of the Public Ministry of Pará informed the author that there had been four disciplinary processes against Josélia Leontina de Barros Lopes: 03340/2003, 36801/2017, 42214/2019, and 17587/2020.
43. Procuração Publica, Cartório Condurú, livr 292, folha 107, 09/12/1992, contained in ITERPA processo 2009.194655, and Procuração, Cartório of Rondon do Pará, libro 028, Folha 168, dated September 9, 1996, and contained in ITERPA public processo 2002.96387.

44. On the jobs of Josélio's daughters, see the self-declaration by each of them in Escritura Pública de nomeação de inventariante, Livro 191, fohas 057/058, ato 095, cartório Travessa Três de Maio n 1503, Belem, Pará. The quote is from a source in Rondon do Pará who prefers to remain anonymous.

45. On the murders attributed by the media to Josélio on May 1995, see *A Província do Pará* and *O Liberal*, May 25, 1995, and *O Liberal* and *Diário do Pará* on June 13, 1995. The quote of Shirley Cristina is from *Tocantins. O Jornal do Carajas*, June 21–27, 1995,

46. Based on documentary evidence in criminal court case 046.1997.2.000029-0, Pará, like motions filed by attorney Vilaça with the letterhead *Barros & Vilaça Lawyers*. The address is the same as the one printed on Josélio's business card. The author attempted to contact Mrs. Vilaça to get her comments, but she never responded.

47. Motion filed on April 29, 1997, in the documents of court case 046.1997-2.000029-0.

48. José Antônio da Silva's statement to the police on February 12, 1997, in court case 046.1997-2.000029-0.

49. Josélio's statement to Detective Moraes, August 18, 1995, in court case 0000272-32.1997.8.14.0046.

50. Documents contained in court case 046.1997-2.000029-0, Pará justice.

51. Rai was freed by order of Judge Amorim on May 2, 1997, according to the documents in court case 046.1997.2.000029-0, Justice of Pará.

10: NOWHERE TO HIDE

1. The statement of eyewitness Jorge Pereira da Silva to the police on February 14, 1995, criminal court case 0000700-89.2000.814.0401.

2. Quote from Josefa de Oliveira Avila statement to police on February 21, 1995, criminal court case 0000700-89.2000.814.0401. The author attempted to verify this information. He contacted Denilson Carlos dos Santos and José Alexander Bastos Dyna, attorneys of the family of Hilário. Olivandro de Oliveira Avila, son of Hilário, was interviewed by the author.

3. Information provided by Josefa de Oliveira Ávila, statement to the police, February 21, 1995.

4. Information provided by Josefa de Oliveira Ávila, statement to the police, February 21, 1995.

5. Author's interview with Elizinete Lopes da Silva.

6. *O Liberal*, February 23, 1995.

7. José Dutra da Costa (Dezinho) statement to the Rondon do Pará police, March 24, 1995, court case 0000700-89.2000.814.0401, Pará criminal justice.

8. Relatório, IPL, 001/95-DPRP, contained in court case 0000700-89.2000.814.0401, Pará criminal justice.

9. On the use of violence, "Violências da PM preocupam deputados," *O Liberal*, March 24, 1995.

10. Author's interview with Sergio Galiza.
11. Raimundo Nonato Rocha statement to the police, March 24, 1995, court case 0000700-89.2000.814.0401.
12. Raimundo Nonato Rocha statement to the police, March 24, 1995, court case 0000700-89.2000.814.0401.
13. Raimundo Nonato Rocha statement to the police, March 24, 1995, court case 0000700-89.2000.814.0401.
14. Elias Dias do Nascimento, mentioned in *Carta Denúncia*, FETAGRI, released on May 18, 1995.
15. Ofício 007/95, FETAGRI, July 10, 1995.
16. Ata do II Congresso dos Trabalhadores Rurais de Rondon do Pará, May 10–11, 1996.
17. Quote from Fetagri-PA/AP and Sindicato dos Trabalhadores Rurais de Rondon do Pará joint communication, undated.
18. Ofício 032/96, Sindicato dos Trabalhadores Rurais de Rondon do Pará, August 11, 1996.
19. Quote from Oficio 032/96, Sindicato dos Trabalhadores Rurais de Rondon do Pará, August 11, 1996.
20. Quotes, descriptions of the scene, and thoughts from author's interviews with José Soares de Brito.
21. José Soares de Brito, interview with the author.
22. Brito statements to police officers in Belém on August 28, 1996.
23. Brito's and Dezinho's statements to police officers in Belém on August 28, 1996.
24. Lawmaker Paulo Rocha, Diário da Câmara dos Deputados, August 28, 1996.
25. FETAGRI communication to Pará's secretary of public security, October 7, 1996.
26. FETAGRI communication to Pará's secretary of public security, October 7, 1996.
27. Luiz Pacheco Neto, "Piaui," statement to the police on October 14, 1996, contained in court case 200620326535, Pará state.
28. Request by officer Raimundo Jose Rodrigues Souza, October 17, 1996.
29. Rule by Judge Margui Gaspar Bittecourt, October 17, 1996.
30. Luiz Pacheco Neto, "Piaui," statement, court case 200620326535, Pará state.
31. Luiz Pacheco Neto, "Piaui," statement, court case 200620326535, Pará state.
32. "Carta Denúncia: Director do STR de Rondon do Pará está ameaçado de morte," February 6, 1997.
33. "Mais um Trabalhador Rural Assassinado no Município de Rondon do Pará," CPT release, March 14, 1997.

11: NOTHING SHINING IN ELDORADO

1. Larry Rohter, "Serra Pelada Journal; Brazilian Miners Wait for Payday After Diet of Bitterness," *New York Times*, August 23, 2004.

2. Numbers of police officers and demonstrators vary greatly among sources. The author has chosen to use the data given in chapter 8 of Miguel Carter, ed., *Challenging Social Inequality: The Landless Rural Workers Movement and Agrarian Reform in Brazil* (Durham, NC: Duke University Press, 2015).
3. *Folha de S. Paulo*, April 17–20, 1996.
4. Larry Rohter, "Acquittals in Massacre Arouse Brazil," *New York Times*, October 29, 1999.
5. *O Globo*, April 19, 1996.
6. Sue Branford and Jan Rocha, *Cutting the Wire*, chapter 7.
7. Carter, ed., *Challenging Social Inequality*, chapter 8. On the name tags, see also Larry Rohter, "Acquittals in Massacre Arouse Brazil," *New York Times*, October 29, 1999, and Matthew Teague, "Police Massacre Case Turns Back Tide of Injustice in Brazil," *Los Angeles Times*, October 6, 2012
8. On police abuses, see James L. Cavallaro and Anna C. Monteiro, *Fighting Violence with Violence: Human Rights Abuse and Criminality in Rio de Janeiro*, Human Rights Watch Short Report no. 2, 1996. On the historical use of violence by elites in Brazil, see R. S. Rose, *The Unpast: Elite Violence and Social Control in Brazil, 1954–2000* (Athens: Ohio University Press, 2005).
9. *O Globo*, April 19, 1996.
10. *O Globo*, April 19, 1996.
11. "Report on the Situation of Human Rights in Brazil," the Inter-American Commission on Human Rights, OEA/Ser.L/V/II.97, Document 29, rev.1, September 29, 1997.
12. Data from *Conflitos no Campo Brasil 1996*, CPT, Goiânia, June 1997.
13. *Conflitos no Campo Brasil 1996*, CPT, Goiânia, June 1997.
14. Rose, *The Unpast*, 226.
15. Diana Jean Schemo, "Violence Growing in Battle Over Brazilian Land," *New York Times*, April 21, 1996.
16. Jeremy M. Campbell, *Conjuring Property: Speculation and Environmental Futures in the Brazilian Amazon*, 60. According to Professor Campbell, "the violence and fraud commonly associated with *grilagem* have delivered much of Brazil's public patrimony into private hands, accentuating the nation's inequality and pushing macroeconomic trends toward greater concentration of wealth."
17. Miguel Carter, "The Landless Rural Workers Movement and Democracy in Brazil," *Latin American Research Review* 45, Special Issue: Living in Actually Existing Democracies (2010):186–217.
18. Carter, ed., *Challenging Social Inequality*, 165–67.
19. See *Folha de S. Paulo*, April 15–20, 1997.
20. *Folha de S. Paulo*, August 16, 1999.
21. See, for instance, *Folha de S. Paulo* on March 30, 1998, and *O Liberal*, April 27, 1998.

22. Manifesto, April 23, 1996, Associação Agropecuária de Rondon do Pará e o Sindicato Patronal de Rondon do Pará.

23. On the reasons causing concern, author's interview with Brito. Data of the MST occupation is from "Sem Terra celebram 30 anos da primeira ocupação realizada pelo MST," MST, October 28, 2015.

24. Data for the context is from *Plano Plurianual—PPA 2002/2005*, Município de Rondon do Pará.

25. On the depletion of reserves, see A. Veríssimo, E. Lima, and M. Lentini, *Pólos Madeireiros do Estado do Pará* (Belém: Imazon, 2002).

26. In 1995, Pará produced 5,375,000 cubic meters of logs for a total of 11,926,000 for Brazil. Source: IBGE, Censo Agropecuario, 1995. On the black market, author's interviews with former Rondon loggers.

27. On the crisis, see Maria Lúcia do Amor Divino Xavier, *O declínio das indústrias madeireiras no município de Rondon do Pará*, dissertation, Universidade Federal do Pará, 2003, 45 and 59. Pará doubled its herd from 1985 to 1995, according to Valdeci Monteiro dos Santos, *A economia do sudeste paraense: fronteira de expansão na periferia brasileira*, dissertation, Instituto de Economia da Universidade Estadual de Campinas, 2011. See tables on 88 and 96.

28. In 1980, the municipality of Rondon do Pará, which includes the town and its interior, had a population of 27,600, and in 1991, of 52,200. By 2000, the population had declined to 39,800. Sources: IBGE Censo Demográfico of 1980, 1991, and 2000.

29. Dezinho's speech contained in footage provided by the CPT. For an overview of the crisis and its consequences, see Valdeci Moneiro dos Santos, "A economia do sudeste paraenese: evidências das transformações estruturais," in Aristedes Monteiro Neto, César Nunes de Castro, and Carlos Antonio Brandão, organizers, *Desenvolvimento Regional no Brasil: políticas, estratégias e perspectivas* (Rio de Janeiro: Ipea, 2017).

30. *O Liberal*, November 20, 1997.

31. Carter, ed., *Challenging Social Inequality*, 274. On the date of the occupation and the motives for it, interview with lawyer Batista and annual reports from CPT for 1997, 1998, 1999.

32. *Conflitos no Brasil 1998*, CPT annual report.

33. CPT communication: "Pistoleiros atiram em trabalhadores rurais da fazenda Jerusalem no município de Rondon do Pará." Marabá, December 16, 1997.

34. Police report by officer James Moreira de Sousa, Delegacia Especializada de Conflitos Agrários (DECA), December 22, 1997.

35. "Queremos Paz em Rondon," pamphlet, unsigned and undated but distributed by late January or early February of 1998. This document was part of the dossier of Petition 1290-04, November 30, 2004, Inter-American Commission on Human Rights, Organization of American States.

36. Quote from *Correio do Tocantins*, January 27–29, 1998; on the accusations about Josélio as the "articulador" of a larger group, see also *Correio do Tocantins*, March 21–26, 1997.
37. Letter addressed February 3, 1998, by the Sindicato, the CPT, and other peasant organizations. Cited in *Petition 1290-04*, November 30, 2004, Inter-American Commission on Human Rights, Organization of American States.
38. On the occupation, *Conflitos no Campo Brasil 1999*, CPT annual report.
39. Denunciation filed by Jucelino Favoreto on February 12, 1999, to the police of Rondon do Pará, case number 59928.
40. Injunction number 063/99, Judge Amorim, March 4, 1999.
41. *Opinião*, May 20–21, 1999.
42. Mandado de prisão, signed by Judge Amorim, May 18, 1999.
43. *Opinião*, May 20–21, 1999.
44. About the number of people in the demonstration and Dezinho's public complaints, see *O Liberal*, June 2 and 3, 1999, and *Correio do Tocantins*, June 4–7, 1999
45. This version of facts was supported by organizations like Justiça Global, Terra de Direitos, and the CPT. The quote is from the brief by the Robert F. Kennedy Memorial Center for Human Rights and Partner Organizations as Amicus Curiae in Support of *Jose Dutra da Costa v. Government of Brazil*— Claim JG/RJ no. 294/04, February 21, 2006.
46. Complaint (FO) no. 460173, June 9, 1999. The author put in requests to both the police and the military police of Pará for access to the complaint but was informed that documents from prior to 2008 had been destroyed. Still, it is mentioned in the official Relatório Fazenda Lacy Barroso (November 25, 1999), found in appendices to the *Comissão Parlamentar Mista de Inquérito da Reforma Agrária e Urbana*. Brasília, November 2005.
47. Complaint (FO) 460178, June 15, 1999.
48. The date of FETAGRI's report on the Fazenda Lacy is October 14, 1999. Source: Relatório Fazenda Lacy Barroso (November 25, 1999), in appendices to the *Comissão Parlamentar Mista de Inquérito da Reforma Agrária e Urbana*. Brasília, November 2005.
49. Processo Administrativo Disciplinar (PAD) 19993006475-1. Also the story "Cooperativa usa títulos podres para pagar dívida," published by *Folha de S. Paulo*, May 14, 2000.
50. About the forced retirement of Mr. Amorim, see Tribunal de Justiça do Estado do Pará, Acórdao 72075, ruled on June 4, 2008.

12: DEATH AND SALVATION

1. In Wellington's statements to the police, the killer referred to the murder as "the service," and so the author uses this word here.

2. Wellington's first statement to police, December 22, 2000, and Wellington's statement on January 25, 2001, both files contained in court case 200620326535, Pará state.

3. Maria Joel's description of the crime is from author's interviews with her.

4. Wellington's first statement to police, November 22, 2000, 9:00 a.m., court case 200620326535, Pará state.

5. *Dezinho: Vida, Sonho, e Luta*, Federação dos Trabalhadores na Agricultura do Estado do Pará (FETAGRI) and Comissão Pastoral da Terra (CPT), 2006.

6. Maria Joel's description of the crime from interviews with her.

7. Dr. Antônio Lopes de Angelo statement to police, December 12, 2000, contained in court case 200620326535, Pará state.

8. Superior Electoral Court, Brazil, *Results of Rondon do Pará, municipal elections, 2000*.

9. Quote from the author's interview with Maria Joel.

10. Quote from the author's interview with Maria Joel.

11. Maria Joel's statement for *Dezinho: Vida, Sonho, e Luta*, Federação dos Trabalhadores na Agricultura do Estado do Pará (FETAGRI) and Comissão Pastoral da Terra (CPT), 2006.

12. Joelson's statement in the documentary *Dezinho: Vida, Sonho, e Luta*, Federação dos Trabalhadores na Agricultura do Estado do Pará (FETAGRI) and Comissão Pastoral da Terra (CPT), 2006.

13. Joelson's account to the author.

14. Joélima's interviews with the author.

15. Statement from *Dezinho: Vida, Sonho, e Luta*, FETAGRI.

16. Shirt description from Laudo de Exame n 018/2001, January 16, 2001, case number 200620326535, Pará state.

17. Gun description given by Raimundo Nonato da Silva, December 27, 2000, court case 200620326535. On the killer's attitude, recollections from Maria Joel's interviews with the author.

18. Maria Joel's interview with the author.

19. Maria Joel's first declaration to police, November 22, 2000, 9:00 a.m., court case 200620326535, Pará state.

20. Quote from Maria Joel's statement in *Dezinho: Vida, Sonho, e Luta*, Federação dos Trabalhadores na Agricultura do Estado do Pará (FETAGRI) and Comissão Pastoral da Terra (CPT), 2006.

21. Maria Joel's first declaration to police, November 22, 2000, 9:00 a.m., court case 200620326535, Pará state.

22. Wellington's first statement to police, November 22, 2000, 9:00 a.m., contained in court case 200620326535, Pará state.

23. Witness Magno Fernandes do Nascimento's statement to the police, November 22, 2000, court case 200620326535, Pará state.

24. *Diário do Pará*, November 25, 2000.

25. Thoughts and unfolding of events prior to the murder conveyed by Wellington in his first statement to police, November 22, 2000, court case 200620326535, Pará state.
26. Maria Joel's interviews with the author.
27. Claudio Marino Ferreira Dias's statement on April 23, 2019, on judgment for court case 0000700-89.2000.814.0401, Pará.
28. Claudio Marino Ferreira Dias's statement on April 23, 2019, court case 0000700-89.2000.814.0401, Pará.
29. Claudio Marino Ferreira Dias's statement on April 23, 2019, court case 0000700-89.2000.814.0401, Pará.
30. Claudio Marino Ferreira Dias's statement on April 23, 2019, court case 0000700-89.2000.814.0401, Pará.
31. Claudio Marino Ferreira Dias's statement on April 23, 2019, court case 0000700-89.2000.814.0401, Pará.
32. Relatório Fazenda Lacy Barros (November 25, 1999), contained in appendices of *Comissão Parlamentar Mista de Inquérito da Reforma Agrária e Urbana*. Brasília, November 2005.
33. Claudio Marino Ferreira Dias statement on April 23, 2019, on judgment for court case 0000700-89.2000.814.0401, Pará.
34. Author's interview with Claudio Marino Ferreira Dias.
35. The certificates or diplomas of the police of Pará awarded to Josélio are found in the court case 0000272-32.1997.8.14.0046 and were presented to the justice by the rancher as a proof of his good behavior.
36. Author's interview with Claudio Marino Ferreira Dias.
37. Claudio Marino Ferreira Dias's statement on April 23, 2019, court case 0000700-89.2000.814.0401, Pará, and author's interview.
38. Joelminha's interviews with the author
39. Joelminha's interviews with the author.
40. Joelminha's interviews with the author.
41. Joelminha's interviews with the author.
42. Joelminha's interviews with the author.
43. Joelminha's interviews with the author.
44. Joelminha's interviews with the author.
45. Autopsy carried out by Instituto Médico-Legal Renato Chaves, November 22, 2000, court case 200620326535, Pará state.
46. Claudio Marino Ferreira Dias's statement on April 23, 2019, on judgment for court case 0000700-89.2000.814.0401, Pará.

13: AN UNUSUAL CASE
1. *O Liberal*, November 22, 2000. Statement by public official Miguel Thomaz.
2. *Diário do Pará*, November 25, 2000.
3. Larry Rohter, "Brazilian Prison Revolt Exposes a Crumbling System," *New York Times*, March 5, 2001.

4. Rohter, "Brazilian Prison Revolt Exposes a Crumbling System."
5. According to a judicial certificate issued on March 9, 2001, Wellington da Silva had no criminal or judicial record. Certidão issued by Maria Naildes Fernandes Chaves, March 9, 2001, contained in court case 200620326535, Pará state.
6. *Diário do Pará*, November 25, 2000.
7. Maria de Jesus Silva's statement to the jury, November 13, 2006, contained in court case 200620326535.
8. According to the police records, Wellington was accompanied by attorney Adriana Andrey Lopes de Lima on November 22, 2000. Quotes are drawn from that statement. Mrs. Adriana Andrey Lopes was contacted by the author but didn't reply for comment.
9. Statement based on interviews with Sergeant Claudio Marino Ferreira Dias.
10. Francisco de Assis Solidade Da Costa, interview with the author.
11. Joelina's TV interview with *Jornal Record*, April 2011.
12. Joelminha's interview with the author.
13. *Diário do Pará*, November 23, 2000. Quote from Airton Luiz Faleiro, presidente of FETAGRI.
14. Statement by Italo Macola, chief of office of Pará state, to the press. *Província do Pará*, November 24, 2000.
15. On minister Raul Jungmann's political career, see Sue Branford and Jan Rocha, *Cutting the Wire: The Story of the Landless Movement in Brazil* (London: Latin America Bureau, 2002). For the quote, see *Diário do Pará*, November 23, 2000, and *O Estado de S. Paulo*, November 24, 2000.
16. *Diário do Pará*, November 23, 2000, and *O Estado de S. Paulo*, November 24, 2000.
17. Wellington de Jesus's statement on November 22, 2000, court case 0000700-89.2000.8.14.0046.
18. *O Liberal*, November 23, 2000.
19. On the number of people, see *O Estado de S. Paulo*, November 24, 2000, and *Jornal do Brasil*, November 24, 2000.
20. Descriptions based on accounts of Dezinho's children, Maria Joel, lawyers Carlos Guedes and José Afonso Batista, photographs and chronicles of newspapers *Província do Pará*, *Diário do Pará*, *O Liberal* on November 24, 2000, and footage from Globo TV stations.
21. *O Liberal*, November 24, 2000.
22. Descriptions based on accounts of Dezinho's children, Maria Joel, lawyers Carlos Guedes and José Afonso Batista, photographs and chronicles of newspapers *Província do Pará*, *Diário do Pará*, *O Liberal* on November 24, 2000, and footage from Globo TV stations.
23. Unless stated otherwise, quotes and statements from detective Walter Resende come from his interviews with the author.

24. On the number of detectives and cops, see *O Liberal*, November 24, 2000.
25. *Conflitos no Campo Brasil, 2000*, CPT, 38.
26. Relatório, Walter Resende de Almeida, December 1, 2000, court case 0000700-89.2000.8.14.0046, Pará state.
27. Fax sent by Shirley Cristina de Barros to detective Resende on December 20, 2000, in court case 0000700-89.2000.8.14.0046.
28. Fax sent by Shirley Cristina de Barros to detective Resende on December 20, 2000, in court case 0000700-89.2000.8.14.0046.
29. On Chico's election, *80 Anos Baixo Guandu. 1935–2015*, undated. Magazine published by the municipal council of Baixo Guandu. Also, Josélio's statement to the police on December 27, 2000, court case 0000700-89.2000.8.14.0046, Pará state.
30. Maria Joel Dias da Costa's statement to the pólice on November 25, 2000, contained in court case 0000700-89.2000.8.14.0046.
31. Claudio Marino Ferreira Dias, interview with the author.
32. About Francisco Martins da Silva Filho's working for Décio José Barroso Nunes and being the brother of the gunman Pedro Alves, see statement given by Francisco Martinson the day of the judgment, June 29, 2014. Video of the decision is available at the website of Pará's tribunal of justice. Concerning Josélio de Barros Carneiro, see his statement to the police on December 27, 2000, court case 0000700-89.2000.8.14.0046.
33. File in court case 046.2000.2.000012-7, Pará justice.

14: THE EVIDENCE MAN

1. Unless otherwise stated, quotes from Francisco Martins da Silva Filho are from interviews with the author and Martins da Silva's statement in court on April 29, 2014, case 0000700-89.2000.8.14.0046, Pará.
2. Court case 046.2000.2.000012-7.
3. Francisco provided this information to the author.
4. Francisco Martins da Silva Filho are from his statement in court on April 29, 2014, case 0000700-89.2000.8.14.0046, Pará. 03:49:00.
5. Quotes from Francisco Martins da Silva Filho are from his statement in court on April 29, 2014, case 0000700-89.2000.8.14.0046, Pará.
6. Detective Resende's interviews with the author.
7. Francisco Martins da Silva Filho's statement on November 27, 2000, court case 200620326535.
8. Statement given by Olávio Silva Rocha to the pólice on December 20, 2000. Court case 0000700-89.2000.8.14.0046.
9. Dezinho referred to Manoel Lopes in *Correio do Tocantins*, January 27–29, 1998. João Lopes de Angelo was from 1972 to at least 1975 mayor of Itanhém, Bahia state, according to official documents.
10. Statement given by Manoel "Duca" Lopes to the police on December 27, 2000. Court case 0000700-89.2000.8.14.0046.

11. Statements provided by João Lopes de Angelo and Antônio Lopes de Angelo on December 20, 2000. Court case 0000700-89.2000.8.14.0046. The author reached out to Dr. Antônio for a comment, contacting him through the cell number and email addresses he had provided to Brazil's electoral court, but Dr. Antônio never responded. João's daughter, lawyer Adriana Andrey Diniz Lopes, didn't reply for comment.

12. Statement given by Manoel "Duca" Lopes to the pólice on December 27, 2000. Court case 0000700-89.2000.8.14.0046.

13. Josélia Leontina married Marcos Vinicius Diniz Lopes on November 12, 1993, according to Escritura Pública, Livro 191, Folhas 057/058/ Ato 095, Cartório Condurú, June 18, 2014.

14. The defense attorney was Adriana Andrey Lopes de Lima, according to court case 0000700-89.2000.8.14.0046.

15. Dr. Angelo had also been elected substitute state legislator in 1998, according to Pará's electoral court.

16. Lourival had attempted to use fake deeds to get a subsidized loan from a state-run bank back in the 1970s, but the ruse hadn't worked because it had been discovered by law enforcement. However, he remained in control of the land. Informação 22/1512-A/75, confidential report, Ministerio do Interior, November 5, 1975.

17. Resende interrogated Mr. Veloso on December 22, 2000. He admitted to having been arrested for violent robbery and for possession without a license, but denied having participated in Pedro's murder, though he admitted he knew Pedro personally. Resende charged him with murder. Years later, prosecutors found no reliable evidence and filed a motion to dismiss. Source is court file for Pedro's murder, number 046.2000.2.000012-7, Pará.

18. Police investigation 037/2000-DMRP, cited in motion of defense attorney Humberto Feio Boulhosa, April 9, 2001, court case 200620326535.

19. Francisco Martins da Silva Filho statement on November 27, 2000, court case 200620326535, and *Correio do Tocantins*, December 5–9, 2000.

20. Francisco Martins da Silva Filho's statement on November 27, 2000, court case 200620326535.

21. Francisco Martins da Silva Filho's statement on November 27, 2000, court case 200620326535.

22. Antônio Lopes de Angelo's statement to the pólice on December 20, 2000, court case 0000700-89.2000.8.14.0046, Pará state.

23. Quotes from Francisco Martins da Silva Filho are from his statement in court on April 29, 2014, case 0000700-89.2000.8.14.0046, Pará.

24. Quotes from Francisco Martins da Silva Filho are from his statement in court on April 29, 2014, case 0000700-89.2000.8.14.0046, Pará.

25. Quotes from Francisco Martins da Silva Filho are from his statement in court on 29/04/2014, case 0000700-89.2000.8.14.0046, Pará.

26. Francisco Martins da Silva Filho's statement on November 27, 2000, contained in court case 200620326535.

27. Sue Branford and Jan Rocha, *Cutting the Wire: The Story of the Landless Movement in Brazil* (London: Latin America Bureau, 2002), 12.

28. According to official information, Francisco joined the witness protection program on January 1, 2001. See Ofício SEJU/SDDH/PROVITA/PA no. 74/2001, September 3, 2001, contained on 388 of court case 0000700-89.2000.814.0401, Pará state.

29. Olivandro de Almeida Ávila's statement to the police on November 27, 2000, court case 200620326535, Pará state. Based also on Olivandro's interview with the author.

30. Carlos Guedes's statement to the police on November 28, 2000, court case 200620326535, Pará state.

31. José Soares de Brito's statement to the police on November 28, 2000, court case 200620326535, Pará state.

32. *A Província do Pará*, December 1, 2000.

33. Autos Cíveis de Interdito Proibitório, requerente: Décio José Barroso Nunes e outros, Requerido: José Dutra da Costa da Silva e José Soares de Brito. Ruling on June 16, 1999, by Judge Paulo Cesar Pedreira Amorim, contained in court case 0000700-89.2000.814.0401, 3307–10.

34. Autos Cíveis de Interdito Proibitório, requerente: Décio José Barroso Nunes e outros, Requerido: José Dutra da Costa da Silva e José Soares de Brito. Ruling on June 16, 1999, by Judge Paulo Cesar Pedreira Amorim, contained in court case 0000700-89.2000.814.0401, 3307–10.

35. Autos Cíveis de Interdito Proibitório, requerente: Décio José Barroso Nunes e outros, Requerido: José Dutra da Costa da Silva e José Soares de Brito. Ruling on June 16, 1999, by Judge Paulo Cesar Pedreira Amorim, contained in court case 0000700-89.2000.814.0401, 3307–10.

36. On the 15,000 reais, see *O Liberal*, November 26, 2000.

37. Detective Walter Resende's report dated November 29, 2000, contained in court case 200620326535, Pará state.

38. Detective Walter Resende's report dated November 29, 2000, contained in court case 200620326535, Pará state.

39. On the details of the arrest, Décio José Barroso Nunes's testimony in the trial, April 29, 2014, court case 0000700-89.2000.814.0401.

40. Termo de ratificação de separação consensual, processo 0460/00, November 17, 2000.

41. Décio José Barroso Nunes's statement to the police, November 30, 2000.

42. *O Liberal*, December 5, 2000.

43. On Brazil's special treatment of college graduates, "Presos com diploma, a elite carcerária do Brasil," *El País*, February 3, 2017. Nunes's level of education was provided by him to Resende.

44. Description and interview based on *O Liberal* photographs published on December 2, 2000.
45. Statement by Claudio Marino Ferreira Dias on April 23, 2019, in the verdict for court case 0000700-89.2000.814.0401, Pará.
46. Claudio Marino Ferreira Dias's interview with the author.
47. Statement by Claudio Marino Ferreira Dias on April 23, 2019, in the verdict for court case 0000700-89.2000.814.0401, Pará.
48. Relatório, detective Walter Resende de Almeida, December 28, 2000. Court case 0000700-89.2000.814.0401.
49. Relatório, detective Walter Resende de Almeida, December 28, 2000. Court case 0000700-89.2000.814.0401.
50. Ruling of Judge Otávio Marcelino Maciel, on Habeas Corpus 7429, December 14, 2000.
51. *Justiça Desmoralizada*, CPT, December 18, 2000.
52. *Correio do Tocantins*, December 5–9, 2000.
53. On the ranchers' association, see Oficio SIRPA no. 06/2000, Rondon do Pará, December 3, 2000. On celebrations of Nunes's release, see *O Liberal*, December 15, 2000.
54. *Correio do Tocantins*, December 5–9, 2000.
55. On the second habeas corpus and the suspension of the investigations, see Oficio no. 261/2001-SCCR-HC, signed by Judge Otávio Marcelino Maciel, April 20, 2001.
56. Ofício no. 072/03-GPT/IC, Centro de Perícia Centífica "Renato Chaves," Belém, March 27, 2003.
57. *O Liberal*, December 17, 2000.

15: A CAUSE LARGER THAN DEATH

1. Rosemary Masters, Lucy N. Friedman, and George Getzel, "Helping Families of Homicide Victims: A Multidimensional Approach," *Journal of Traumatic Stress* 1, no. 1 (1988): 109–25. For a study of Brazilian victims, see Daniella Harth da Costa, Kathie Njaine, and Miriam Schenker, "Repercussions of Homicide on Victims' Families: A Literature Review," *Ciência & Saúde Coletiva* 22, no. 9 (2017): 3087–97.
2. Joelminha's interviews with the author.
3. Joelminha's interviews with the author.
4. Sue Branford and Jan Rocha, *Cutting the Wire*, 131.
5. *Jornal do Brasil*, December 13, 2001.
6. On Nunes's claims, author's interview with Humberto Feio Bulhousa, Nunes's defense attorney at the time.
7. On the confronted demonstrations, see *Jornal do Brasil*, December 13, 2001. Also based on author's interviews with Maria Joel, de Assis, and Batista.
8. Author's interview with Francisco de Assis Solidade.

9. *Nota a Imprensa. Assassinado mais um sindicalista em Rondon do Pará.* Signed by Sindicato dos Trabalhadores Rurais de Rondon do Pará, CPT, FETAGRI, and CUT. February 7, 2004.

10. Data from Oficio JG/RJ 294/04, November 30, 2004, contained in the file of the case of Dezinho at the Inter-American Commission on Human Rights, Organization of American States. This issue will be discussed later in the book.

11. For instance, labor court cases 107-JCJ-00627/99. A judge would say that in Josélio's fazendas, workers were treated "like animals." Rule by Judge Raimundo Itamar Lemos Fernandes Junior, June 25, 1999.

12. *Jornal do Brasil*, December 11, 2001.

13. Josélio de Barros Carneiro's statement to the police on December 27, 2000, court case 0000700-89.2000.8.14.0046.

14. Imposto sobre propiedad territorial rural 2001, Fazenda Serra Morena, Ministerio da Fazenda, contained in ITERPA administrative processo 2002.96387.

15. Author's interview with Sergeant Claudio Marino Ferreira Dias. Mr. Marino also provided this account in court on August 13, 2019, court case 0000700-89.2000.814.0401, Pará.

16. Author's interview with Francisco Martins da Silva Filho.

17. *Jornal do Brasil*, December 11, 2001.

18. *Jornal do Brasil*, December 11, 2001.

19. *Jornal do Brasil*, December 11, 2001.

20. Idelma Santiago da Silva et al., *Mulheres em perspectiva: trajetórias, saberes e resistências na Amazônia Oriental* (Belém: Paka-Tatu, 2017).

21. Santiago da Silva et al., *Mulheres em perspectiva*, 76 and 96.

22. Vanessa Barbara, "Life as a Brazilian Woman," *New York Times*, April 23, 2014. Femicide rates from Shasta Darlington, "Domestic Abuse, Shown Blow by Blow, Shocks Brazil," *New York Times*, August 7, 2018.

23. In 2014, for instance, twenty-one out of twenty-seven rural unions in southern Pará were led by men, according to Santiago da Silva et al., *Mulheres em perspectiva*, 76 and 96. Also based on interviews with Maria de Jesus Silva Moura, former Workers' Party leader and a victim of gender discrimination.

24. According to Ata da Assembléia Geral Extraordinaria, STR, August 9, 2002, this was a mandate for six months, but Maria Joel would be reelected in January 31, 2003, for a four-year term, according to Ata do IV Congresso do S.T.R. de Rondon do Pará.

25. Santiago da Silva et al., *Mulheres em perspectiva*, 101.

16: THE LAW OF THE GUN

1. See Atas 19, 22, and 23 of the reunião do comité de Decisão Regional do Incra/SR(27)MB, September 17 and November 25, 2002. Darwin Boerner Jr., at the time superintendent of INCRA in the region, declined to be interviewed for this book.

2. SEI SR-27/MBA 54600.002803/2002-39, INCRA.

3. On the INCRA funds, "Notícias do Sindicato," *Informativo do STTR de Rondon do Pará*, Ano 1, no. 1, March 2008.

4. Interview with Eliane dos Santos, born in Bahia and forty-seven years old in 2018, Projeto de Assentamento Unidos Para Vencer.

5. Number of members published by *O Estado de S. Paulo*, February 21, 2005.

6. Bianca Pyl, "Libertados 'coavam' insetos e girinos para poderem beber água," *Repórter Brasil*, August 29, 2008.

7. *Nota a Imprensa: Assassinado mais um sindicalista em Rondon do Pará*. Signed by Sindicato dos Trabalhadores Rurais de Rondon do Pará, CPT, FETAGRI, and CUT, February 7, 2004.

8. On polls, *Jornal do Brasil*, January 28, April 14, and April 23, 2002.

9. Letter addressed by Shirley Cristina to ITERPA's president on March 14, 2002, in ITERPA processo 2002/60712.

10. The issue of the fake deed was published in Pará's official gazette on March 15, 1979. It included Josélio's full name.

11. Ofício 553/2002-PG, ITERPA, in processo 2002/60712.

12. Shirley Cristina's letter to ITERPA's president dated April 15, 2002, processo 2002/96387.

13. Court case number 0000805-91.2002.8.14.0046, Pará state.

14. Court case number 0000805-91.2002.8.14.0046, Pará state.

15. Statement given to the police by Luiza da Silva Barros, mother of the alleged killer Walisson Geronimo Rodrigues da Silva, contained in court case number 0000805-91.2002.8.14.0046, Pará state.

16. Statement given to the police by Adelina Fernandes do Nascimento, sister of the victim, contained in court case number 0000805-91.2002.8.14.0046, Pará state.

17. Ordem de Missão, Polícia Civil, dated on February 10, 2004, and contained in court case 0000526-73.2004.8.14.0046, Pará state. Also in *Diário do Pará*, February 10, 2004.

18. On the UN officer in Marabá, see UN Commission on Human Rights, *Civil and Political Rights, Including the Question of Disappearances and Summary Executions: Extrajudicial, Summary or Arbitrary Executions*; Report of the Special Rapporteur Asma Jahangir, January 28, 2004, E/CN.4/2004/7/Add.3. Also verbal note E/CN.4/2004/G/33. Although Maria Joel was not mentioned by name in the report, her meeting with Jahangir was confirmed to the author by activist Sandra Carvalho, who was also present in the meeting. Quote is from Maria Joel's interview with the author.

19. Court case 0000526-73.2004.8.14.0046, Pará state.

20. Maria Joel Dias da Costa statement on February 8, 2004, court case 0000526-73.2004.8.14.0046, Pará state.

21. Statement given by the union's director Cordiolino José de Andrade on February 11, 2004, court case 0000526-73.2004.8.14.0046, Pará state.

22. Josélio de Barros Carneiro's statement to the police on December 27, 2000, court case 200620326535.
23. Josélio de Barros Carneiro's statement to the police on February 16, 2004, court case 0000526-73.2004.8.14.0046.
24. Josélio's struggle to keep control of the land, including his eventual hiring of lawyers, is from documents found in ITERPA processos 2002/96387 and 2002/60712.
25. According to the office of the state attorney general of Pará, in communication with the author, the deed of the Fazenda Serra Morena had been declared a fake. Ofício no. 001242/2021-PGE-PFAM, 11/11/2021.
26. INCRA SR-27 communication to Ouvidoria Agrária Nacional, signed by Celso Aparecido Florêncio, January 31, 2005.
27. INCRA SR-27 communication to Ouvidoria Agrária Nacional, signed by Celso Aparecido Florêncio, January 31, 2005.
28. On the use of lawyers, see attorney João Sá's motion addressed to the president of ITERPA, October 9, 2007, in processo 2002/60712.
29. From a total budget of some 101,000 reais, Josélio de Barros Carneiro donated 30,000 reais to the campaign, according to the data published by Tribunal Superior Eleitoral.
30. Settlement (Projeto de Assentamento) Rainha da Paz, Gavião, and Mantenha. Data from Superintendência Regional do Estado do Pará, SR 27, INCRA.
31. *Carta Maior*, April 11, 2005.
32. On threats and slave labor, see annual reports from the CPT, *Conflitos no Campo Brasil*, for 2004 and 2005.
33. Quote from Report no. 71/08, Petition 1290-04, José Dutra da Costa Admissibility, October 16, 2008, Inter-American Commission on Human Rights, Organization of American States. The date of the litigation is from the original complaint, Ofício JG/RJ 294/04, November 30, 2004. For some reason never fully explained and based on often contradictory arguments, Sandra Carvalho, from the NGO Justiça Global, refused to grant the author access to the original denunciation and the supporting files. The Brazilian Ministry of Foreign Affairs, responding to the several formal requests by the author, granted access to them.
34. Darci Frigo, 2001, Brazil, https://rfkhumanrights.org/people/darci-frigo. Also, Frigo's interview with the author.
35. Brief by the Robert F. Kennedy Memorial Center for Human Rights and Partner Organizations as Amicus Curiae in Support of *José Dutra da Costa v. Government of Brazil*—Claim JG/RJ no. 294/04, February 21, 2006. Supplied by the organization at the request of the author.
36. Letter to Nilmário Miranda, January 5, 2005.
37. *O Globo*, January 16, 2005.
38. *O Globo*, January 16, 2005.
39. Author's interview with Nilmário Miranda.

40. Author's interview with Nilmário Miranda, and Carlos Eduardo Gaio et al., *Na linha de frente: defensores de direitos humanos no Brasil, 2002–2005,* Justiça Global, 2006.

41. On Dorothy and Fontanella, author interview with Dorothy's colleague during her time in Vila Rondon, Rebecca Spires (Roberta Lee Spires).

42. Soybeans are typically grown in temperate climates, but new varieties of seeds and improvements in the often poor Amazon soil, such as the use of lime, made the crop resilient against the tropical heat. See, for instance, A. M. Buainain et al., *O mundo rural no Brasil do século 21. A formação de um novo padrão agrário e agrícola* (Brasília: Embrapa, 2014).

17: LAND OR WE BURN THE JUNGLE

1. "The Great Brazilian Land Grab," *Forbes*, July 25, 2005.

2. For a story of the colonization process of the BR-163, see Mauricio Torres, Juan Doblas, and Daniela Fernandes Alarcon, *"Dono é quem desmata"* (Altamira: Instituto Agronômico da Amazônia, 2017), and Jeremy M. Campbell, *Conjuring Property: Speculation and Environmental Futures in the Brazilian Amazon* (Seattle: University of Washington Press, 2015).

3. The story that better illustrated the lawlessness around the BR-163 was that of Márcio Martins da Costa, a man who in the late 1980s and early 1990s left his own mark under the name "Pará's Rambo." His story is briefly presented in *Conjuring Property*. Also, *Veja*, October 10, 1991.

4. Daniel C. Nepstad, Claudia M. Stickler, and Oriana T. Almeida, "Globalization of the Amazon Soy and Beef Industries: Opportunities for Conservation," *Conservation Biology* 20, no. 6 (2006):1595–603.

5. *Eating Up the Amazon*, Greenpeace International, April 2006. On Mato Grosso's geography, Alexandre José Cattelan and Amélio Dall'Agnol, "The Rapid Soybean Growth in Brazil," *OCL* 25, no. 1 (2018): D102. https://www.ocl-journal.org/articles/ocl/pdf/2018/01/ocl170039.pdf.

6. Nepstad, Stickler, and Almeida, "Globalization of the Amazon Soy and Beef Industries: Opportunities for Conservation."

7. "In 2003 China imported 21 million tons of soybeans, 10% of world production and 83% more than it imported in 2002; 29% of this soy came from Brazil." Nepstad, Stickler, and Almeida, "Globalization of the Amazon Soy and Beef Industries: Opportunities for Conservation." On droughts in the United States, Ralph Mondesir, "A Historical Look at Soybean Price Increases: What Happened Since the Year 2000?" *Beyond the Numbers: Prices & Spending* 9, no. 4 (U.S. Bureau of Labor Statistics, March 2020).

8. Nepstad, Stickler, and Almeida, "Globalization of the Amazon Soy and Beef Industries: Opportunities for Conservation."

9. S. L. Woodgate and R. G. Wilkinson, "The Role of Rendering in Relation to the Bovine Spongiform Encephalopathy Epidemic, the Development of EU Animal By-Product Legislation and the Reintroduction of Rendered Products

into Animal Feeds," *Annals of Applied Biology* 178, no. 3 (2021): 430–41. Also, information from the Center for Food Safety, https://www.centerforfood safety.org/issues/1040/mad-cow-disease/mad-cow-disease-q-and-a.

10. Sandra Blakeslee and Marian Burros, "Danger to Public is Low, Experts on Disease Say," *New York Times*, December 24, 2003.

11. On being undetectable, "BSE Frequently Asked Questions," U.S. Department of Agriculture, https://www.usda.gov/topics/animals/bse-surveillance -information-center/bse-frequently-asked questions. Also, Woodgate and Wilkinson, "The Role of Rendering in Relation to the Bovine Spongiform Encephalopathy Epidemic, the Development of EU Animal By-Product Legislation and the Reintroduction of Rendered Products into Animal Feeds."

12. Suzanne Daley, "Europe Takes Toughest Steps to Fight Mad Cow Disease," *New York Times*, December 5, 2000. Quote is from Woodgate and Wilkinson, "The Role of Rendering in Relation to the Bovine Spongiform Encephalopathy Epidemic, the Development of EU Animal By-Product Legislation and the Reintroduction of Rendered Products into Animal Feeds."

13. Nepstad, Stickler, and Almeida, "Globalization of the Amazon Soy and Beef Industries: Opportunities for Conservation."

14. "The Great Brazilian Land Grab," *Forbes*, July 25, 2005, and "Why Soybeans Are the Crop of the Century," *Financial Times*, June 20, 2017.

15. Nepstad, Stickler, and Almeida, "Globalization of the Amazon Soy and Beef Industries: Opportunities for Conservation."

16. *Jornal do Brasil*, January 24, 2005.

17. Quote is from Miguel Carter, "The Landless Rural Workers Movement and Democracy in Brazil," *Latin American Research Review* 45, Special Issue: Living in Actually Existing Democracies (2010): 186–217.

18. On the 500 percent increase, "Amazônia Ganha Unidades de Conservação e 10. Distrito Florestal Sustentável," Ministry of Environment of Brazil, press release, February 12, 2006.

19. The quote is from Scott Wallace, "Last of the Amazon," *National Geographic Magazine*, January 2007.

20. Susanna B. Hecht, "From Eco-Catastrophe to Zero Deforestation? Interdisciplinarities, Politics, Environmentalisms, and Reduced Clearing in Amazonia," *Environmental Conservation* 39, no. 1 (2012): 4–19.

21. *Jornal do Commercio*, Manaus, February 4, 2005.

22. *Jornal do Commercio*, Manaus, February 2, 2005.

23. Campbell, *Conjuring Property.*

24. Larry Rohter, "Brazil, Bowing to Protests, Reopens Logging in Amazon," *New York Times*, February 13, 2005. On the arms, *Jornal do Commercio*, Manaus, February 1, 2005.

25. On "blood will flow" see *O Globo*, February 4, 2005. The quote is from Luciano de Meneses Evaristo, former head of the department of environmental protection at IBAMA, in interview with the author.

26. "Last of the Amazon," *National Geographic Magazine*, January 2007.
27. On the kidnapping of the head of Brazil's National Fund for the Environment, see Philip M. Fearnside, "Brazil's Cuiabá-Santarém (BR-163) Highway: The Environmental Cost of Paving a Soybean Corridor Through the Amazon," *Environmental Management* 39, no. 5 (2007): 601–14.
28. On Senator Flexa, see *O Globo*, February 4, 2005; the quote from the *Times* is from Rohter, "Brazil, Bowing to Protests, Reopens Logging in Amazon."
29. Rohter, "Brazil, Bowing to Protests, Reopens Logging in Amazon."
30. Roseanne Murphy, *Martyr of the Amazon: The Life of Sister Dorothy Stang* (Ossining, NY: Orbis Books, 2007).
31. Murphy, *Martyr of the Amazon*.
32. Murphy, *Martyr of the Amazon*.
33. Larry Rohter, "Brazil Promises Crackdown After Nun's Shooting Death," *New York Times*, February 14, 2005. Also, author's interviews with Tom Stang, one of the brothers of Sister Dorothy.
34. Murphy, *Martyr of the Amazon*, 78.
35. Murphy, *Martyr of the Amazon*, 119.
36. Murphy, *Martyr of the Amazon*, 122.
37. Author's interview with Nilmário Miranda.
38. "Sister Dorothy's Killers," *New York Times*, March 2, 2005.
39. On the number of troops sent by Lula, Stephan Schwartzman et al., "Social Movements and Large-Scale Tropical Forest Protection on the Amazon Frontier: Conservation from Chaos," *Journal of Environment & Development* 19, no. 3 (2010): 274–99.
40. "Governo conclui obras de pavimentação da BR-163, no Pará," Agência Brasil, November 28, 2019.
41. On the size of the protected areas and quote, see Marina T. Campos and Daniel C. Nepstad, "Smallholders, the Amazon's New Conservationists," *Conservation Biology* 20, no. 5 (2006): 1553–56.
42. Campos and Nepstad, "Smallholders, the Amazon's New Conservationists."
43. On the size of the reserves, "Amazônia Ganha Unidades de Conservação e 10. Distrito Florestal Sustentável," press release from the Ministry of Environment, February 12, 2006. Quote from Schwartzman et al., "Social Movements and Large-Scale Tropical Forest Protection on the Amazon Frontier: Conservation from Chaos."
44. Doug Boucher et al., *Deforestation Success Stories: Tropical Nations Where Forest Protection and Reforestation Policies Have Worked*, Union of Concerned Scientists, June 2014, chapters 1 and 2.
45. H. K. Gibbs et al., "Brazil's Soy Moratorium," *Science* 347, no. 6220 (2015): 377–78.
46. On the impacts, see Marina Piatto and Lisandro Inakake de Souza, *10 anos da moratória da soja na Amazônia: História, impactos e a expansão para o Cerrado* (Piracicaba, SP: Imaflora, 2017). Also Lisa Rausch and Holly K.

Gibbs, "The Low Opportunity Costs of the Amazon Soy Moratorium," *Frontiers in Forests and Global Change* 4 (March 2021): 621685. On the role of Greenpeace Brasil, see the press release "Moratória da Soja na Amazônia: dez anos de resultados," Greenpeace Brasil, November 19, 2016.

47. Official data provided by INPE. http://www.obt.inpe.br/OBT/assuntos /programas/amazonia/prodes. See also "10 Years Ago the Amazon Was Being Bulldozed for Soy—Then Everything Changed," Greenpeace USA press release, 2016.

48. Official data provided by INPE. http://www.obt.inpe.br/OBT/assuntos /programas/amazonia/prodes Also, Boucher et al., *Deforestation Success Stories*, chapters 1 and 2.

49. Quote is from Boucher et al., *Deforestation Success Stories*, 5.

50. *O Globo*, February 14, 2005.

51. *O Estado de S. Paulo*, February 21, 2005.

52. "Só uma pessoa tem proteção policial no Pará," *Folha de S. Paulo*, February 14, 2005.

53. Henry Chu, "Light Shines on State Called Brazil's Heart of Darkness," *Los Angeles Times*, April 6, 2005.

54. Quote and data on attendees from the account of Renato Pita, published by Movimento Direitos Humanos, http://www.humanosdireitos.org/atividades /historico/117-Ato-Cultural-em-Rondon-do-Para--.htm)

55. *Opinião*, April 12-13, 2005.

56. Author's interview with Mrs. Pitanga.

57. Account of the meeting from Ricardo Rezende, who that day was referring to Father Josimo Morais Tavares, killed in 1986.

58. President Lula's answer was confirmed by eyewitnesses Father Rezende and Maria Joel in interviews with the author.

18: AMAZONIAN JUSTICE

1. Descriptions based on *Diário do Pará* and *O Liberal*, November 11, 2006.

2. Forensic report in criminal court case number 200620326535, Pará.

3. Author's interview with Américo Lins da Silva Leal in Belém.

4. Quote from documentary *They Killed Sister Dorothy* (2008) by Daniel Junge.

5. On the obscene gestures, see *Jornal do Brasil*, August 24 and 25, 1999; on the acquittal, *Folha de S. Paulo*, August 19, 1999.

6. "Coronel que liderou massacre de Carajás é preso após 16 anos," *Folha de S. Paulo*, May 8, 2012. See also Matthew Teague, "Police Massacre Turns Tide of Injustice in Brazil," *Los Angeles Times*, October 6, 2012.

7. *O Liberal*, October 11, 2006.

8. For pictures of the demonstration, see *Amazônia Hoje*, *O Liberal* and *Diário do Pará* from November 13 and 14, 2006.

9. *Amazônia Hoje*, November 14, 2006.

10. *Jornal Amazônia*, November 14, 2006.

11. Wellington de Jesus Silva's statement to the court on November 13, 2006, court case number 200620326535.
12. *O Liberal*, November 14, 2006.
13. Wellington de Jesus Silva's statement to court on November 13, 2006, court case number 200620326535.
14. Maria de Jesus Silva's statement in court, November 13, 2006, court case number 200620326535.
15. Sentence signed by Judge Raimundo Mosés Alves Flexa on November 13, 2006, court case 200620326535.
16. *O Liberal*, November 14, 2006.
17. *Amazônia Hoje*, November 14, 2006.
18. *Amazônia Hoje*, November 14, 2006.
19. *Amazônia Hoje*, November 14, 2006.
20. Lawyer Américo Leal's interview with the author.
21. Leal's interview with the author.
22. Description of Wellington's clothing based on photographs published by *Amazônia Hoje* on November 14, 2006.
23. Maria Joel interviews with the author.
24. *Amazônia Hoje*, November 14, 2006.
25. On the district attorney's request to the judge, see Alegações finais, signed by Promotoria de Justiça da Comarca de Rondon do Pará on November 14, 2006, court case 0000700-89.2000.8.14.0046. Contacted by the author multiple times for comment, the prosecutor never replied.
26. On the district attorney's request to the judge, see Alegações finais, signed by Promotoria de Justiça da Comarca de Rondon do Pará on November 14, 2006, court case 0000700-89.2000.8.14.0046.
27. On the federal police report, see the report by Luiz Eduardos Navajas Telles Pereira signed on August 9, 2005.
28. In their statements, Nunes's brother-in-law, Nilvan Moreira Costa, and his father-in-law, Manoel Ferreira Costa, as well as the brother of the latter, Geovano Ferreira da Costa, supported the supposition that Hilário's family and Lourival were the masterminds. Statements of the forementioned are found in court case 0000700-89.2000.814.0401, Pará state.
29. Alegações finais, December 4, 2006.
30. OMP no. 1699/2011—DPF/MBA/PA, MJ—Departamento de Polícia Federal—SR/PA.
31. Comissão Parlamentar Mista de Inquérito criada através do requerimento no. 13, de 2003-CN. Ata da 28a Reunião realizada em 6 de abril de 2005, Diário do Senado Federal, Suplemento. August 12, 2006.
32. OMP no. 1699/2011—DPF/MBA/PA, MJ—Departamento de Polícia Federal—SR/PA.
33. The list of assets and when they were acquired is found in OMP no. 1699/2011—DPF/MBA/PA, MJ—Departamento de Polícia Federal—SR/PA.

34. Nunes declared to the justice that the combined value of the Fazenda Lacy, including machinery, houses, the land, and the timber was 107 million reais. See 566–80 of federal court case 0006677-95.2012.4.01.3901/PA, TRF1.

35. Comissão Parlamentar Mista de Inquérito criada através do requerimento no. 13, de 2003-CN. Ata da 28a Reunião realizada em 6 de abril de 2005, Diário do Senado Federal—Suplemento. August 12, 2006.

36. For a list of alleged wrongdoings of Nunes during those years, see Sérgio Sauer, *Violação dos direitos humanos na Amazônia: conflito e violência na fronteira paraense* (Rio de Janeiro: Heinrich Böll Stiftung, 2005), 70–80.

37. OMP no. 1699/2011—DPF/MBA/PA, MJ—Departamento de Polícia Federal—SR/PA.

38. Interview under the condition of anonymity.

39. On the dangers endured by civil servants, see, for instance, "Fiscalização de rotina vira inferno," *O Liberal*, August 20, 2006, and "Ibama destaca as irregularidades," *Diário do Pará*, October 21, 2006.

40. Termo de Ajustamento de Conduta no. 79/2005, Ministério Público do Trabalho, Procuradoria do Trabalho do Munícipio de Marabá.

41. For a list of labor lawsuits against Nunes covering years from 1995 to 2013, see pages 3023–3032 of court case 0000700-89.2000.814.0401, Pará. For a detailed official account of some of these suspected wrongdoings, see, for instance, Auto de Infraçao no. 01927129-8, Ministério Público do Trabalho, Procuradoria do Trabalho do Munícipio de Marabá, 2010.

42. Comissão Parlamentar Mista de Inquérito criada através do requerimento no. 13, de 2003-CN. Ata da 28a Reunião realizada em 6 de abril de 2005, Diário do Senado Federal, Suplemento, August 12, 2006.

43. Report on Nunes's businesses by attorney Maurel Mamede Selares, April 18, 2013, Public Labor Prosecution Office of Brazil.

44. Carvoaria Chapadao Ltda was added in the "dirty list" in 2011, according to data provided to the author by the Ministry of Economy (Brazil's freedom of information act request number 03005.037879/2021-26). According to police files (OMP no. 1699/2011—DPF/MBA/PA, MJ—Departamento de Polícia Federal—SR/PA), Carvoaria Chapadao was located within the Fazenda Lacy and was owned by Nunes's second wife.

45. "Brazil to issue 'dirty list' of employers using slave labor based on court findings," Reuters, July 3, 2019.

46. *Senzalas na Amazônia*, Correio Braziliense, May 2, 2006.

47. *Senzalas na Amazônia*, Correio Braziliense, May 2, 2006.

48. The CPT reported in 2004 some 274 cases of slave-like labor conditions on eleven ranches and for charcoal kilns in Rondon; in 2005, the number was 149. Data from annual reports from CPT, *Conflitos no Campo Brasil*, 2004, 2005, and 2006.

49. *O Liberal*, December 14, 2006, and "Premiação sem discurso de Lula não mereceu atenção da mídia," *Carta Maior*, December 15, 2006.

50. Maria Joel's interviews with the author.
51. Sentença, March 26, 2007, contained in court case files 0000700-89.2000.8.14.0046.
52. Appeal filed on April 9, 2007, court case 046.2000.2.000004-4.
53. Press release from Terra de Direitos and CPT, April 24, 2007, and "Presos pistoleiros contratados para matar sindicalista em Rondon do Pará," *Repórter Brasil*, April 26, 2007.
54. Motion filed on April 25, 2007, Lucinery Helena Resende Ferreira, court case files 0000700-89.2000.8.14.0046.
55. The author contacted Mrs. Resende Ferreira for comment, but never received an answer.
56. Report no. 71/08, Petition 1290-04, José Dutra da Costa Admissibility, October 16, 2008, Inter-American Commission on Human Rights, Organization of American States.
57. Maria Joel's interview with the author.
58. *Diário do Pará*, January 2008.
59. Certidão carcerária issued by Luiz Correa Jr. on April 11, 2006, in court case 200620326535, Pará state.
60. Certidão carcerária issued by Ruberval Lopes da Silva on October 5, 2006, in court case 200620326535, Pará state.
61. *O Liberal*, June 6, 2011.
62. On jobs lost, see "O aumento do desemprego no setor formal na indústria da madeira no estado do Pará—2008," Departamento Intersindical de Estatística e Estudos Socioeconômicos (DIEESE), January 30, 2009. On population estimates, IBGE "População residente estimada—Rondon do Pará," 2008.
63. In 2001, Rondon do Pará had a budget of 14.5 million reais; in 2010, it was 44.6 million. About 5 percent of that amount was generated by local taxes, the remaining 95 percent coming from transfers from either the federal or the state government. Plano Plurianual—PPA 2006/2009 (lei n.478/2005, December 27, 2005) and 2014/2017 (lei n.665/2013-PE, December 12, 2013). In 2001, Rondon do Pará spent 7.3 million reais on staff expenses; by 2010, that number had climbed to 23.8 million. Plano Plurianual—PPA 2006/2009 (lei n.478/2005, December 27, 2005) and 2014/2017 (lei n.665/2013-PE, 02/12/2013).

19: SINK OR SWIM

1. Author's interview with Francisco de Assis Soledade da Costa.
2. Author's interview with Francisco de Assis Soledade da Costa and federal lawmaker Beto Faro.
3. Author's interview with Maria de Jesus Silva Moura, former leader of the Workers' Party in Rondon do Pará.
4. Author's interview with Rondon-based reporter Ivan Santana.

5. Divulgação de Candidaturas e Contas, Tribunal Superior Eleitoral, Rondon do Pará, 2008.

6. Divulgação de Candidaturas e Contas, Tribunal Superior Eleitoral, Rondon do Pará, 2008.

7. Judge Therezinha Martins da Fonseca ruling, May 29, 2008, court case files 0000700-89.2000.8.14.0046.

8. "Jornal de Campanha. Informativo da Campanha da 'Frente Popular Transforma Rondon,'" August 2008.

9. For a socioeconomic overview of some of these areas, see "1.3. Aspectos económicos," Plano Plurianual—PPA 2010/2013, Município de Rondon do Pará.

10. On her campaign, "Jornal de Campanha. Informativo da Campanha da 'Frente Popular Transforma Rondon,'" August 2008. Also, interviews with eyewitnesses.

11. Julio Jacobo Waiselfisz, *Mapa da Violência 2013. Homicídios e Juventude no Brasil* (Rio de Janeiro: Centro Brasileiro de Estudos Latino-Americanos, 2013). Data on homicides in Rondon do Pará for 2009, 2010, and 2011 is on 62.

12. Results provided by Brazil's top electoral court (www.tse.jus.br).

13. "TRE casa mandato de prefeito de Rondon do Pará," *Diário do Pará*, August 18, 2010.

14. Recurso especial eleitoral n. 4851-74.2009.6.14.0000, Tribunal Superior Eleitoral, May 8, 2012.

15. "Um prefeito é cassado a cada 16 horas no país," *Época*, May 5, 2009.

16. F. Daniel Hidalgo and Simeon Nichter, "Voter Buying: Shaping the Electorate Through Clientelism," *American Journal of Political Science* 60, no. 2 (2016): 436–55.

17. Simeon Nichter, "Vote Buying in Brazil: From Impunity to Prosecution," *Latin American Research Review* 56, no. 1 (2021): 3–19.

18. *Opinião*, April 12–13, 2005.

19. *Opinião*, April 12–13, 2005.

20. Agribusiness exports data provided by the Ministry of Agriculture, years 2002 and 2010, respectively. Quote from Mr. Powell is from Larry Rohter, "Brazil Is Ready to Use Clout to Get a Fair Deal," *New York Times*, December 12, 2005.

21. Wilder Robles and Henry Veltmeyer, *The Politics of Agrarian Reform in Brazil: The Landless Rural Workers Movement* (New York: Palgrave Macmillan, 2015), p. 4–6. Also, author's interview with minister of agrarian development Guilherme Cassel, and data quoted by him in Marco Weissheimer, "'Um dos objetivos do golpe é quebrar a agricultura familiar,' diz Guilherme Cassel," *Sul 21*, June 21, 2017. https://www.brasildefato.com.br/2017/06/21/um-dos-objetivos-do-golpe-e-quebrar-a-agricultura-familiar-diz-guilherme-cassel/.

22. Robles and Veltmeyer, *The Politics of Agrarian Reform in Brazil*, 4–6.
23. Miguel Carter, "The Landless Rural Workers Movement and Democracy in Brazil," *Latin American Research Review* 45, Special Issue: Living in Actually Existing Democracies (2010): 186–217.
24. Censo agropecuário 2017: resultados definitivos, IBGE, 2019.
25. Ata da Eleiçao e Posse da Diretoria e Conselho Fiscal da FETAGRI, March 28, 2009.
26. *O Estado de S. Paulo*, June 11, 2011.

20: THE WIDOW MUST FALL

1. Author's interviews with Zudemir Santos Jesus, and *Diagnóstico sobre as situações de ameaças de morte contra trabalhadores e trabalhadoras rurais do sul e sudeste do Pará*, CPT, 2012.
2. Author's interviews with Zudemir Santos Jesus and her daughters now living in the United States.
3. Data from "Carta à população de Marabá," June 17, 2011, FETAGRI, MST e FETRAF, in *Pastoral da Terra* 36, no. 204 (April–June 2011).
4. Author's interview with Antonio Gomes, nicknamed Pipira, a land activist under threat of death and FETAGRI coordinator.
5. See, for instance, "Um exemplo do descaso com a Reforma Agrária," in *Pastoral da Terra* 36, no. 204 (April–June 2011).
6. Felipe Milanez, "The Death of Zé Cláudio and Maria," *Vice*, November 6, 2011.
7. Tom Phillips, "Amazon Rainforest Activist Shot Dead," *The Guardian*, May 24, 2011.
8. *O Estado de S. Paulo*, May 25, 2011.
9. The Forest Code was one of the most controversial pieces of environmental legislation in Brazil during these years. One of the main requirements that the Forest Code established to tame deforestation was that landowners in the Amazon would have to set aside no less than "80 percent of their property (land area) in native vegetation," which caused disappointment among many fazendeiros. Still, the moratorium was perceived by campaigners as a way to provide impunity to environmental offenders and foster further grilagem. Andrea A. Azevedo et al., "Limits of Brazil's Forest Code as a Means to End Illegal Deforestation," *Proceedings of the National Academy of Sciences* 114, no. 29 (2017): 7653–58.
10. Congressman José Sarney Filho. "Eles morreram pela floresta," *Época*, May 27, 2011.
11. "Faroeste brasileiro," *Folha de S. Paulo*, June 1, 2011.
12. Sue Branford, "Brazil high court Forest Code ruling largely bad for environment, Amazon: NGOs," Mongabay, March 1, 2018.
13. "Cronologia da violência," *Pastoral da Terra* 36, no. 204 (April–June 2011). Also, "Líder camponês é assassinado a tiros em Rondônia," *G1*, May 27, 2011.

14. The name is Almirandi Pereira Costa. "Cronologia da violência," *Pastoral da Terra* 36, no. 204 (April–June 2011), and "Liderança do Quilombo Charco sofre tentativa de assassinato," *CPT Maranhão*, May 30, 2011.

15. The name of the victim is Herenilton Pereira. "Cronologia da violência," *Pastoral da Terra* 36, no. 204 (April–June 2011).

16. "Trabalhador é encontrado morto no mesmo assentamento de José Cláudio e Maria do Espírito Santo," CPT, May 30, 2011.

17. "Cronologia da violência," *Pastoral da Terra* 36, no. 204 (April–June 2011).

18. The victim was Obede Loyola Souza. "Cronologia da violência," *Pastoral da Terra* 36, no. 204 (April–June 2011), and "Quinto trabalhador rural é assassinado em 20 dias no Pará," *Folha de S. Paulo*, June 15, 2011.

19. *O Liberal*, June 16, 2011.

20. "Manifestação de acampados do Pará dura 46 dias," CPT, *Conflitos no Campo 2011*.

21. Events in May and June 2011 were described to the author by Maria Joel, Joelson, lawyer Batista, lawmaker Beto Faro, Zudemir, and a policeman who requested to remain anonymous.

22. Camila Penna de Castro, *Conexões e controvérsias no INCRA de Marabá: o Estado como um ator heterogêneo*, dissertation, Universidade de Brasília, 2013, 48–50.

23. Author's interview with lawmaker José Roberto Oliveira Faro.

24. Motions by Nunes's defense attorney Miguel Szaroas Neto in Supreme Court case HC 103867.

25. Documentary *SMS para Brasília SOS para Dona Joelma*, Movimento Humanos Direitos, September 2011.

26. *Fantástico*, TV Globo, July 31, 2011.

27. Moura's interview in *Conversa com Bial*, TV Globo, July 11, 2017.

28. Mr. Moura's communications with the author.

29. "Brazilian actor Wagner Moura joins campaign to end modern slavery," International Labor Organization, press release, August 4, 2015.

30. On the meeting, see "Brasília: Encontro com Ministra dos Direitos Humanos e Ministro da Justiça—filme," Movimento Humanos Direitos, July 7, 2011.

31. Quotes from Minister Cardozo and Maria Joel extracted from the videos recorded during the meeting.

32. "Rondon do Pará—PA: O MHuD esteve presente em Ato Público por Dezinho—filme," Movimento Humanos Direitos, November 26, 2011.

33. Report no. 71/08, Petition 1290-04, Jose Dutra da Costa Admissibility, October 16, 2008, Inter-American Commission on Human Rights, Organization of American States.

34. Documents of the case were shared by the Ministry of Foreign Affairs after submitting a FOIA request.

35. Tabela de Análise das Demandas Oriundas da Justiça Global, Sindicato dos

Trabalhadores Rurais de Rondon do Pará e Comissão Pastoral da Terra. Obtained by the author through Maria Joel's lawyers.

36. Tabela de Análise das Demandas Oriundas do Justiça Global, Sindicato dos Trabalhadores Rurais de Rondon do Pará E Comissão Pastoral da Terra.

21: "LOAD THE TRUCKS"

1. *Acordão*, August 9, 2011, Supreme Court case HC 103.867.

2. In 2010, officials from the Ministry of Labor had inspected Nunes's sawmills for nine days and found 171 workers in an irregular situation. Some didn't receive payment on time, while others didn't have the right to paid vacation or had been terminated without receiving their paychecks. Relatório de Fiscalização, Madereira Urubú LTDA-EPP, March 10–19, 2010, Ministério do Trabalho e Emprego.

3. Ruling by Judge Jônatas dos Santos Andrade on February 17, 2012, Labor Court case number 0001053-26.3011.5.08.0117.

4. The alleged use of Nunes's children to hide assets was raised by Judge Abeilar dos Santos Soares Jr. in his ruling of May 7, 2012, in Ação de Execução 0001053.26.2011.5.08.0117 and Embargos de Terceiros 0000820-92-2012.5.08.0117;0000819-10.2012.5.08.0117; 0000801-86.2012.5.08.0117 0000798-34.2012.5.08.0117; 0000803-56.2012.5.08.0117.

5. Serviço Publico Federal, Departamento de Polícia Federal-SR/PA, OMP N. 1699/2011-DPF/MBA/PA.

6. Author's interview with R.O.F. Except for Judge Jônatas dos Santos Andrade, the real name of the officers working at the labor court of Marabá have been changed to avoid potential retaliation.

7. Ação Publica Civil 0006677-95.2012.4.01.3901, Tribunal Regional Federal 1.

8. Ação Publica Civil 0006677-95.2012.4.01.3901, Tribunal Regional Federal 1.

9. Nunes's statement presented to the jury on August 14, 2019, court case 0000700-89.2000.814.0401.

10. Serviço Publico Federal, Departamento de Polícia Federal-SR/PA, OMP N. 1699/2011-DPF/MBA/PA.

11. Data on number of people and trucks, see *O Liberal*, January 20, 2012.

12. Author's interview with F.O.J.

13. Author's interview with S.O.J.

14. On the quotes about the cattle, http://www.braziliancattle.com.br/english/ On the herd owned by Mr. Nunes, see Serviço Público Federal, Departamento de Polícia Federal-SR/PA, OMP N. 1699/2011-DPF/MBA/PA.

15. See, for instance, Natacha Simei Leal, "About Zebus and Zebuzeiros: Value and Price, Influences and Substances in Elite Cattle Auctions," *Vibrant* 13, no. 2 (2016): 95–109.

16. Natacha Simei Leal, "About Zebus and Zebuzeiros."

17. "Vaca é vendida por quase R$3 milhoes em Minas," *O Estado de S. Paulo*, May 12, 2010.

18. On the Parla appreciation, see "Vaca nelore alcança valorização de R$ 7 milhões," *CanalRural*, October 31, 2016.

19. Case number 2881-85.2015.4.01.3903, Federal Justice, Altamira, Pará, Brazil. Also based on author's interviews with Luciano Evaristo, former head of IBAMA's environmental protection department.

20. Quote is from "MP denuncia novamente o maior desmatador da Amazônia à Justiça," *G1*, December 7, 2016.

21. On prosecution, see Ação Civil Pública 001503-60.2016.4.01.3903, Tribunal Regional Federal da 1a região. Contacted by the author in 2019 and 2022, defense attorney Rubens Silvera Neto didn't reply to requests for comment.

22. Nunes's statement to the judge, April 29, 2014, court case 0000700-89.2000.814.0401.

23. On Vale's size, see "Petrobras foi a nona empresa mais valiosa do mundo em 2009," *O Estado de S. Paulo*, January 9, 2010; on the ruling of Judge Jônatas against Vale, see "Vale faz acordo para não pagar R$300 mi no Pará," *Folha de S. Paulo*, July 24, 2010.

24. Certidão, signed by Rodrigo Xavier de Mendonça on February 16, 2012, and contained in court case 0001053-26.2011.5.08.0117, Tribunal Regional do Trabalho da 8a Regiao.

25. Declaração, Associação Agropecuaria Rondonense, March 9, 2009, in court case 0000700-89.2000.814.0046, Pará.

26. "Nota de repúdio da OAB—Subseção de Rondon do Pará," January 12, 2011.

27. Certidão, signed by Rodrigo Xavier de Mendonça on February 16, 2012, and contained in court case 0001053-26.2011.5.08.0117, Tribunal Regional do Trabalho da 8a Região.

28. Author's interview with officers from the Tribunal Regional do Trabalho da 8a Região.

29. Auto de arrematação em Leilao no. 03/2012 and 04/2012, court case 0001053-26.2011.5.08.0117, TRT 8a Região.

30. Inventory made by Mr. Nunes's lawyer, Marli Siqueira Fronchetti, court case 0001053-26.2011.5.08.0117, TRT 8a Região.

31. Appeal by lawyer Marli Siqueira Fronchetti, signed on February 18, 2012, court case 0001053-26.2011.5.08.0117, TRT 8a Região. Also based on author's interview with attorney Fronchetti.

32. Appeal by lawyer Marli Siqueira Fronchetti, February 18, 2012, court case 0001053-26.2011.5.08.0117, TRT 8a Região.

33. Appeal by lawyer Marli Siqueira Fronchetti, signed on February 18, 2012.

34. According to Nunes's lawyers, the value of the 948 cows seized was 2.08 million reais.

35. Author's interview with Nunes's attorney, Marli Siqueira Fronchetti.

36. *O Liberal*, February 20, 2012.

37. Author's interview with F.O.J.

38. Report from Prosecutor Maurel Mamede Selares, April 18, 2013, Procuradoria Regional do Trabalho da 8a Região.
39. Report from Prosecutor Maurel Mamede Selares, April 18, 2013.
40. *O Liberal*, August 21, 2013.
41. Labor court cases 00001673-67.2013.5.08.0117 and 0000262-18.2015.5.08.0117, 2a Vara do Trabalho de Maraba.
42. See, for instance, federal labor cases 0001396-81.2013.5.08.0107, 0001342-85.2013.5.08.0117, 0001330-71.2013.5.08.0117, TRT 8a Região.
43. Statistics and information dashboard of labor inspection in Brazil, Radar SIT, 2012.
44. See "Ministra Eleonora participa da entrega do Prêmio Direitos Humanos 2012," press release, Secretaria de Direitos Humanos da Presidência da República. Quote is from "Entrevista Juiz Jônatas Andrade," Tribunal de Justiça de Santa Catarina (TJSC), November 11, 2016.
45. Quote from court case 0000067-97.2016.5.08.0117, TRT 8a Região.
46. Statement of Nunes's former lawyer quoted in ACP 0000067-96.2016.5.08.117, Procuradoria do Trabalho da 8a Região, 2016.
47. Author's interview with lawyer O.R. At the request of the source, his name has been changed to guarantee anonymity.
48. Certidão, 2a Vara do Trabalho de Marabá, undated court document.
49. Certidão, 2a Vara do Trabalho de Marabá, undated court document.
50. *O Liberal*, July 25, 2013.
51. "Juiz denuncia que está sendo ameaçado de morte em Marabá, PA," *G1*, July 25, 2013.
52. *O Liberal*, July 24–30, 2013.
53. Relatório, Inquérito Policial n.0208/2013-4-DPF/MBA/PA, October 9, 2013, court case 0000700-89.2000.8.14.0046, Pará State.
54. Nunes's statement to the judge on August 14, 2014, criminal court case 0000700-89.2000.814.0401, Pará state.
55. Author's interview with attorney Marli Siqueira Fronchetti.

22: SHE IS OUT

1. Her program is available at Brazil's top electoral court: https://divulgacand contas.tse.jus.br/divulga/#/candidato/2012/1699/05738/140000021219.
2. Official data from Tribunal Superior Eleitoral for the 2012 mayoral election.
3. Official data from Tribunal Superior Eleitoral for the 2012 mayoral election.
4. Detalhamento dos Bens, Brazil's top electoral court: https://divulga candcontas.tse.jus.br/divulga/#/candidato/2012/1699/05738/140000021219/bens.
5. Detalhamento dos Bens, Brazil's top electoral court: https://divulgac andcontas.tse.jus.br/divulga/#/candidato/2012/1699/05738/140000021219/bens.

6. https://www.tse.jus.br/hotsites/estatistica2012/quadro-votacao.html.

7. *Diário do Pará*, January 4, 2013.

8. Descriptions based on photographs of the event and interviews with eyewitnesses Joelminha and Maria Joel.

9. Termo de posse da prefeita municipal, registered at the Cartório Elciria Oliveira, January 7, 2013.

10. *Diário do Pará*, January 4, 2013.

11. Unless stated otherwise, all quotes are from the author's interviews with Judge Gabriel Costa Ribeiro.

12. Court case 0001093-59.2009.8.14.0046, TJPA, Pará State.

13. Court case 0001093-59.2009.8.14.0046, TJPA, Pará State.

14. Report from Judge Gabriel Costa Ribeiro to the president of the judges of Pará State, Dr. Paulo Roberto Ferreira Vieira, September 23, 2009.

15. Civil court case 0000741-10.2006.8.14.0046 (2006.1.000865-6), Pará justice. Rule published by the official gazette of the justice of Pará (Diário da Justiça TJPA) no. 5123/2012, October 2, 2012.

16. Civil court case 0000741-10.2006.8.14.0046 (2006.1.000865-6), Pará justice.

17. Civil court case 0000741-10.2006.8.14.0046 (2006.1.000865-6), Pará justice.

18. Civil court case 0000737-30.2006.8.14.0046 (2006.1.000862-6), Pará justice.

19. Civil court case 0000737-30.2006.8.14.0046 (2006.1.000862-6), Pará justice.

20. Request published by the official gazette of the justice of Pará (Diário da Justiça TJPA) no. 5123/2012, 02/10/2012. Also, based on Banco da Amazônia's answers to the author, who filed a request through Brazil's freedom of information act.

21. Author's interview with attorney André Souza.

22. According to lawyer Patrícia Severo, who represented Shirley Cristina and some of her sisters in a civil case, and to the official communication of Banco da Amazônia to the civil court of Rondon do Pará on February 18, 2020, the unpaid debts at that point amounted to 18.2 million reais. Documents contained in court case number 0010474-32.2017.8.14.0046.

23. On the pirate radio station and the suspected use of illegal propaganda, see processo administrativo no. 408-19.2012.6.14.0051. 405-64.2012.6.14.0051, Ação Cautelar no. 404-79.2012.6.14.0051; Ação de propaganda irregular no. 578-88.2012.6.14.0051; AIJE no. 416-93.2012.6.14.0051 and AIJE no. 417-78.2012.6.14.0051. On the use of public resources, Acordão 27.266 and Recurso Eleitoral 418-63.2012.6.14.0051, Tribunal Regional Eleitoral do Pará. Also, *Jornal Opinião*, September 15–17, 2012.

24. *Jornal Opinião*, September 18, 2012.

25. Acordão 27.266 and Recurso Eleitoral 418-63.2012.6.14.0051, Tribunal Regional Eleitoral do Pará. Also, author's interviews with Judge Gabriel.

26. On potential violations and the removal from office of Mayor Shirley Cristina

de Barros Malcher, see processo administrativo no. 408-19.2012.6.14.0051. 405-64.2012.6.14.0051, Ação Cautelar no. 404-79.2012.6.14.0051; Ação de propaganda irregular no. 578-88.2012.6.14.0051; AIJE no. 416-93.2012.6.14.0051 and AIJE no. 417-78.2012.6.14.0051. On Mayor Shirley Cristina's claims, see lawyer Marcelo Araujo de Alburquerque's appeal to the judge, electoral court case 418-63.2012.6.14.0051, Pará state.

27. See, for instance, court cases 0001442-05.2012.814.0046, 0000200-28.2012.814.0046 (exceção de suspeição 2013.3.007333-3), 0000149.89.2012.814.0046 (exceção de suspeição 2013.3.007348-2), 2013.7.001617-9 (reclamação disciplinar c/c pedido de afastamento liminar), 0001442-05.2012.814.0046.

28. Court case 0001489-50-2012.814.0046 (ação de suspeição 2013.3.017073-3).

29. Reclamação disciplinar c/c pedido de afastamento liminar, corregedoria do tribunal de justiça do estado do Pará, 2013.7.001617-9.

30. Reclamação disciplinar c/c pedido de afastamento liminar, corregedoria do tribunal de justiça do estado do Pará, 2013.7.001617-9.

31. Processo 2013.7.001617-9, sindicancia administrativa 005/2013.

32. Judge Gabriel's motion, May 24, 2013, in processo 2013.7.001617-9, sindicancia administrativa 005/2013.

33. Decision by Judge Maria de Nazaré Saavedra Guimaraes, July 31, 2013, processo 2013.7.001617-9, sindicancia administrativa 005/2013.

34. See, for instance, Exeção de suspeição 2013.3.007333-3, Câmaras Civis Reunidas, TJPA.

35. See, for instance, Exeção de suspeição 2013.3.007348-2, Câmaras Civis Reunidas, TJPA.

36. Notas Taquigraficas da Exceçoes de Impedimento n.s 83-10, 82-25, Sessão Ordinária de Julgamento do Tribunal Regional Eleitoral do Pará, May 16, 2013. Quote from Judge Leonardo de Noronha Tavares.

37. Notas Taquigraficas da Exceçoes de Impedimento n.s 83-10, 82-25, Sessão Ordinária de Julgamento do Tribunal Regional Eleitoral do Pará, May 16, 2013. Quote from Judge Leonardo de Noronha Tavares. Contacted by the author, the office of Judge Noronha declined to comment.

38. Contacted by the author, former senator Flexa didn't reply to requests for comment.

39. Court case 83-10.2013.614.0051, Agravo de Instrumento, September 30, 2014. Gilmar Ferreira Mendes, Tribunal Superior Eleitoral.

40. Ruling by Judge Gabriel Costa Ribeiro, October 16, 2014, case (AIJE) 418-63.2012.6.14.0051, 51 Zona Eleitoral, TRE. Also, Ação Cautelar 404-79.2012.6.14.0051, Ação por propaganda Eleitoral irregular 578-88. 2012.6.14.0051, Ação de Investigação Judicial Eleitoral 416-93.2012.6.14.0051 and 417-78.2012.6.14.0051.

41. Description and quotes based on a video of the event.

42. AIJE 577-06.2012.6.14.0051, 51 Zona Eleitoral, TRE.

43. REsp no. 418-63.2012.6. 14.0051/PA, Judge Gilmar Mendes, Superior Electoral Court of Brazil.
44. Electoral court case 0600947-79.2018.6.14.0000, 0601356-45.2018.6.00.0000/PA and Recurso Ordinario no. 0600947-79, Superior Electoral Court of Brazil.

23: THE TRIAL

1. "Assento de óbito," Cartório Cerqueira Cesar, São Paulo, Livro C-0071, no. 41926, Folha 035V, 29/04/2014.
2. José de Barros Neto was elected in 2012. Tribunal Superior Eleitoral, eleição municipal 2012.
3. "Morre Josélio Barros (1933–2014)," Rogério Medeiros, *Século Diário*, April 29, 2014.
4. "Morre Josélio Barros (1933–2014)," Rogério Medeiros, *Século Diário*, April 29, 2014.
5. *Século Diário*, April 29, 2014.
6. *Século Diário*, April 29, 2014.
7. Author's interview with Paulo Orlando Rodrigues de Mattos, retired police detective of Minas Gerais.
8. "Indulto a condenado a 229 anos," *O Tempo*, September 11, 2013.
9. *Estado de Minas*, July 17, 2020.
10. Requerimento no. 451, 2014, Senado Federal, published in the Diário do Senado Federal, 30/04/2014.
11. Requerimento, Wandenkolk Gonçalves, May 13, 2014, Câmara dos Deputados.
12. On the transfer of land to his children, see ITERPA processos 2017/189100, 2017/285945, 2017/250779, and 2017/228200.
13. On the will or testament, see Assento de óbito, Cartório Cerqueira Cesar, São Paulo, Livro C-0071, no. 41926, Folha 035V, April 29, 2014.
14. Civil court case 0010474-32.2017.8.14.0046, Rondon do Pará.
15. On the problems of the deeds, see ITERPA processos 2017/189100, 2017/285945, 2017/250779 and 2017/228200.
16. See ITERPA processos 2017/189100, 2017/285945, 2017/250779, and 2017/228200.
17. Case number 0010474-32.2017.8.14.0046 and 0800845-93.2020.8.14.0046, civil court of Rondon do Pará.
18. Planilha, Laudo de Avaliação do Processo 192015730001091-0, Shirley Cristina de Barros, contained in civil court case 0010474-32.2017.8.14.0046.
19. Ruling by Judge Tainá Monteiro da Costa, July 30, 2021, court case 0010474-32.2017.8.14.0046, Pará.
20. Ata do 70 Congresso Geral Ordinária do Sindicato dos Trabalhadores Rurais de Rondon do Pará, 30-31/01/2014.

21. Court case 0000711-34.2000.814.004601, trial jury celebrated on October 24, 2013.
22. *Jornal Liberal*, April 30, 2014
23. Attorney Roberto Lauria declined to be interviewed for this book, but attorney Marcos Noboru, who was also in the courtroom and had represented Nunes since early in the case, was interviewed by the author for purposes of context and to provide the reader with the defendant's perspective of events.
24. Prosecutor Franklin interview with the author.
25. "Temido latifundiário e madeireiro do Sudeste do Pará é condenado a 12 anos de prisão pelo homicidio do sindicalista Dezinho," press release by Justiça Global, May 2, 2014.
26. *Jornal Liberal*, Globo TV, April 29, 2014.
27. About the use of hearsay evidence in Brazil, see Recurso Especial no. 1.373.356–BA (2013/0097292-2), Superior Court of Justice.
28. Video recording of the trial jury, court case 0000700-89.2000.814.0401.
29. The outcome of Neto's case is mentioned in documents of court case 0000700-89.2000.814.0401, Pará state.
30. Geovano Ferreira da Costa's statement on June 15, 2004, court case 0000700-89.2000.814.0401, Pará state.
31. Video recording of the trial jury, court case 0000700-89.2000.814.0401, time 02:09:00 to 02:15:00.
32. Video recording of Nunes's second trial jury, celebrated on August 13 and 14 of 2019, time code 04:33:00 and onward.
33. Francisco Martins da Silva's interview with the author.

24: A CERTAIN SENSE OF JUSTICE

1. See Tribunal de Justiça do Estado do Rio de Janeiro, 36 Vara Civel, court case 19.404/95, and announcements published in *Jornal do Commercio*, December 12, 1995, A8.
2. Several witnesses testified to the police that Pedro worked for Nunes, for instance, Daniel Matciulevicz, Luiz Carlos Aviz de Oliveira, and Maria Regina Rodrigues da Silva, court case 0000421-29.2000.8.14.004, Pará state.
3. Video recording of the trial jury, August 13, 2019, court case 0000700-89.2000.814.0401, time 03:58:00–04:00:00.
4. Police investigation 037/2000-DMRP, cited in motion of defense attorney Humberto Feio Boulhosa, April 9, 2001, court case 0000700-89.2000.814.0401, Pará state.
5. Luiz Carlos Aviz de Oliveira admitted in a statement provided to the police on December 15, 2000, that he was with Pedro the night of the murder. He denied any involvement in the killing. Court case 0000421-29.2000.8.14.004, Pará state.

6. José Lopes de Angelo's statement to the police on December 20, 2000. Court case 0000700-89.2000.814.0401, Pará state.

7. Statement of Luiz Carlos Aviz de Oliveira, December 15, 2000, in court case 0000421-29.2000.8.14.004, Pará state.

8. Statement of Daniel Matciulevicz, November 28, 2000, court case 0000421-29.2000.8.14.004, Pará state.

9. Dr. Joaquim statement to Detective Resende on December 20, 2000, in court case 0000421-29.2000.8.14.004.

10. Antônio Lopes de Angelo, statement to Detective Resende on December 20, 2000, court case 0000700-89.2000.8.14.0046. Dr. Antônio didn't respond to requests for comment.

11. Report by Detective Resende, December 27, 2000, in court case number 0000421-29.2000.8.14.004.

12. Laudo de Exame de Corpo de Delito: Exumação/Necropsia, Liv 1138 Fls 074, June 8, 2005, court case 0000421-29.2000.8.14.004.

13. Motion by prosecutor Mauro José Mendes de Almeida, September 22, 2005, court case 0000421-29.2000.8.14.004, Pará.

14. Transcription of the audiotapes on Favoreto are on 790–96 of criminal court case 0000700-89.2000.814.0401, Pará state.

15. Video recording of the trial jury held on April 29, 2014, court case 0000700-89.2000.814.0401, time 06:14:00 to 06:30:00.

16. Video recording of the trial jury, April 29, 2014, court case 0000700-89.2000.814.0401, time 06:14:00 to 06:30:00.

17. Video recording of the trial jury, April 29, 2014, court case 0000700-89.2000.814.0401, time 09:11:00 onwards.

18. *Jornal Liberal*, 2nd edition, TV Globo Pará, April 30, 2014.

EPILOGUE

1. Acórdao 20160445432208, 1a Camara criminal isolada, Tribunal de Justiça do Estado do Pará.

2. "Tribunal de Justiça do Pará apura supostas transgressões de juízes," *O Liberal*, July 17, 2019.

3. Data on the number of jury trials provided by attorney Lauria in the closing arguments presented to Pará's court on October 13, 2021.

4. *Ofício* 728/2019-MP/PGJ, by attorney general Gilberto Valente Martins, July 16, 2019.

5. Representação, protocolo 2015.6.003753-9, Tribunal de Justiça do Estado do Pará.

6. Video recording of the jury trial, April 23, 2019, case number 0000700-89.2000.814.0046.

7. For the defense of Mr. Lauria, see the video recording of the 46 sessao ordinária, Tribunal Pleno, Tribunal de Justiça do Estado do Pará, December 4, 2019.

8. Pedido de Provicências 0005099-14.2019.2.00.0000, Corregedoria da Região Metropolitana de Belém.

9. Pedido de providências case number 0005099-14.2019.2.00.0000, Pará justice.

10. Pedido de providências case number 0005099-14.2019.2.00.0000, Pará justice. Data on the number of jury trials provided by attorney Lauria in the closing arguments presented to Pará's court on October 13, 2021.

11. Revisão Disciplinar 0008381-89.2021.2.00.0000, CNJ.

12. Video recording of the jury trial is available at the TJPA's oficial website: https://webcast.overseebrasil.com.br/tjpa/#ondemand//watch/?v=242807 8d9fc58416aa36. The incident described takes place at time 03:24:00.

13. Video recording of the jury trial, August 13, 2019, case number 0000700-89.2000.814.0046.

14. State prosecutor Franklin interview with the author, and "Viúva de líder da luta contra grileiros vive há quase 20 anos sob proteção policial," *Record TV*, November 16, 2020.

15. Ruling by Judge Ronaldo Marques Valle, February 14, 2022, court case 0000700-71.2000.8.14.0046, Pará.

16. Alex Cuadros, *Brazillionaires: The Godfathers of Modern Brazil* (London: Profile Books, 2016). Quote from chapter 2.

17. *Conflitos no Campo Brasil 2020*, CPT, May 2021.

18. On his unpaid taxes, see the federal public database (https://www.listade vedores.pgfn.gov.br/) and Pará's Secretaria de Estado da Fazenda (https://app.sefa.pa.gov.br/consulta-divida-ativa/#/). On the Fazenda Lacy, federal case 0006677-95.2012.4.01.3901, TRF1.

19. The environmental fines refer to IBAMA's Auto de Infração ZAG04W12, GS3VJQ9I, JH40Y3XD, BWIF70UB. On the labor case, ruling by Judge Amanda Cristhian Mileo Gomes Mondença, November 8, 2021, ACPCiv 0000301-15.2015.5.08.0117, TRT 8a Região.

20. On the transfer of ownership to R. C. Moreira Costa, the author cites information provided by the Ministry of the Economy of Brazil via Brazil's freedom of information law (case number 13035.100497/2021-42). Data on exports from Brazilian Beef Exporters Association (ABIEC). http://abiec .com.br/en/exports-consult/#.

21. On the investigations on the meat plant, the Ministry of Agriculture claimed the "right to privacy" in denying the author access to the investigations Procedimentos de fiscalização e inspeção no. 21000.003278/2021-41, 21000.003304/2021-31, 21000.044832/2020-60, 21000.065973/2020-16; Relatório de fiscalização no. 21000.003291/2021-09; Auto de Infração no. 21000.009034/2021-72, 21000.043115/2021-00, 21000.017251/2021-36, 21000.080633/2020-15, 21000.076914/2020-73, 21000.060051/2020-12, 21000.030523/2020-11, 21000.033280/2020-64, 21000.049513/2020-41, 21000.027148/2020-13, 21000.024236/2020-63, 21000.007363/2020-06,

21000.079465/2019-81, 21000.042280/2019-11, 21000.048499/2018-43, 21000.048499/2018-43. On the tax incentives from Pará state to the meat plant, see *Diário Oficial* April 12, 2019, number 33.850 and June 8, 2020, number 34.247.

22. Statement contained in the court case 0002843-13.2012.8.14.0046, Pará state.

23. Ruling by Judge Tainá Monteiro da Costa, March 17, 2022. Court case 0800422-02.2021.0.14.0046, Pará state

24. Data from CPT compiled by the author, 2017–2022.

25. On the environmental activists, see "IACHR and UN Human Rights Condemn Murders of Environmental Activists and Quilombolas in Brazil," release 017/22, January 24, 2022.

26. *Brasil de Fato*, February 11, 2022.

27. Nathalie Butt et al., "The Supply Chain of Violence," *Nature Sustainability* 2, no. 886 (2019): 742–47.

28. Data from Brazil's top electoral court. https://divulgacandcontas.tse.jus.br /divulga/#/candidato/2018/2022802018/PA/140000625297.

29. Data on foreign trade, both general and agribusiness, compiled by the author from the databases of Comex Stat (comexstat.mdic.gov.br/pt/comex-vis) and the Ministry of Agriculture of Brazil (http://indicadores.agricultura.gov.br /agrostat/index.htm).

A NOTE ON SOURCES

1. In a signed statement addressed to the court in Espírito Santo, Josélio had admitted to the murder of the two alleged hitmen about to kill his father in Baixo Guandu's train station and the murder of Major Cavalcanti. The signed statement and the motion by his defense attorney, Mario Rocha, is in the files of court case 1.277/67, Vitória, Espírito Santo. According to the report of detective Adonias Marques dos Santos, December 1975, in classified file ACE 3582/83, Josélio also admitted to have hired out the murders of Reis and Vieira, in the so-called chainsaw murder.

Index

activists
countries with tropical forests
and, 303–4
exposing lawlessness, 303–4
murder statistics, 303
murder victim examples, 303–4
See also specific individuals
Agribusiness Association of Rondon,
112, 133
agricultural production
funds/settlements from land
occupation, 236
land inequality and, 230–31
large-scale landowners vs. small
landowners, 230–31
Lula's presidential terms and, 230
state support and, 230
See also colonization of Amazon;
soybean production/Amazon;
specific individuals
Alcântara, Cândido Afonso de, 26
Alves de Silva, Pedro
background/crime, 168, 169, 170,
174, 280, 282, 283–85
descriptions, 168, 169
Dezinho murder and, 167–68, 175,
176, 180, 285–86
investigation into his murder, 168,
173–74, 288, 361n17

Josélio's warning and, 189
murder of/motive, 168, 170, 173,
174, 176, 286, 292
Nunes and, 168, 170, 174, 175, 176,
181, 215, 216, 280, 282–86, 287,
291–92, 300, 383n2
Regina (wife/widow), 171, 177, 284,
287
secret evidence of, 174–75, 176,
183, 215, 216, 288–89
See also Silva Filho, Francisco
Martins da
Alves dos Santos, Pedro
background/homicide and, 65
cattle ranch/homesteaders and,
60–61
chainsaw murders/murder
suspect, 61, 65, 336n18
gunmen of, 61–62
land/land-grabbing, 60–61, 62,
65, 66
selling land, 65, 66, 194
Alves Fagundes, Alfim, 115
Alves Tuma, Angela Alice, 299, 300
America westward expansion/
strategies, 92, 101
Amorim, Paulo Cesar Pedreira
corruption/removal, 147, 170
Dezinho incarceration, 145, 200

Amorim, Paulo Cesar Pedreira (*cont.*)
    Nunes/occupation and, 179–80
    Rai/Josélio and, 128, 131
Andrade, Jônatas dos Santos
    background/views, 250–51
    descriptions, 245, 248, 250–51
    Human Rights Prize, 254–55
    on Nunes, 257
    Nunes's assets seizure decision,
      247
    Nunes's threat against, 256
    Nunes's violations and, 245
    security, 256
    undercover Nunes investigation,
      245
    Vale mining company and, 250,
      257
    See also Nunes, Décio José
      Barroso ("Delsão") debt/assets
      seizure
Andreazza, Mário David, 57
Anti-Slavery International, 13
Arc of Deforestation, 90, 91
Armstrong, Neil, 39
Assis, Francisco de, 145, 163, 187
Associated Press, 75
Ávila, Josefa, 132–33
Ávila, José Hilário
    background/Rondon business, 114
    Dezinho/Dezinho murder and,
      215, 221
    land claims, 114–15, 117, 118
    land occupation, 114–15, 116–18
    violence and, 114, 117, 118
    See also Vila Gavião land
      occupations
Ávila, José Hilário's murder
    Brito as suspect, 133, 137
    description, 132
    Dezinho as suspect/defense,
      133–34, 137
    Josefa on, 132–33
    Josélio and, 135–36

    law enforcement and, 133–34, 136,
      137
    retaliation and, 137, 166
    Sindicato and, 133
Ávila, Olivandro, 137, 179, 216, 277
Aviz de Oliveira, Luiz Carlos, 286, 287,
    288, 383n5

Barros Carneiro, Josélio de
    Agribusiness Association of
      Rondon, 112
    arrest/sentence (Cavalcanti
      assassination), 35–36, 39, 69
    authorities support of, 66
    background, 26–27, 33
    buying his claimed land and, 195
    buying land from Alves, 65,
      73–74
    Cavalcanti assassination and,
      21–22, 23–24, 25, 35–36, 259
    "chainsawing" peasants by, 68
    children, 25, 66
    death and leaving no will/
      consequences, 269–70
    death/cause, 268
    debt cases against, 263–64
    descriptions, 5, 17, 21, 25, 27, 35,
      112, 188, 189, 198, 217, 259
    evicting settlers, 66, 74, 103–4
    Faria raid and, 127
    Fazenda Serra Morena/
      controversies, 66, 69, 72,
      73–74, 81, 103, 104, 105, 127,
      188, 195, 239, 347n54, 366n25
    Flexa Ribeiro, Fernando de Souza
      and, 205
    funeral/wake, 268–69
    gang denial, 188
    impunity, 25–26
    INCRA officials/GPS coordinates
      of land, 198
    Jornal do Brasil land-related
      conflicts, 188

as land grabber, 103–5, 112, 195,
197–98, 269–70, 347n54,
365n10
land occupations/reactions, 117,
118, 135–36, 144, 147, 195
move to Amazon, 36–37
murder of men/defending father,
25–26, 326n33
murders/murder rumors and, 5,
6–7, 17, 25–26, 68, 105, 127–28,
347–48nn57–58
reputation, 112, 198, 268
Ribamar murder and, 197
settling in Paragominas, 66
Sindicato denouncing, 144–45
as target (in Baixo Guandu), 34–35
workers/working conditions, 4, 5,
7–8, 10, 11, 12, 13–20, 188
See also chainsaw murders;
Dezinho/Josélio investigation;
specific events/individuals;
Te-Chaga-U ranch
Barros Carneiro, Josélio de/Dezinho
murder
admitted contacts, 188
Alves de Silva, Pedro and, 168
as suspect, 167, 179, 188–89, 200
visit to father, 167, 188, 261
warnings before, 189
Barros, Jorio de, 26–27
Barros, José de
background, 26–27
Cavalcanti/assassination and, 23,
34–35
Josélio moving to Amazon, 36
Olga (wife) and children, 26–27
politics/enemies, 27, 29, 34
ranch/business of, 21, 23
Barros, José Eloy, 130, 270
Barros, José Francisco de ("Chico"),
26–27, 33, 34–35, 167, 268
Barros, Josélia, 26–27
Barros, Josil, 26–27

Barros Lopes, Josélia Leontina de
helping father, 125, 129, 130
husband, 173, 361n13
investigations on, 129–30
lawsuits and, 129
as prosecutor/career, 9, 13, 125
Te-Chaga-U ranch and, 13–14
Barros Malcher, Shirley Cristina de
Costa Ribeiro, Gabriel and,
252–53, 262, 263, 264–67
debts/Fazenda Nova Delhi
Agropecuaria, 263–64,
380n22
defending/supporting father,
17, 18, 130–31, 167, 195, 198,
260–61
descriptions, 130, 252, 259
ExpoRondon and, xvi
husband, 130, 252, 259, 305
Joelma ("Joelminha") and, 260–61,
305
Josélio views/support of, 130, 198,
226–27, 258, 366n29
on land occupation, 230
law firm, 130
as lawyer, 17
at Nunes's raid, 252
politics following ban, 306
Sindicato/landless movement and,
230, 235, 243
Te-Chaga-U ranch and, 17
Barros Malcher, Shirley Cristina de/
mayor elections
finances/asset report (2012), 258
first attempt, 198
inauguration ceremony (2013),
259–60
legal battle/outcome (2008),
229–30, 231
removal as mayor/ban, 266–67
2008, 224, 226–27, 228
vote (2012), 258–59
Barros Neto, José de, 268, 382n2

Batista Gonçalves Afonso, José
  Brazilian celebrities/Maria Joel
    and, 208
  as CPT lawyer/activism, 143, 187,
    191, 194
  description/background, 143
  Dezinho and, 146–47
  Maria Joel safety and, 240
  Nunes and, 216, 220, 221, 227
  Nunes's trials and, 298, 299
  Wellington trial and, 210, 212
Batista, Vicente, 43
Belem, Rita, 53, 59
Bernstein, Carl, xx
"Bimbim, Colonel" (Cipriano da Silva,
    Secundino), 32–34
Bolsonaro, Jair Messias, xi, 84, 206,
    303–4
Bolsonaro, Percy Geraldo, 84
bovine spongiform encephalopathy
    (BSE), 203–4
BR-163, 202–3, 204, 205, 206–7, 250,
    309–10
BR-222, xv, 39–40, 165, 174, 175, 285
Brasília inauguration, 41
Brazil
  Amazon drug trafficking/gang
    violence beginnings, 228
  Confederates (US Civil War)
    settling in, 31–32
  constitution (1988), 95
  coup (1964), 29
  criticism/response to activist
    murders, 164
  democracy transition/
    consequences, 95–96, 97, 141,
    197, 257
  "economic miracle" (1968–1974), 56
  economic problems (1980s), 83
  Eldorado Massacre, 140–42
  4000 BP local societies, 46
  hearsay evidence, 275
  immigration late 1800s, 31–32

  independence, 31
  inequalities, 83, 90, 97, 98, 106
  judges' role, 298
  jury trial differences with US, 272
  labor rights/laws, 246
  lawlessness, 200–201, 204–5
  pre-Columbian societies, 46
  reputation in reducing
    deforestation, 207
  sexism/discrimination against
    women, 191
  soil/agriculture, 46
  turmoil in 1950s, 27–29
  voter buying, 228, 229
  See also specific events/issues;
    specific individuals/locations
Brito, José Soares de
  activism views/actions, 11, 98–100,
    105, 106
  background, 11
  descriptions, 98, 135
  on Dezinho, 107, 108, 145
  Dezinho/Dezinho family
    relationship, 98, 99, 107, 108
  Dezinho murder and, 164–65, 179
  Dezinho relationship strains,
    145–46, 165
  Hilário murder and, 133, 137
  home arson, 11, 291
  Josélio/Josélio investigation, 11, 15,
    128, 290–91
  kidnapping attempt of, 134–35,
    137, 291
  on land-grabbing schemes/fraud,
    99–100, 103
  land occupations and, 132, 142,
    179
  Maria Joel relationship/Dezinho
    murder, 165
  as marked man/enemies, 11, 131,
    134–36, 137, 291
  Nunes's trial and, 290–91
  safety concerns, 134–36

Sindicato and, 11, 105, 107, 134, 145–46, 190
Sindicato position/Dezinho and, 145–46
stroke/consequences, 290
BSE (bovine spongiform encephalopathy), 203–4
bullets with cyanide, 22

Câmara, Paulo Sette
Josélio investigation, 12, 13, 14, 16, 128–29
position, 12, 13
Campbell, Jeremy M., 57–59, 100, 101, 102
Canuto, Sebastião Batista
background, 65
chainsaw murders and, 65, 67, 69
Cardoso, Dirceu, 66
Cardoso, Fernando Henrique, 13, 16–17, 141, 142, 224
Cardozo, José Eduardo, 237, 242–43
Carepa, Ana Júlia, 212
Carvalho Neto, Gil Bonifácio
Cavalcantes and, 6, 7, 19
description, 3
escape plans/escape, 9–10, 11, 118
following investigation, 19
Josélio murders/investigation, 11, 14–15, 16, 119
killing field, 4, 5–6, 7
Manuel and, 3–4, 5
Te-Chaga-U ranch, 3–6, 118
castanheira trees, xiv
Castelli, Giuseppe, Father, 59
Castro, Fidel, 37
Catholic Church
of Brazil, 72–73
homesteaders/peasants and, 61, 72–73
cattle
embryo sales, 249
Guzerá cattle, 269

Nelore cattle, xiv, 3, 100–101, 248, 249
price factors, 249
Tabapuã cattle, 248
Cavalcante family
Carvalho Neto, Gil Bonifácio and, 6, 7, 19
escape plans/escape from Te-Chaga-U ranch, 7, 9–10, 11, 118, 127
following escape, 14
following investigation, 19
Te-Chaga-U ranch/Josélio and working conditions, 6–8, 11
Cavalcante, Luiz Bezerra
background, 6
description, 6
Te-Chaga-U ranch/Josélio and, 6, 14, 16, 118, 127
See also Cavalcante family
Cavalcante, Sueny Feitosa
Dezinho and, 8, 9–10
Josélio investigation, 16
Te-Chaga-U ranch/Josélio and, 6
See also Cavalcante family
Cavalcanti da Silva, Orlando
assassination/Josélio and, 21–22, 23–24, 25, 35–36, 259
assassination of, 21–23
background to assassination, 29, 34–35
position, 21
Ceará
background/description, 4–5
murder/killing field and, 4, 7, 11, 19, 119
See also Dezinho/Josélio investigation
chainsaw murders
anonymous note, 63–64
descriptions, 66–69
disappearances/dead bodies, 63–64

chainsaw murders (*cont.*)
lure used, 63, 67
media/reporters, 61–63, 64
See also specific individuals
chainsaw murders/Josélio
attorney, 69–70
buying Alves land, 65
disappearance, 69–70
hired assassins, 66–67
Lincoln and, 70–71, 73
parole/conditions, 69
prison/release, 70–71, 73
victims/bodies, 67–68
character witness meaning, 291
Chico (Te-Chaga-U ranch worker),
4, 5, 18
children of Maria Joel/Dezinho. *See*
Dias da Costa, Maria Joel/
Dezinho children; *specific
issues/individuals*
Chu, Henry, 208
Cipriano da Silva, Secundino
("Colonel Bimbim"), 32–34
Civil War (US), 31
Cockburn, Alexander, 37
Collor de Mello, Fernando, 122
colonization of Amazon
boom-and-bust models, 114
BR-163/settlements and, 202
conflicting stories on, xix
devaluation of nature/indigenous
people, xvii
fazendeiro "trophies," 65, 236
financial incentives, 339n82
frontier concept and, 8
government policy shift/
agribusiness (beginning in
1960s), xvii, 4, 72–73
indigenous people and, xvii, 39,
46, 49–50, 52, 84
inequalities and, xii, xvi–xvii, xix, 73
land conflict/fraud, 49–50, 59–60,
65, 71–73, 99–105

Operation Amazonia/
consequences, 37–38, 45, 49,
53, 56, 99
PA-70 road and, 39–40, 41, 45, 46,
49, 50, 59, 60, 72, 165, 202
propaganda and, 57–58
road construction/PIN, 56–57
soybean/agribusiness and, 202–3
Vila Rondon transformation,
54–56
See also land-grabbing issue;
*specific events; specific
individuals/groups*
colonizers slaughter by Gavião-
Kyikatêjê
DER/request following, 40–41, 59
descriptions, 39, 40, 43, 50
group hiking to/from scene,
42–44
peasants after/Vila Rondon,
39–40
possible counteroffensive, 41, 42
reasons for, 39
Confederates (US Civil War) settling
in Brazil, 31–32
Conjuring Property (Campbell), 57–58
Contestado region
description/lawlessness, 29–30,
32–35, 120, 268
economic development/
consequences, 32
gold and, 30–31
Minas Gerais/Espírito Santo
dispute, 29–30, 32
Ouro Preto, 30
strongmen, 32–35
trails/roads and, 30
Cordeiro family
Leite family feud, 121–22, 123
slaughter of, 121–22
Cordeiro, Helvécio Augusto, 123
Cordeiro, Maria Luiza, 121–22
Costa, Joaquim Carlos, 288

Costa Ribeiro, Gabriel
  background, 261
  case against Barros family
    members/debt, 263–64
  case against policemen/threats,
    262–63
  defense against Shirley Cristina's
    accusations, 265
  descriptions, 252, 261
  modernizing local justice system,
    262
  position, 252
  Shirley Cristina conflicts and,
    252–53, 262, 263, 264–67
  Shirley Cristina's campaign
    wrongdoings and, 264–67
Costa Ribeiro, Raquel, 265
Cotrim Soares, Antonio
  background, 45
  FUNAI hiring, 41
  resignation as *sertanista*, 53
Cotrim Soares, Antonio/
    Gavião-Kyikatêjê
  attitude change towards, 50
  colonizers conflict, 41–45, 46,
    49–50, 52–53
  first contact, 45, 46–48
  ultimatum, 49–50
Coutinho de Queiroz, José, 55–56
CPT (Pastoral Land Commission), 73,
    89, 98, 105, 106, 107, 108, 109,
    111, 117, 138, 141, 143, 144, 146,
    179, 186–87, 188, 190, 191, 193,
    194, 199, 200, 208, 209, 210,
    219, 221, 238, 240, 243, 301
Creutzfeldt-Jakob disease, 203
Criminal Syndicate, 33
Curionópolis beginnings, 85

Dário Godinho Rodrigues, José
  Cordeiro family/slaughter and, 123
  Faria and, 123
  murder of, 127, 128

Davis, John Weaver/sons and family,
    74, 98
debt bondage/slavery, 7, 8, 10, 12–13,
    16–17, 18, 19, 31, 60, 97–98,
    147, 188, 200, 218–19, 242,
    254–55, 273–74
deforestation in Amazon
  activists/fate of Amazon, 304
  Arc of Deforestation, 90, 91
  black market, 143
  BR-163 road/settlements, 202–3,
    205–7
  charcoal production, 91
  claiming/clearing land, 103
  climate change and, 304
  criticism, 143
  drivers in Rondon, 90–91
  employment/jobs, 223–24
  family as "biggest destroyer,"
    249–50
  fazendeiros, 66
  Forest Code, 237, 375n9
  government monitoring, 207
  government restrictions on, 143,
    207
  impunity, 302
  indigenous groups defending
    forests, 96
  international community/China
    and, 309–10
  landowner threats/government
    responses, 205
  logging boom end and, 143
  logging trucks descriptions, xvi
  Lula and, 204, 205, 206
  murders of forest defenders (2011),
    236–38
  parks banning, 204–5
  slash-and-burn activities/
    superfires, 4, 55, 91, 92
  slave use in, 60
  soybean production, 203, 204,
    207

deforestation in Amazon (*cont.*)
    timber industry, 90–91
    See also colonization of Amazon;
        specific activists; *specific
        individuals*
Demick, Barbara, xix
Dezinho
    activism views/actions, 8, 8p,
        93–94, 97, 105, 106–7
    childhood/family, 88–89
    descriptions, 8, 88, 89, 107, 133,
        177, 194
    full name, 8
    honors/settlements, 192–93
    jobs in timber industry, 92–94
    on Josélio/land occupiers, 135–36
    leadership style, 194
    Lula admiration by, 96, 97, 117,
        220
    mayoral election/council member,
        147, 149, 225, 228
    Moisés possible appointment,
        149–50
    murder threat incident, 151–52
    on Rondon socioeconomic
        situation, 143–44
    security/law enforcement and, 116,
        151–52
    settling in Rondon do Pará, 75
    Sindicato/causes, 8–10, 98, 105,
        106–7, 134, 145–46, 173, 192
    Sindicato/personal reasons, 146
    Sindicato position change,
        145–47
    threats/as marked man, 9, 115–16,
        117, 131, 135–36, 149, 151–52,
        154–55, 164, 176, 200
    uniting people/visiting people,
        107–8
    Vila Gavião land occupation,
        114–15, 116–18, 132, 134
    Workers' Party/reasons for
        joining, 147

    See also Dias da Costa, Maria Joel;
        Dias da Costa, Maria Joel/
        Dezinho; *specific individuals/
        events*
Dezinho/Josélio investigation
    description, 14–17, 18–19, 118, 119,
        128, 130
    following investigation, 20
    Josélio defense/retaliation, 17, 18,
        19–20
    planning, 9–10, 12
Dezinho's murder
    autopsy report, 157
    description, 153–54
    family grieving/funeral, 163–64,
        165–66, 184
    federal government/government
        and, 164
    funeral/procession, 165–66
    Joélima and, 156, 156–57, 304–5
    Joelina and, 149, 152, 153, 154,
        163–64, 222
    Joelma ("Joelminha") and, 156–57,
        164
    Joelson and, 151, 154, 275
    Maria Joel and, 149, 150, 152, 153,
        154–55, 164, 185
    media, 164
    motive and, 178
    public response to, 161, 164
    retaliatory riot possibility, 157, 161
    See also Nunes, Décio José
        Barroso ("Delsão") trial for
        Dezinho's murder; Silva,
        Wellington de Jesus
Dezinho's murder investigation
    delay in, 183, 200, 208
    early theories/suspects, 166–67
    human resources, 166
    Josélio as suspect, 164–65
    justice system corruption and,
        182–83
    Lula and, 208–9, 215

Maria Joel on Nunes, 183
Maria Joel statement, 167–68, 171,
    175, 176
See also Resende de Almeida,
    Walter; specific individuals
Diário of Pará, 7
Dias da Costa, Maria Joel
    Araujo and, xvii–xix
    childhood/background, 86, 185
    descriptions, xvii–xviii, 8, 86, 89,
        139, 189–90
    Dezinho's brother's visit/request,
        183
    Dezinho's risks/support, 108–9,
        116, 137–38, 139, 146, 147
    Eva relationship (sister), 81, 187
    extended family/mother and, 79,
        81–82, 93–94
    Josélio's death and, 270
    Rondo/first impressions, 79–82,
        86–87, 185
    Sindicato and, xvii–xviii, 146
    See also Dezinho; specific
        individuals/events
Dias da Costa, Maria Joel after
    Dezinho's murder
    extended family/mother and,
        184–85
    Joélima on remaining, 185
    Joelma ("Joelminha") eating
        disorder, 184, 190
    justice/suspected murderers, 185,
        187, 188–89, 199, 200, 208–9,
        215–16, 270, 295, 296, 297
    kindergarten job, 186
    remaining in Rondon/risks,
        185–86
    sending children away, 186
    views of, 304
Dias da Costa, Maria Joel after
    Dezinho's murder/activism
    actors helping/public campaign,
        208–9, 210, 212, 242–43

Brazil's government/deal over
    Dezinho's murder, 244–45
children and, 190, 195
civil disobedience, 238
criticism of her/family, 222–23, 243
descriptions/reputation, 192,
    193–95, 241
Dezinho's murder anniversary/
    demonstrations, 187, 243
disappointment in Lula, 220
encouragement and, 186
extended family/mother and,
    186–87
FETGRI position, 231
her potential murder and, 205–6
Human Rights Prize, 219, 254–55
incident of hired men/murder of,
    220–21
Inter-American Commission on
    Human Rights, 199–200, 242,
    243–44
intimidation attempts, 196
Joélima and, 222, 241
Joelma ("Joelminha") and, 187, 190,
    223
Joelson and, 190, 238–40
landless movement/protest
    against Rousseff government,
    236, 238
land occupations/new settlements,
    192–95
landowners and, 187, 196–98
on land reform, 199
leadership style, 194
Lula/Lula meetings, 207–9, 215,
    219–20
Marabá move, 231
Nunes report/after Lula meeting,
    215–16
propaganda campaigns, 193–94
security, 199, 208, 220, 222
security/threats and hiding (2011),
    238–41

Dias da Costa, Maria Joel after
Dezinho's murder/
activism (*cont.*)
Sindicato election/position, 191
Sindicato membership increase,
194
Stang's murder and, 207–8
threats, 196–97, 199, 201, 207–8,
209, 220–21, 238–41, 243
UN representative meeting,
196–97
Wellington's escape/vanishing
and, 221, 222, 226, 238
*See also specific events; specific
individuals/locations*
Dias da Costa, Maria Joel/Dezinho
children's health problems/
pollution, 90
finances, 82–83, 94, 109, 139
house of, 94, 148
Joélima/security and, 151
Joelma ("Joelminha") and security,
138–39
Joelson/security and, 138, 151, 152
Josélio investigation and, 129
meeting, 89
relationship, 94
in São Luis, 82–83
security/family security, 9, 116,
138–39, 150, 151–52
settling in Rondo, 75, 88
threats to family/as outcasts,
137–38
See also specific family members
Dias da Costa, Maria Joel/Dezinho
children
childhood (Joelma/"Joelminha",
Joelson, Joélima, Joelina),
9, 81, 82, 86, 90, 92, 94, 108,
116, 138–39, 149, 150, 151, 152,
154–55, 156–57
Joelma ("Joelminha") marriage/as
councilwoman, 260–61

Joelma ("Joelminha") political
background/shift, 260–61,
305–7
See also specific events/individuals
Dias Soares, Alfonso, 17
"dirty list" of employers, 218–19
Dutra da Costa, José. *See* Dezinho
Dutra da Costa, Valdemir, 88–89, 96,
183

Eldorado do Carajás massacre. *See*
Eldorado Massacre
Eldorado Massacre
background causes, 141–42
description, 140–41
Leal and, 211
state police and, 140–41
environment/Amazon destruction
boomtowns/local governments
and, 223–24
Brazilian Catholic Church
recognition of, 73
Brazilian government views/
defense, 91–92
devaluation of nature/indigenous
people, xvii
examples in other nations, 92
IBAMA (environmental
protection agency), 202, 205,
207, 218, 250, 309
pioneers views, 91–92
See also deforestation; pollution;
specific mining/locations
Escobar, Pablo, 242
Espírito Santo descriptions, 22, 30
See also Contestado
Espírito Santo, Maria do, 236, 237
European Union, 203
ExpoRondon/center, xvi, 112, 251–52,
268

Fantástico (TV news program), 242
Faria, Antônio Carlos Corrêa de

Adélcio initial arrest, 122
background, 123
description, 120
Leite family crimes and, 120–21
recapture of Adélcio, 122–27
Faro, Beto, 240–41
Fate of the Forest, The (Hecht and
Cockburn), 37
Favoreto, Jucelino, 145
Favoreto, Vicente Paulo, 181, 282, 283,
289, 292
Fazenda Fé em Deus, 194
Fazenda Lacy, 114, 147, 156, 179, 200,
217, 218, 246–47, 274, 277, 293,
301, 349n21, 372n34
Fazenda Nova Delhi Agropecuaria
debt case, 263
Fazenda Primavera II, 145, 166, 288
Fazenda Serra Morena, 66, 69, 72,
73–74, 81, 103, 104, 105, 127,
188, 195, 239, 347n54, 366n25
Fazenda Tulipa Negra
land-grabbing and, 179, 276
land occupations, 166–67, 177, 285
Lopes, Kyume Mendes and, 179,
192–93, 276
renaming for Dezinho, 192–93
fazendeiros
definition, xiii
"trophies" of, 65, 236
See also specific individuals
Ferraz Rodrigues, Paulo, 288
Ferreira dos Santos, Fausto
Cavalcanti assassination and,
21–25, 35, 36
description, 21
Ferreira, Durval, 145
Ferreira, Lucinery Helena Resende
on Dezinho/Maria Joel, 221
Lourival and, 221
Nunes case/dismissal and, 215,
221, 227
position, 215

Ferreira Rocha, Arnaldo, 305
FETAGRI, 8–9, 12, 14, 16, 19, 20, 107,
108, 110, 115, 117, 133, 134, 140,
144, 145, 146, 147, 152, 163, 164,
175, 186–87, 191, 193, 198, 200,
210, 231, 235–36, 240, 270, 293,
351n35
Figueiredo Correia, Jader de, 51
Figueiredo Report, 51–52
Financial Times, 204
fire in Amazon (summary), 55
Flexa, Raimundo Moisés Alves
bribery scandal/consequences,
297–99
description/background, 210, 273,
297
Nunes's trial, 271, 273, 277–78, 281,
290, 291, 292, 295
Wellington trial/sentence, 210,
211–12, 213, 271
Flexa Ribeiro, Fernando de Souza,
205, 266, 269
Folha de S. Paulo, 16, 60, 106, 142, 208,
283
Fontanella, Giuseppe, Father
background/description, 61
Davis family murders and,
74–75
investigation of/deportation, 72,
74–75, 340n96
peasants/land and, 61, 72, 73, 74,
89, 106
Fonteles de Lima, Paulo, 105
Food First, California-based,
199–200
Forbes, 204
Forest Code, 237, 375n9
Francisco, José, 43
Freitas, Francisco Veloso de, 173–74,
288, 361n17
Freitas Leite, Antônio Maria Jr.,
299–300
Frigo, Darci, 199–200

frontier concept, 8
FUNAI (National Indian Foundation),
    41, 45, 52, 53, 309

Gavião-Kyikatêjê Indigenous group
    arrows/other weapons, 39, 43, 47
    location/description, 39, 40
    mayor of Imperatriz/raid, 51, 52
    metal access and, 47
    removal/reservations, 52–54,
        80, 99
    traditions/way of living, 48–49
    See also colonizers slaughter by
        Gavião-Kyikatêjê; Cotrim
        Soares, Antonio
Gazeta, A, 25
Geisel, Ernesto, 71
Global Witness, 303
gold mining
    Contestado region, 30–31
    Serra Pelada, 84–86, 110, 140
Gomes da Silva, Juracy ("Bodão"), 174,
    283, 292
Gomes da Silva, Juracy/father, 183
Gomes de Almeida, Lincoln
    Josélio/chainsaw murders and,
        70–71, 73
    position/background, 70
Gonçalves da Silva, Luiz, 220–21
Gonçalves, Wandenkolk, 269
Good, the Bad and the Ugly, The, 80
Goulart, João, 29
Greater Carajás Program, 83–84
grief
    Dezinho murder/family, 163–64,
        166, 184
    homicide/effects on relatives,
        164
grilagem/grileiro
    meaning/origin, 99
    See also land-grabbing issue
Guardian, The, 236–37
Guedes, Carlos, 117

Guerrilha do Araguaia (1974), 74
Guzerá cattle, 269

Hart to Hart (American TV series), 94
hearsay evidence, 274–75
Hecht, Susanna, 37, 99
Hemming, John, 47, 91
Homestead Act (US/1862), 57
howler monkeys' hoots, 44
human rights abuses (Amazon)
    children, 98, 242
    debt bondage/slavery, 7, 8, 10,
        12–13, 16–17, 18, 19, 31, 60,
        97–98, 147, 188, 200, 218–19,
        242, 254–55, 273–74
    "dirty list" of employers, 218–19
    hinterland, 10–11, 12
    international community/China,
        309–10
    perpetrators' impunity, 141
    reasons for, 254, 255
    See also specific events/locations;
        specific individuals

IBAMA (environmental protection
    agency), 202, 205, 207, 218,
    250, 309
impunity
    causes/consequences (summary),
        302
    crime/lawlessness, xii, 10, 35, 141,
        172, 188, 214, 221, 241, 301, 302
    deforestation, 302
    money/status and, 25–26, 35, 141,
        172, 214, 220, 294, 301, 302–3,
        304, 335n4
    See also specific individuals
INCRA (Instituto Nacional de
    Colonizaçãoe Reforma
    Agrária), 98, 100, 102, 103–4,
    114–15, 117, 192, 193, 194, 197,
    198–99, 236, 238, 240, 244, 276
Indian Museum of Rio de Janeiro, 52

Indian Protection Service (SPI), 45, 47, 48, 52
indigenous groups
constitution/land rights and, 95–96
contact/gifts, 47, 48
epidemics, 46, 49
federal trespassers ban/enforcement, 49–50
Figueiredo Report/abuses, 51–52
fire use by, 55
as forest protectors, 96
FUNAI changing policy, 45
gardens of, 42, 49
pre-Columbian societies, 46
removal/reservations, 52–54
See also colonization of Amazon; specific groups
Instituto Nacional de Colonizaçãoe Reforma Agrária (INCRA), 98, 100, 102, 103–4, 114–15, 117, 192, 193, 194, 197, 198–99, 236, 238, 240, 244, 276
Inter-American Commission on Human Rights, 10, 141, 199–200, 242, 243
International Committee of the Red Cross, 53–54
International Labor Organization (ILO), 12, 13, 242
iron ore deposits/consequences, southern Pará, 83–84
Ismael dos Santos, Maria, 62, 104
Itacaiúna, 46–47, 48
ITERPA (Land Institute of Pará state), 102, 104, 195, 198

Jornal do Brasil, 34, 49–50, 60, 129, 201
Jornal do Brasil series/land-related conflicts
Josélio and, 188
Maria Joel photo, 189–90, 201

Nunes and, 189
overview, 188–90
José de Oliveira, Artur and Gavião-Kyikatêjê, 40, 50
Juarez (nickname/hired assassin); 66–67
Jungmann, Raul, 164

Kayapó Indigenous group, 202, 203, 206–7, 310
King, Martin Luther, Jr., 295
Kubitschek, Juscelino, 28–29

Lacerda, Carlos, 28
land
"social function" of property, 111
subsistence peasants and, 310
See also agricultural production; colonization of Amazon; land-grabbing issue; specific individuals/organizations
land conflict
terror campaign/murders (2011), 236–38
"trophies," 65, 236
See also specific individuals/issues
land distribution
Cardoso/federal government views, 142, 164
family farmer funds, 236
inequity/statistics, 141–42
landowners responses, 142
See also land occupation
land-grabbing issue
crickets and, 99
examples, 103–5, 249–50, 335n4
fraud scale, 100–101
grilagem/grileiro meaning/origin, 99
inequity/statistics, 141–42, 354n16
investigation request, 244
lack of land registry system, 59–60, 101–2, 346n34

land-grabbing issue (*cont.*)
   land conflicts/violence, 102, 105
   land occupation threats, 117–18
   landowners response to MST,
      111–12
   Lulu and, 206
   MST and, 109–11
   notary publics and, 101–2
   overview, 99–105
   Rondon area, 117
   Rousseff government and, 236
   scheme descriptions, 99–101
   See also colonization of Amazon;
      specific individuals
Land Institute of Pará state (ITERPA),
   102, 104, 195, 198
landless movement
   Eldorado Massacre and, 142–43
   poll on, 142
   See also specific events/issues;
      specific individuals/
      organizations
land occupations
   Josélio's land and, 147
   landowners response, 142, 145,
      194–95
   law enforcement and, 115, 140–41,
      144, 145, 146
   as model strategy, 117
   MST and, 110–11, 140, 177–78
   seizure of land/new settlements,
      192–93, 198–99, 239
   See also specific individuals/
      locations
land occupations/Dezinho
   denouncing Josélio, 144–45
   Fazenda Jerusalem, 144, 166
   Fazenda Tulipa Negra, 166–67,
      177, 285
   ideas on, 143–44
   incarceration/Amorim and,
      146–47
   Primavera II ranch, 145, 166

   protest camp, Marabá, 144
   testimony/Nunes's trial, 274
landowners
   employment and, 223
   Leal defending, 211
   See also specific individuals
Lauria, Roberto
   description, 272
   as Flexa's defense lawyer, 298,
      299
   Nunes's appeal/right to retrial,
      297
   Nunes's trial, 272, 275–77, 279–81,
      289–91, 294–95
Leal, Américo Lins da Silva
   defense strategies, 211
   description/background, 211
   people hiring and, 212, 221, 224
   Stang and, 211
   Wellington trial and, 211–13, 214
Leite, Adélcio Nunes
   arrest, 122
   Cordeiro family slaughter and,
      121, 122
   death, 269
   description, 120
   escape/escape route, 122, 126–27
   Eva Nilma (wife), 123, 125, 126
   false identity/Souza and, 124,
      126–27
   homicides, 120, 269
   Josélio and, 5, 119, 125, 126, 127,
      269
   recapture of, 122–26, 269
   sentences and, 127, 269
   See also Souza
Leite, Alírio, 121
Leite family
   Cordeiro family feud, 121–22, 123
   Cordeiro family slaughter/
      investigation, 121–22
   description, 120–21
   goals, 120

Lobato Prato, Franklin
  description, 272
  Nunes's trials, 272–74, 275,
    280–81, 287, 290, 292–93, 294,
    299, 300–301
  threat to, 300–301
Lopes, Adriana, 173
Lopes, de Angelo, Antônio
  background, 173
  mayoral election (2008), 224, 225,
    227
  Pedro's death/hospital, 173,
    286–87
  politics, 173, 176, 224, 225, 227,
    361n15
Lopes family, 173, 179, 181, 188, 286
Lopes, João, 173
Lopes, José Dias, 24, 25, 35
Lopes, Kyume Mendes, 179, 192–93,
    276
  See also Fazenda Tulipa Negra
Lopes, Manoel ("Duca"), 145, 173
Lopes, Marcos, 173
Los Angeles Times, 52, 208
Lourdes Ribeiro da Silva, Maria de, 8
Lula da Silva, Luiz Inácio
  as activist/views, 97, 236
  agricultural production and, 230
  arrest, 253–54
  BR-163/settlements and, 204, 205,
    206–7
  childhood, 96
  deaths of first wife/son, 96
  deforestation issue, 204, 205, 206
  descriptions, 97
  "dirty list" of employers, 218–19
  hopes for, 96, 97
  landowners and, 194, 197–98
  Maria Joel and, 200, 207–9, 215,
    219–20
  presidency and, 106, 194
  remarriage, 97
  Sindicato/Maria Joel and, 200

social programs, 224
Workers' Party founding by, 9
Lynch, Ana Lúcia Bentes, 16

Machado da Cunha, Antônio Fernando
  Josélio and, 65, 66, 70, 103
  land-grabbing and, 65, 103
  mad cow disease, 203–4
Malcher Dias Neto, João
  description, 259
  Ferreira Rocha, Arnaldo and, 305
  Nunes's raid and, 252
  Shirley Cristina and, 130, 252, 259,
    305
Maniçoba de Moura, Wagner, 242
Manuel (Te-Chaga-U ranch worker),
    3–4, 5
Mariano da Silva Rondon, Cândido, 48
Marino Ferreira Dias, Claudio
  background/investigations, 156
  Dezinho's murder/investigation,
    155, 158, 168, 182
  Josélio's warnings, 189
  Josélio/views of, 156, 182
  Nunes/views of, 182, 293
  Wellington and, 155, 157–58, 182
  Ygoismar and, 158
mayoral elections (1988)
  democracy and, 95
  results, 228
  Workers' Party and, 96, 106
mayoral elections (2008)
  candidates, 224
  hostility towards Sindicato, 224
  legal battle over, 229–30
  polls, 226, 228
  vote buying and, 229
mayoral elections (2008) and Maria
    Joel
  campaign/promises, 226, 227–28
  children and, 225–26, 228, 229
  decision to run, 225–26
  finances, 236

mayoral elections (2008) and Maria
    Joel (*cont.*)
    Joelma ("Joelminha") on, 228, 229
    polls, 226, 228
    risks/security, 225–26
    running mate, 226
    Workers' Party, 225, 226, 227, 229
mayoral elections (2012), 258–59
Medeiros, Carlos, 103
Médici, Emílio Garrastazu, 56, 58
Mendes, Chico/murder, 116, 236
Miranda, Nilmário, 200, 201
Miranda, Pedro
    Nunes/Nunes's workers and,
      255–56
    Nunes's threats and, 256–57
    security, 256
    work of, 255
Monde, Le, 52
Montgomery, Paul, 51
moon landing, first, 39
Moraes, João Nazareno Nascimento/
    Josélio investigation, 14, 15, 17,
      18–19
Moreira Costa, R. C., 301
Moreira, Odeb, 112
Morricone, Ennio, 80
Mother Maria Reservation/
    conditions, 52–54
Movimento Humanos Direitos, 208
MST (Landless Workers Movement),
    109–10, 111, 114, 117, 140, 141,
    142, 143, 144, 177–78, 198,
    235–36, 348n9
MST (Landless Workers Movement)
    founding/goals and strategies,
    109–11

Narcos (Netflix series), 242
Nascimento, Magno Fernandes do
    Dezinho's murder/Wellington and,
      155, 195, 196
    as eyewitness/Wellington and, 195

    investigation into murder/
      motives, 195–96, 197, 200
    murder of, 195–96
National Geographic, 53, 204
National Indian Foundation (FUNAI),
    41, 45, 52, 53, 309
Nelore cattle, xiv, 3, 100–101, 248, 249
Neves, Tancredo, 95
New York Times, 51, 52, 60, 75, 84, 97,
    161, 206
Nova Almeida, 21
Nunes, Décio José Barroso ("Delsão")
    Agribusiness Association of
      Rondon, 112–13
    Alves de Silva, Pedro and, 168,
      170, 174, 175, 176, 181, 215, 216,
      280, 282–86, 287, 291–92, 300,
      383n2
    arrest/release, 180–83
    business (2000–2006), 216–17
    business diversification, 114
    case against/appeals (2008), 227
    case against/dismissal (2006),
      215–16, 220, 236
    cattle, 114, 216–17, 248, 250,
      251–52, 253, 293, 301, 378n34
    conflicts history, 215, 217
    criminal gang/"Judge, the" and,
      172, 174, 176, 215, 274, 293
    descriptions, 112–13, 170, 200,
      217–18, 255, 256, 257, 272, 301
    Dezinho/murder and, 114, 175,
      180–82, 200, 209, 215–16, 217,
      261, 291
    "dirty list" of employers, 218
    divorce/consequences, 180–81
    environmental offenses fines, 301
    Fazenda Lacy/land-grabbing
      controversy, 114, 147, 156, 179,
      200, 217, 218, 246–47, 274, 277,
      293, 301, 349n21, 372n34
    FETAGRI accusations against, 147
    Hilário and, 118

illegal deforestation, 200, 217, 293
impunity, 294, 301–2
investigation/FETAGRI
    accusations, 156
Joelma ("Joelminha") with son
    encounter, 307–8
Jornal do Brasil land-related
    conflicts, 189
King of Wood, the (nickname), 113
land-grabbing and, 147, 179–80,
    189, 215, 246–47, 274
land occupations/landless people
    and, 147, 179–80, 189
lawsuits against/response, 254,
    255, 256, 292, 301
Lopes family and, 181
Magno murder and, 196
meat plant problems/
    investigations, 301
men hired to kill Maria Joel and,
    220–21
murder rumors/accusations
    against, 174, 196, 283–85
O Globo lawlessness series, 200
reforestation, 114
report on (after Maria Joel/Lula
    meeting), 215–16
reputation/status with
    landowners, 217
Rondon economy and, 247
Silva Filho, Francisco Martins da
    and, 170, 172, 174, 189, 215, 216,
    221
tax debt, 301
timber business/strategies, 113–14,
    218, 246–47
workers' treatment/murders, 113,
    216, 218, 245, 254, 255, 274,
    283–85, 292, 301, 377n2
See also specific events/individuals
Nunes, Décio José Barroso ("Delsão")
    debt/assets seizure
    Alma/risks and, 248, 249, 250

assets in other names/front men,
    246, 377n4
cows, 248, 250, 251–52, 253, 378n34
ExpoRondon and, 251, 253
Frank and, 248–49, 250
Jarbas and, 248, 250
judge joining asset seizure, 247,
    248, 251, 252, 253
land-grabbing and, 246, 247
media/townspeople and, 253–54
morning of asset seizure, 247–48
Nunes lawyers/appeal, 253, 256–57
Nunes responses, 253, 256–58
repeated violations/strategies, 245
undercover investigation, 245, 247
vehicles, 248–49, 251, 253
See also Andrade, Jônatas dos
    Santos; *specific individuals*
Nunes, Décio José Barroso ("Delsão")
    trial for Dezinho's murder
    appeal/right to retrial (2016), 297
background, 270–71
description, 271–72, 273–95
difficulties in getting state
    attorney, 272–73
Joelma ("Joelminha") and, 273
Josélio mentions, 276–77, 293,
    294–95
Maria Joel at retrial (2019), 299–300
Maria Joel/family following trial,
    295–96, 306
Maria Joel's children and, 273, 295
Maria Joel suspicions on Flexa, 298
Maria Joel testimony/problems,
    273–77
Nunes continuing appeals, 301
Nunes leaving room, 273
Nunes testimonies, 291–94, 300
retrial (2019), 298, 299–300
verdicts/sentences and "appeal at
    liberty," 295, 298, 300
See also specific individuals
Nut Road descriptions, xv, 39, 40–41

Oakland Institute, 199–200
O Estado de S. Paulo, 60, 207–8
O Globo, 121, 200–201, 207, 229
O Globo series on Brazil's lawlessness
   Maria Joel photo, 201
   Nunes, 200
   overview, 200–201
O Liberal, 253
Oliveira, Moisés Soares de
   Dezinho and, 149–50, 175
   election/landowners and, 149,
      164–65, 173, 175, 178
   Maria Joel and, 186
Oliveira Pereira, Edilson
   becoming mayor, 267
   mayoral vote (2012), 258
Operation Amazonia/consequences,
   37–38, 45, 49, 53, 56, 99
Organization of American States
   (OAS), 10

PA-70 road
   colonization and, 39–40, 41, 45,
      46, 49, 50, 59, 60, 72, 165, 202
   descriptions/dangers, 54, 57
Pacheco Neto, Luiz/Hilário murder
   implicating Dezinho/Brito, 136–37
   murder/escape, 132
   Neuza (fiancée) and, 134
   reasons for murder, 134, 136–37
Parla FIV AJ (Nelore cow), 249
Pastoral Land Commission (CPT), 73,
   89, 98, 105, 106, 107, 108, 109,
   111, 117, 138, 141, 143, 144, 146,
   179, 186–87, 188, 190, 191, 193,
   194, 199, 200, 208, 209, 210,
   219, 221, 238, 240, 243, 301
Pedro II, Emperor, 31
Pereira, Bruno, 303
Phillips, Dom/murder, 303
Phillips, Tom, 236–37
Piaui family, 285
Piaui/murder, 283–85, 292

pistoleiros description, xii
Pitanga, Camila
   description, 208
   helping Maria Joel, 208–9, 242, 243
pollution
   charcoal kilns, 91
   children of Dezinho/Maria Joel,
      90, 92
   gold mines/Serra Pelada, 85
   slash-and-burn activities/
      superfires, 4, 55, 91, 92
   See also deforestation in Amazon
Possuelo, Sydney, 45
Powell, Colin, 230
pre-Columbian societies, Brazil, 46
presidential elections (2022)
   Bolsonaro, 308–9
   Lula, 308, 309
"preventive habeas corpus," 69–70, 197
prison conditions, Brazil/Latin
   America, 161–62
Programa de Integração Nacional
   (PIN), 56–57
Provincia do Pará, A, 14, 61, 64

Rai
   defense/imprisonment, 130–31
   as José Antônio da Silva, 130
   Josélio and, 5, 119, 128, 130–31, 145
   Josélio/Ceará and, 5, 119, 128,
      130–31
Ramos, Adelino, 237
Red Cross, International Committee
   of the Red Cross, 53–54
Reis, Conceição dos, 63, 68, 73
Reis, Eni, 64
Reis family
   move from Vila Rondon, 73
   See also specific individuals
Reis, Geraldo dos, 63, 64, 65, 71
Reis Silva, Antônio dos
   disappearance, 63
   See also chainsaw murders

Resende de Almeida, Walter
  description, 166
  Dezinho's murder investigation,
    166, 167, 171–77, 178–79, 180,
    181, 188, 189, 215, 288
  Francisco's interrogation/
    statements, 171–79, 278, 283,
    285, 289
  Josélio and, 167, 188
  Nunes's investigation, 216
  Pedro's murder investigation,
    173–74, 288, 361n17
Rezende, Ricardo, Father, 208, 209,
  242
Ribeiro da Silva, José Cláudio/murder,
  236–37
River of Doubt expedition, 48
Road of the Jaguar, 41
Robert F. Kennedy Memorial Center
  for Human Rights,
  199–200
Rocha, Anterino Pereira, 144
Rocha, Janilton Silva, 226
Rocha, Olávio Silva
  Francisco on, 172, 173, 188
  mayoral elections, 106, 176
  mayoral elections (2008), 224, 225,
    226, 227, 228
  mayoral elections (2008) legal
    battle/outcome, 229–30, 259
  Nunes and, 189
Rondon do Pará (summary)
  activists, xii
  budgets, 224, 373n63
  colonists, xiii–xvi
  descriptions/crime, xi–xii,
    xiv–xvi, 75, 79–81, 83–87,
    141, 235
  elite and city hall, 223, 224
  inequalities, xii, xvi–xvii, xix
  lumber industry/sawmills
    descriptions, 79–80, 81, 143
  medical system and, 81

population changes, 80, 84, 143,
  355n28
power system, xii, xiii
sexism of, 190–91
unemployment/population
  decrease, 143–44
See also specific events/
  individuals; Vila Rondon
Roosevelt, Teddy, 48
Rousseff, Dilma, 236, 237, 238, 242,
  243, 272

Sabatella, Letícia
  description, 208
  helping Maria Joel, 208–9, 210,
    212, 242, 243
St. Joseph Hospital, Rondon, 173,
  286–87, 288
Salgado, Sebastião, 85
Salgado Vieira dos Santo, Iacy, 180
Santos, Alberto Nogueira dos, 291
Santos de Jesus, Zudemir dos
  description, 191
  Maria Joel and, 191, 222
  on Nunez/landowners, 218
  positions, 191, 218
  Sindicato and, 235
Santos Dios, Maria Eva dos
  daughters' escape from Rondon,
    235
  on Dezinho, 93–94, 285
  health problems, 235
  on Maria Joel, 186, 192
  Maria Joel relationship (sister),
    81, 187
  politics, 225, 228
  quitting Sindicato/leaving
    Rondon, 235
  selling clothes, 109
  settling in Rondon, 82
  Shirley Cristina and, 235
  Sindicato, 108–9, 191, 235
  threats/security, 235

Santos, Eliane dos, 193
Santos, Ribamar Francisco dos
  description/Sindicato, 196
  murder as intimidation, 196, 199
  murder of/investigation, 196, 197,
    198, 200
Sarney, José, 95
Seattle University School of Law,
  199–200
Silva, Antônio Gregório da. *See*
  Toninho
Silva Filho, Francisco Martins da
  background/parents, 169–70
  case against Nunes and (2006), 215
  on criminal gang, 172–75, 280,
    286, 287, 289
  description, 169
  Dezinho's murder and, 285
  Dezinho threat tip, 167–68, 176
  fear/threats and attempt to flee,
    170–71, 177
  land occupation/Dezinho and,
    167, 177
  Lucas (brother), 286–87
  Maria (wife), 170, 171, 172, 177, 178,
    284
  mistrust of police, 172
  Nunes as employer, 283
  Nunes's bribes and, 279, 300
  Nunes's trials/safety concerns,
    278, 279–90, 300
  Pedro relationship (brother),
    167–71, 178, 280
  Pedro's murder/warning, 189, 280,
    286–88
  Resende interrogating, 171–79,
    278, 283, 285, 289
  statements/Dezinho's murder
    investigation, 171–77
  witness protection plan/family,
    178, 278–79, 300, 362n28
  See also Alves de Silva, Pedro
Silva, José Antônio da. *See* Rai

Silva Menezes, Agamenon da, 309–10
Silva, Rogério, 148, 162, 180, 216, 271
Silva, Wellington de Jesus
  background/Rondon, 148–49, 162,
    163
  descriptions, 150, 163, 211
  mother/family members and, 213,
    214
  traffic stop, 158
Silva, Wellington de Jesus/Dezinho's
  murder
  Adriana (defense attorney), 162,
    173
  bounty, 153, 163
  crowd beating/injuries, 154, 155,
    157, 161, 185
  custody/questioning, 157, 162–63,
    182
  description, 148–49, 150–54
  Nunes link, 179
  preparation/cover story, 148, 150,
    153–54
  professional killer question, 163
  weapon used, 152, 162, 182, 271
Silva, Wellington de Jesus/Dezinho's
  murder trial
  activist supporters/protests, 210,
    212
  description, 210–14
  lawyer change/defense, 211–13, 214
  Maria Joel/children and, 210, 212,
    213, 214–15
  Nunes mentions/Nunes's defense
    attorneys, 212, 213, 214
  permission to leave jailhouse/
    disappearance, 221–22, 238,
    271, 302
  sentence, 213, 298
  See also specific individuals
Silva, Ygoismar Mariano da
  Dezinho murder and, 148–49, 153,
    154, 162, 180, 216, 271
  traffic stop, 158

Sindicato
Araujo's visit to, xvii–xviii
headquarters description, xvii
history/purpose, 8
land-grabbers/land occupation
and, 114, 117, 143, 144–45
leader following Maria Joel,
235
women leaders, 191
See also specific individuals
Sister Dorothy. See Stang,
Dorothy Mae
slash-and-burn activities, 4, 55, 91
slavery/debt bondage, 7, 8, 10, 12–13,
16–17, 18, 19, 31, 60, 97–98,
147, 188, 200, 218–19, 242,
254–55, 273–74
SMS for Brasília, SOS for Dona Joelma
(documentary), 242
Socrates, 29
Souza
descriptions, 5–6, 118–19
Te-Chaga-U ranch/killings, 5–6,
14, 118–19
See also Leite, Adélcio Nunes
Souza Barros, Sirley de, 259, 263
Souza, Clovis, 42
Souza Costa, Lourival de
criminal gang and, 173
Dezinho's murder and, 179,
182, 194, 215, 216, 261, 271,
276
Dezinho's murder trial, 271, 276
Fazenda Santa Monica, 173, 182,
194, 196, 197, 215, 261, 276
Ferreira and, 221
land-grabbing and, 173, 194,
361n17
land occupation, 196, 197
Ribamar murder and, 197
Souza Neto, Domicio de ("Raul"), 173,
182, 271
"soybean highway," 204

soybean production/Amazon
beginnings, 201, 202–3
BR-163 settlements and, 202–3,
204
deforestation and, 203, 204, 207
growth/reasons, 201, 203–4, 367n7
innovations needed, 367n42
moratorium on, 207
Stang, Dorothy Mae
activism, 201, 206, 219
background, 201
descriptions, 201, 205
enemies on, 206
Human Rights Prize, 219
Maria Joel and, 207
Miranda and, 201, 206
murderers, 209, 271
murder trial, 211, 271, 297
threats/murder, 201, 205–6, 236
Sunday Times, 52
Survival International, 52

Tabapuã cattle, 248
Tadeu Dalla Sily, Renato/murder, 127,
128
Te-Chaga-U ranch
Barros family/Josélio and, 3–12,
13–20
descriptions, 4
Josélio buying, 104
killing field, 3–4
land occupation and, 147
name meaning, 4
See also specific individuals
terra preta soil, 46
Toninho
Cavalcanti assassination, 21–25,
35, 36
description, 21
Torres da Silva, José Geraldo. See Zé
Geraldo
Trans-Amazonian Highway, 56–57
Turner, Frederick Jackson, 58

Vargas, Getúlio, 27, 28, 226
Vice, 236
Vicente, Alvaristo
  Cavalcanti assassination, 21–25,
    35, 36
  description, 21–22
Vieira de Rezende, Eurico, 66
Vieira Ramos, Honório
  disappearance, 63
  See also chainsaw murders
Vilaça de Lima, Eliana, 130
Vila Gavião land occupations
  descriptions, 114–15, 116–18, 132,
    276
  following Hilário murder, 134,
    138
  settlements, 239
  See also Ávila, José Hilário; *specific
    individuals*
Vila Rondon
  descriptions/lawlessness,
    59–60, 75
  land conflicts (summary), 59–60,
    71–73

transformation, 54–56
  See also Rondon do Pará
    (summary); *specific events/
    individuals*

Wałęsa, Lech, 97
Wallace, Scott, 204
Washington Post, 60
Workers' Party, 8–9, 11, 12, 17, 96, 98,
    106, 107, 109, 115, 129, 130, 134,
    146, 147, 150, 154, 194, 212, 223,
    225, 226, 227, 229, 240–41,
    260, 306
Workers' Party description/founding,
    8–9

Zé Geraldo
  background/description, 12
  Dezinho/Brito and Hilário's
    murder, 133
  Josélio investigation, 12, 13–14, 15,
    19, 128, 129
Zezinho (nickname/hired assassin),
    66–67